Andrea Klinglmair

Exploring the Public Value of Renewable Energy in Austria

AF099272

Andrea Klinglmair

Exploring the Public Value of Renewable Energy in Austria

A Stated Preference Study focussing on Hydropower

Südwestdeutscher Verlag für Hochschulschriften

Impressum / Imprint
Bibliografische Information der Deutschen Nationalbibliothek: Die Deutsche Nationalbibliothek verzeichnet diese Publikation in der Deutschen Nationalbibliografie; detaillierte bibliografische Daten sind im Internet über http://dnb.d-nb.de abrufbar.
Alle in diesem Buch genannten Marken und Produktnamen unterliegen warenzeichen-, marken- oder patentrechtlichem Schutz bzw. sind Warenzeichen oder eingetragene Warenzeichen der jeweiligen Inhaber. Die Wiedergabe von Marken, Produktnamen, Gebrauchsnamen, Handelsnamen, Warenbezeichnungen u.s.w. in diesem Werk berechtigt auch ohne besondere Kennzeichnung nicht zu der Annahme, dass solche Namen im Sinne der Warenzeichen- und Markenschutzgesetzgebung als frei zu betrachten wären und daher von jedermann benutzt werden dürften.

Bibliographic information published by the Deutsche Nationalbibliothek: The Deutsche Nationalbibliothek lists this publication in the Deutsche Nationalbibliografie; detailed bibliographic data are available in the Internet at http://dnb.d-nb.de.
Any brand names and product names mentioned in this book are subject to trademark, brand or patent protection and are trademarks or registered trademarks of their respective holders. The use of brand names, product names, common names, trade names, product descriptions etc. even without a particular marking in this works is in no way to be construed to mean that such names may be regarded as unrestricted in respect of trademark and brand protection legislation and could thus be used by anyone.

Coverbild / Cover image: www.ingimage.com

Verlag / Publisher:
Südwestdeutscher Verlag für Hochschulschriften
ist ein Imprint der / is a trademark of
AV Akademikerverlag GmbH & Co. KG
Heinrich-Böcking-Str. 6-8, 66121 Saarbrücken, Deutschland / Germany
Email: info@svh-verlag.de

Herstellung: siehe letzte Seite /
Printed at: see last page
ISBN: 978-3-8381-3701-8

Zugl. / Approved by: Klagenfurt, Alpen-Adria-Universität, Diss., 2013

Copyright © 2013 AV Akademikerverlag GmbH & Co. KG
Alle Rechte vorbehalten. / All rights reserved. Saarbrücken 2013

TABLE OF CONTENTS

1 Introduction and problem definition 1

2 Theoretical background: externalities and public goods 4

3 The significant role of renewable energy sources 13
    3.1 Renewable energy investments 13
    3.2 Global climate and energy goals 16
        3.2.1 International agreements 17
        3.2.2 European progress towards the climate and energy goals 22
        3.2.3 Austrian progress towards the stipulated climate and energy goals 25
    3.3 Austrian climate and energy policy 31
    3.4 The structure of the Austrian electricity sector 34
    3.5 Green electricity funding in Austria 42
    3.6 Renewable energy potentials 46
    3.7 Assets and drawbacks of renewable energy expansion 55
    3.8 Research question 67

4 Economic valuation with Stated Preference techniques 71
    4.1 General remarks 71
    4.2 Contingent valuation 74
    4.3 Choice modelling 80
    4.4 Comparison of the two methods and general limitations 82

5 Previous research 85
    5.1 Renewable energy studies 85
    5.2 Wind power studies 104
    5.3 Hydropower studies 114
    5.4 Comparison of previous findings 118

6 Study design and implementation 123
    6.1 The general approach 123
    6.2 The choice experiments 128

	6.2.1	Choice experiment for hydropower expansion........................... 128
	6.2.2	Choice experiment for renewable energy expansion.................. 134
	6.2.3	Choice experiment for a specific hydropower project................ 137
6.3	Experimental design 141	
6.4	Contingent valuation exercise 150	
6.5	The questionnaire 151	
6.6	Pre-Testing 155	
6.7	Data collection... 157	
6.8	Identifying non-valid responses 160	
6.9	Representativeness of the samples 163	

7 Public attitude towards hydropower in Austria 171
   7.1 Descriptive analysis of public attitude 171
   7.2 Determinants of public attitude towards hydropower expansion.... 186
      7.2.1 Results of the estimated models.. 190
      7.2.2 Comparison of the models 193
      7.2.3 Odds-Ratios and probabilities... 197

8 Willingness to pay for hydropower expansion 202
   8.1 Descriptive statistical analysis of WTP..................................... 202
   8.2 Determinants of stated WTP 205

9 Preferences for the attributes of hydropower expansion 214
   9.1 The econometric framework 214
   9.2 Model results – hydropower... 220
   9.3 Willingness to pay – hydropower... 229
   9.4 Welfare analysis – hydropower.. 233
   9.5 The impact of current electricity bill on WTP 238

10 Preferences for a specific hydropower project 247
   10.1 Attitude towards the planned hydropower project.................... 249
   10.2 Model results – regional case study 254
   10.3 Willingness to pay – regional case study 261
   10.4 Welfare analysis – regional case study..................................... 263

11	Preferences for different renewable technologies	267
	11.1 Attitude towards renewable energy	267
	11.2 Model results – renewable energy	270
	11.2.1 Generic parameter estimates	270
	11.2.2 Alternative-specific parameter estimates	276
12	Concluding remarks	284
References		290
Appendix A: Tables		313
Appendix B: Questionnaire – Hydropower		322
Appendix C: Questionnaire – Regional hydropower case study		333
Appendix D: Questionnaire – Renewable energy		344

LIST OF TABLES

Table 1:	Global investments in renewable energy by sectors, 2011	14
Table 2:	GHG emission reduction targets by 2020 for emissions beyond the EU-ETS	20
Table 3:	Targets for the share of renewable energy sources in gross final consumption by 2020	21
Table 4:	Progress in the field of renewable energy in the EU-27 countries, 2006-2010	23
Table 5:	Shares of hydropower and thermal power plant production (in %), 1970 and 2011	37
Table 6:	Power plants and installed capacity (in MW) in Austria, 2011	37
Table 7:	Electricity generation by size and technology (in GWh), 2011	40
Table 8:	Feed-in tariffs for new green power facilities (in cent/kWh), 2012	44
Table 9:	Investment grants for small- and medium-scale hydropower plants	44
Table 10:	Installed capacity of Austrian green electricity facilities (in MW), 2006 and 2011	45
Table 11:	Additional renewable energy capacities in a future scenario for 2020 (in GWh)	49
Table 12:	Proposed power plant investments in Austria until 2020 (in billion €)	50
Table 13:	Hydropower plants in the process of construction by size and technology	52
Table 14:	Hydropower plants in the planning stage by size and technology	53
Table 15:	Renewable energy power plants in the planning stage	54
Table 16:	Quality elements for the determination of the water body status in rivers	64
Table 17:	The categorisation of TEV into use and non-use values	72
Table 18:	Elicitation techniques for contingent valuation studies	75
Table 19:	Overview of existing stated preference applications – renewable energy	101
Table 20:	Overview of existing stated preference applications – wind power	112

Table 21:	Overview of existing stated preference applications – hydropower	117
Table 22:	Sample sizes and response rates of previous CE and CV studies	121
Table 23:	Attributes and levels used in the CE for hydropower expansion	130
Table 24:	Directly stated WTP for hydropower in the pre-test	133
Table 27:	Components of the randomized CE designs	142
Table 28:	Efficiency of the CE design for hydropower expansion	147
Table 29:	Efficiency of the CE design for renewable energy expansion	148
Table 30:	Efficiency of the CE design for the regional hydro power case study	149
Table 31:	Extent of attitudinal section of the questionnaire	152
Table 32:	Extent of the designed questionnaires	155
Table 33:	Conducted pre-tests of the questionnaires	156
Table 34:	Required sample size depending on error probability and level of precision	159
Table 35:	The different samples of the online survey	160
Table 36:	Categorisation of protest responses	161
Table 37:	Sample sizes after adjustment for protest responses	162
Table 38:	Gender or respondents compared to total population	163
Table 39:	Age of the respondents compared to total population	165
Table 40:	Geographical distribution of the samples on hydro power and renewable energy	166
Table 41:	Geographical distribution of the regional hydropower sample	167
Table 42:	Educational level of respondents compared to total population	169
Table 43:	Professional situation of respondents compared to total population	169
Table 44:	Distribution of disposable monthly household income in the samples	170
Table 45:	Perceived state of information regarding hydropower use in Austria	175
Table 46:	Recreational activities along Austrian rivers	176
Table 47:	Reasons for the individual concernment imposed by a hydropower plant	180

Table 48:	Perceived impact of hydropower on recreational activities by attitude	183
Table 49:	Description of the variables used in the Logit model	189
Table 50:	Logit models explaining people's attitude towards hydropower expansion	191
Table 51:	Results of the Likelihood Ratio Tests	195
Table 52:	Statistically best fit model – Odds-Ratios	199
Table 53:	Probability of a positive attitude for two different individuals	201
Table 54:	Stated WTP for the expansion of hydropower and current electricity bill (per month)	203
Table 55:	WTP for hydropower expansion according to individual characteristics	204
Table 56:	Description of the variables used in the Tobit and Hurdle model	207
Table 57:	Results of the Tobit and Hurdle model	210
Table 58:	Likelihood ratio test comparing the Tobit and Hurdle model	213
Table 59:	Description of the variables used in the choice model on hydropower	221
Table 60:	Model estimates – hydropower	223
Table 61:	Hausman tests for IIA – hydropower	224
Table 62:	Result of the likelihood ratio test – hydropower	225
Table 63:	Importance of the attributes – hydropower	231
Table 64:	Estimates of marginal WTP – hydropower	232
Table 65:	EWF for different scenarios (per household/month) – hydropower	235
Table 66:	Aggregation of economic welfare (in million € per year) – hydropower	237
Table 68:	Separate EC models for sample A and B	244
Table 69:	Estimates of MWTP for sample A and B	246
Table 70:	The hydropower project Graz-Puntigam	247
Table 71:	Recreational activities along the Mur	251
Table 72:	Distance between project location and respondents' home	253
Table 73:	Description of the variables used in the choice model – regional case study	255
Table 74:	Hausman tests for IIA – regional case study	256
Table 75:	Model estimates – regional case study	257

Table 76:	Result of the likelihood ratio test – regional case study	258
Table 77:	Estimates of marginal WTP –regional case study	262
Table 78:	Importance of the attributes – regional case study	262
Table 79:	EWF for different scenarios (per household/month) – regional case study	264
Table 80:	Aggregation of economic welfare (in million € per year) – regional case study	265
Table 81:	Model estimates using generic parameters – renewable energy	272
Table 82:	Comparison structure between technologies	273
Table 83:	Wald tests comparing alternative specific constants	274
Table 84:	Comparison of MWTP from the unlabelled and labelled CE (with generic parameters)	275
Table 85:	Model estimates alternative-specific parameters – renewable energy	277
Table 86:	Wald tests comparing technology-specific parameter estimates	279
Table 87:	MWTP across renewable technologies	281
Table 88:	MWTP for hydropower expansion from unlabelled and labelled CE	283
Table A1:	Conversion of national WTP values from previous research	314
Table A2:	Comparison between Logit and Probit model	319
Table A3:	MWTP from simple calculation – hydropower	320
Table A4:	MWTP from simple calculation – partitioned sample hydropower	320
Table A5:	MWTP from simple calculation – regional case study	320
Table A6:	MWTP from simple calculation – renewable energy (generic estimates)	321
Table A7:	MWTP from simple calculation – renewable energy (alternative-specific estimates)	321

LIST OF FIGURES

Figure 1:	Market inefficiency in the presence of an external cost	6
Figure 2:	Market inefficiency in the presence of an external benefit	7
Figure 3:	Optimal provision of a public good	11
Figure 4:	Global investments in renewable energy (in billion € at current prices), 2004-2011	13
Figure 5:	European investments in renewable energy (in billion € at current prices), 2004-2011	15
Figure 6:	Investments in renewable power plants in Austria (in million €), 2008 and 2010	16
Figure 7:	The major targets of the EU climate and energy package until 2020	19
Figure 8:	Total GHG emissions in the EU-15 (in Mt CO_2 equivalents), 1990-2010	24
Figure 9:	Total GHG emissions in the EU-27 (in Mt CO_2 equivalents), 1990-2010	25
Figure 10:	Austrian share of renewable energy sources in gross final consumption, 2006-2020	26
Figure 11:	Austrian share of renewable energy sources in the transport sector, 2005-2020	27
Figure 12:	GHG emissions in Austria (in Mt CO_2 equivalents), 1990-2010	28
Figure 13:	GHG emissions in Austria by sectors (in %), 2010	29
Figure 14:	GHG emissions in Austria by sectors (in Mt CO_2 equivalents), 1990-2010	30
Figure 15:	The main initiatives of the Austrian climate and energy policy	31
Figure 16:	The three pillars of the Austrian energy strategy	32
Figure 17:	Gross domestic electricity production (in %), 2011	35
Figure 18:	Gross domestic electricity production (in GWh), 1970-2011	36
Figure 19:	Hydropower plants in Austria by technology, 2011	38
Figure 20:	Hydropower plants in Austria by size, 2011	39
Figure 21:	Thermal power plants and renewable energy facilities in Austria, 2011	41
Figure 22:	The basic system of green electricity funding in Austria	43
Figure 23:	Hydropower potential in Austria (in GWh)	47
Figure 24:	Hydropower potentials across Austrian federal states (in GWh)	48

Figure 25:	Expansion targets for renewable energy imposed by the green electricity act	49
Figure 26:	Hydropower plants in the process of construction among Austrian federal states	51
Figure 27:	Hydropower plants in the planning stage among Austrian federal states	53
Figure 28:	Electricity production costs of renewable energy sources in Austria (in cent/kWh)	56
Figure 29:	CO_2 emissions from different technologies (in g CO_2 equivalent per kWh)	57
Figure 30:	The main environmental impacts associated with the use of hydropower	62
Figure 31:	Current ecological status of Austrian rivers (in %), 2009	65
Figure 32:	Conflict of interests and ecological trade-offs of hydropower use	68
Figure 33:	Basic research questions of the dissertation	69
Figure 34:	Economic valuation techniques in relation to the concept of total economic value	73
Figure 35:	Basic idea of implementing a choice experiment	81
Figure 36:	Multi-stage procedure of a stated preference study	124
Figure 37:	Distribution of directly stated WTP for hydropower in the pre-test (in %)	134
Table 25:	Attributes and levels in the CE for renewable energy expansion	136
Table 26:	Attributes and levels in the CE for the specific hydro power project	140
Figure 38:	Choice set example for hydropower	143
Figure 39:	Choice set example for renewable energy	144
Figure 40:	Choice set example for the specific hydropower project	145
Figure 41:	General structure of the questionnaire	151
Figure 42:	Number of protest responses in the different samples	162
Figure 43:	Perceived importance of future renewable energy expansion	172
Figure 44:	Preferred energy sources for future electricity generation in Austria	173
Figure 45:	Preferred renewable energy sources for future expansion in Austria	174
Figure 46:	Attitude towards hydropower use and the construction of new plants in Austria	175

Figure 47:	Perceived number of hydropower plants in respondents' environment	177
Figure 48:	Stated distance between the nearest hydropower station and respondents' home	178
Figure 49:	Individual concernment of people by the closest hydropower station	179
Figure 50:	Respondent's knowledge of hydropower construction plans near their home	181
Figure 51:	Perceived impact of hydropower on recreational activities by concernment	182
Figure 52:	Respondents' agreement to several statements on intensified hydropower use	184
Figure 53:	Agreement to landscape and wildlife impacts by attitude	185
Figure 54:	ROC-curve of model specification 1	196
Figure 55:	ROC-curve of model specification 2	197
Table 67:	Procedural test results for preference and scale parameter equality	242
Figure 56:	Result of the optimization exercise searching for optimal relative scale	243
Figure 57:	Design of the hydropower plant in Graz-Puntigam	248
Figure 58:	Attitude towards hydropower and its expansion along the Mur	250
Figure 59:	Perceived impact of the new hydropower station on recreational activities	252
Figure 60:	Individual concernment of people by the new hydropower plant	254
Figure 61:	Perceived importance of future renewable energy expansion (2)	268
Figure 62:	Preferred renewable energy sources for future expansion in Austria (2)	269
Figure 63:	Respondents' agreement to several statements on renewable energy	270

LIST OF ABBREVIATIONS

ASC	Alternative Specific Constant
BGBL	Bundesgesetzblatt
BRM	Binary Response Model
CBA	Cost Benefit Analysis
CHP	Combined Heat and Power
CO_2	Carbon Dioxide
DB	Double Bounded
DC	Dichotomous Choice
df	Degrees of Freedom
EC	European Commission
EC	Error Component
EEA	European Environment Agency
EPC	European Parliament and Council
ETS	Emissions Trading System
EU	European Union
EV1	Extreme Value Type 1
EWF	Economic Welfare
G	Gramm
G	Graz
GDP	Gross Domestic Product
GHG	Greenhouse Gases
GU	Graz-Umgebung
GWh	Gigawatt hour
HMWB	Heavily Modified Water Body
IEA	International Energy Agency
IIA	Independence of Irrelevant Alternatives
IID	Independently and Identically Distributed
IR	Interval Regression
IRENA	International Renewable Energy Agency
JRC	Joint Research Centre
KRW	Korean Won
kWh	Kilowatt hour
LC	Latent Class
LL	Log Likelihood
LPM	Linear Probability Model

LR	Likelihood Ratio
LULUCF	Land Use, Land Use Change and Forestry
MB	Marginal Benefit
MC	Marginal Cost
MEB	Marginal External Benefit
MEC	Marginal External Cost
MLE	Maximum Likelihood Estimator
MNL	Multinomial Logit
MRS	Marginal Rate of Substitution
MSB	Marginal Social Benefit
MSC	Marginal Social Cost
Mt	Million tonnes
MW	Megawatt
MWTP	Marginal Willingness to Pay
MU	Marginal Utility
MXL	Mixed Logit
NIMBY	Not In My Backyard
NH	Nordhessen
NL	Nested Logit
NOK	Norwegian Kroner
NREAP	National Renewable Energy Action Plan
OLS	Ordinary Least Squares
PJ	Petajoule
PPP	Purchasing Power Parity
PRO	Probit
PV	Photovoltaics
RBMP	River Basin Management Plan
REL	Random Effects Logit
REP	Random Effects Probit
RES	Renewable Energy Sources
ROC	Receiver Operating Characteristic
RPL	Random Parameter Logit
SB	Single Bounded
SBF	Statistically Best Fit
SDC	Socio-Demographic Characteristics
SEK	Swedish Krona
SP	Stated Preferences

TEV	Total Economic Value
UNEP	United Nations Environment Programme
UNFCCC	United Nations Framework Convention on Climate Change
USD	US-Dollar
WFD	Water Framework Directive
WS	Westsachsen
WTA	Willingness to Accept
WTP	Willingness to Pay

1 Introduction and problem definition

Renewable energy sources play a significant role in the Austrian energy sector. Due to the natural and environmental conditions, Austrian electricity generation is largely based on hydropower. In particular, more than half (57.4 %) of the domestically generated electricity currently comes from hydropower installations. In spite of this large proportion, there is still substantial potential for new hydropower facilities, especially for small-scaled ones. Indeed, it is planned to open up the remaining potentials. According to the 20/20/20 targets of the European Union (EU), Austria is bound to reduce its greenhouse gas (GHG) emissions, increase the share of renewable energy sources in gross final consumption, and improve energy efficiency until 2020. Additionally, Austria is committed to reduce its GHG emissions within the Kyoto Protocol. The further expansion of hydropower utilisation may contribute significantly to achieve these global climate and energy targets. Consequently, the intensified use of hydropower is an integral part of the Austrian energy strategy. However, hydropower use creates multiple impacts. Although investments in hydropower are undisputed due to security of supply issues, climate change and dependency concerns, the technology is subject to some disadvantages. Hydropower plants are often seen as a blot on the landscape and a threat for the ecosystem. Thus, the expansion of hydropower is principally in conflict with the objectives of nature and water conservation as for instance the European Water Framework Directive (WFD). Accordingly, future hydropower investments are associated with an "energy-water trade-off" between economic and climate-related advantages (such as the emission-free generation of electricity) and the negative environmental side effects. These positive and negative impacts can be seen as external effects imposed by the intensified use of hydropower.

There is a general agreement upon the need to consider all costs and benefits, including external ones, when political decisions, such as the expansion of hydroelectric power, are to be made. Moreover, an increasing tendency to legitimate political decisions with figures has been observed in the past. However, this is often fraught with difficulties since externalities like reduced air emissions or environmental impacts are usually not reflected in market prices. Hence, the assignment of monetary values to these effects represents a way to integrate them into public decision-making (SUNDQVIST AND SÖDERHOLM, n.d.:1). This is the aim of the present work: examining public preferences for the multiple impacts of an expansion of hydroelectric power in Austria in order to broaden the strategic basis of public decision making. More precisely, the dissertation aims to assess people's general attitude towards renewable energy and hydropower use in Austria. The main part of the current work refers to the monetary valuation of the multiple impacts (trade-off) associated with future hydropower investments. For this purpose, direct evaluation methods, also known as "Stated Preference Techniques" have to be primarily used, since the monetary values of the positive and negative externalities of increased hydropower cannot be derived from actual market behaviour. Finally, the dissertation deals with the quantification of public preferences for different renewable technologies beside hydropower (biomass, photovoltaics and wind power).

Beside the general examination of the role renewable energy sources play in the Austrian energy sector and the analysis of existing literature, a comprehensive online-survey of approximately 1,400 Austrian citizens has been carried out within the scope of this work. Based on the statistical evaluations, it was shown that people have in principle a positive attitude towards hydropower use, as well as the construction of new plants along Austrian rivers. The Austrian population is generally willing to pay for the positive impacts imposed by the expansion of hydroelectric power, but wish to be compensated for the loss of nature new hydropower plants are associated with.

The first part, chapter 2, provides the theoretical foundation of this work and includes a brief review of the economic theory behind externalities and public goods. Chapter 3 deals with the role of renewable energy sources against the backdrop of global climate and energy goals like the reduction of GHG emissions or the increase of the share of renewable energy sources. Furthermore, chapter 3 contains a detailed description of the Austrian electricity sector, as well as a presentation of future renewable energy potentials and the assets and drawbacks of renewable energy investments. The theoretical background on stated preference techniques is covered in chapter 4 of this work. This is followed up by a summary of previous economic valuation studies on the topic of renewable energy, wind and hydropower using stated preference methods (chapter 5). Chapter 6 refers to the study design and survey implementation. The main part of the current doctoral thesis is made up of the statistical evaluations that are based on the implemented surveys. First, people's general attitude towards increased hydropower use in Austria is analysed in chapter 7. This part of the work includes a logistic regression model explaining the determinants of public attitude towards hydropower. Chapter 8 comprises the analysis of directly stated willingness to pay (WTP) for the expansion of hydropower. By means of a Tobit and two-step Hurdle model, determinants of stated WTP are explained. The econometric framework used to quantify public preferences and the results of the choice experiments are presented in chapters 9 to 11. In particular, chapter 9 refers to the evaluation of future hydropower expansion, while chapter 10 addresses a regional hydropower case study in the province of Styria. In chapter 11, preferences for the multiple impacts of different renewable energy sources are examined. Chapter 12 concludes the thesis and contains relevant policy implications drawn from the results of this work.

2 Theoretical background: externalities and public goods

Economic and political decisions are often associated with consequences for second or third parties that decision makers do not consider. Such consequences are called externalities. Generally speaking, an externality is a cost or benefit resulting from some action or decision that is imposed or bestowed upon parties not involved in the transaction (CASE ET AL., 1999:363). If an external cost is caused by an economic or political decision, or generally speaking a market transaction, one speaks of a negative externality. Conversely, an external benefit corresponds to a positive externality (PINDYCK AND RUBINFELD, 2003:872ff).[1]

From a microeconomic perspective, marginal cost is the cost of producing an additional unit of a good or service that goes at the expense of the producer of that good or service. In the presence of a negative externality, the production of a good or service causes an external cost that is not borne by the producer but borne by second or third parties. Hence, "a marginal external cost is the cost of producing an additional unit of a good or service that falls on people other than the producer of the good or service" (PARKIN ET AL., 2005:333). Following this, marginal social cost (MSC) accrued to the entire society is composed of marginal cost (MC) and marginal external cost (MEC). That is:

$$MSC = MC + MEC \qquad (2.1)$$

Analogous to the definition of marginal cost, the marginal benefit can be interpreted as the benefit the consumer obtains from an additional unit of a good or service. In case of a positive externality the consumption of a good or service creates an external benefit to someone other than the consumer of the good or service. Accordingly, "a marginal external benefit is the benefit from an additional unit of a good or service that people other than

---

[1] Additionally, externalities are often defined according to whether the cost or benefit arises from the consumption or production of a good or service. This is why they are also referred to as production or consumption externalities (PARKIN ET AL., 2005:330).

the consumer enjoy" (PARKIN ET AL., 2005:337). The sum of the marginal benefit (MB) and the marginal external benefit (MEB) is the total benefit enjoyed by society called marginal social benefit (MSB). That is:

$$MSB = MB + MEB \qquad (2.2)$$

If decision makers fail to consider social costs and benefits, inefficient decisions will result (CASE ET AL., 1999:363). This can be illustrated for both cases, the presence of an external cost and an external benefit. Let us start with the external cost example. Figure 1 shows a simple market diagram with quantity on the horizontal axis and price on the vertical axis. The demand curve D is represented by the marginal benefit (MB) to the buyers of the corresponding good. The supply curve S measures marginal cost (MC) of the producers. Market equilibrium is given at quantity q* and price P* where supply equals demand (S=D), or equivalently, marginal cost is equal to marginal benefit (MC=MB). However, this equilibrium is inefficient since not all costs – private and external – are considered.[2] An efficient allocation of resources, however, requires that marginal benefit is equal to marginal social cost (MSC). With this equilibrium condition, negative externalities, i.e. external costs, are taken into account. Accordingly, the efficient equilibrium is given at quantity q_{eff} and price P_{eff}. Comparing the efficient outcome with the general market equilibrium, it can be concluded that – in the presence of external costs – the market equilibrium leads to an overproduction of the good or service. This overproduction creates a deadweight loss shown by the triangle between MC and MSC in Figure 1 (PARKIN ET AL., 2005:334; PINDYCK AND RUBINFELD, 2003:873).

---

[2] In the market equilibrium q*, the marginal social cost exceeds marginal benefit representing an inefficient outcome.

Figure 1: Market inefficiency in the presence of an external cost

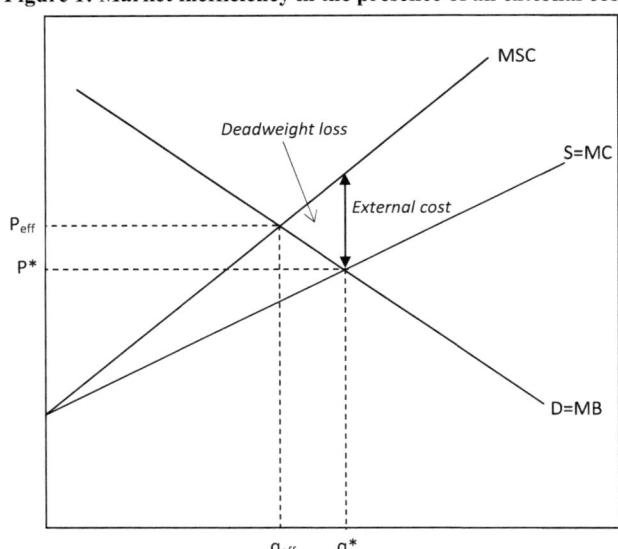

Source: PARKIN ET AL. (2005:334); OWN REPRESENTATION

Market inefficiency can also be shown for a positive externality, i.e. the presence of an external benefit. This is illustrated in Figure 2. As before, the supply curve corresponds to the marginal cost curve (S=MC) and demand equals marginal benefit (D=MB). Generally, market equilibrium occurs at the intersection of supply and demand curve (S=D), or more precisely, where marginal benefit equals marginal cost (MB=MC). This point is given at quantity q* and price P*. However, this allocation of resources is inefficient due to the fact that not all benefits are taken into account.[3] In the presence of a positive externality, marginal social benefit (MSB) accounting for the external benefit should be taken into consideration. An efficient resource allocation is given at q_{eff} and P_{eff} where marginal cost is equal to the marginal social benefit (MC=MSB). Hence, the non-consideration of the external benefit would lead to an underprovision of the good or service that creates the externality. This undersupply creates a deadweight loss represented by the triangle shown in Figure 2 (PARKIN ET AL., 2005:338; PINDYCK AND RUBINFELD, 2003:875).

---

[3] At quantity q*, marginal social benefit is higher than marginal cost indicating that the market equilibrium is associated with an inefficient allocation of resources.

Figure 2: Market inefficiency in the presence of an external benefit

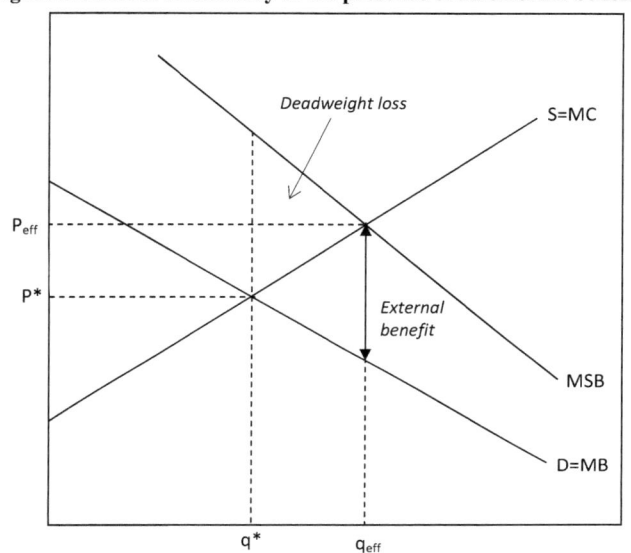

Source: PARKIN ET AL. (2005:338); OWN REPRESENTATION

To conclude, the presence of externalities provokes market inefficiencies. The non-consideration of external costs leads to an oversupply of the good that creates the externality, while ignored external benefits cause an under-provision of the good. Thus, external costs and benefits should be factored into decisions in order to arrive at an efficient allocation of resources (CASE ET AL., 1999:366). The question is how producers can be induced to reduce or increase the quantity of the corresponding good to the efficient amount. Generally, there are five different approaches to solve the problem of externalities: government-imposed taxes or subsidies, bargaining and negotiation (Coase theorem), legal rules and procedures, selling or auctioning rights to impose externalities, and finally, direct government regulations (CASE ET AL., 1999:370).

The first approach refers to the idea that marginal taxes and subsidies represent a direct way to force decision makers to consider external costs or benefits in their decisions. Thereby, the imposed tax should equal exactly the marginal external cost. Conversely, a subsidy should be granted in the amount of the marginal external benefit in order to arrive at an efficient solution (CASE ET AL., 1999:370; PARKIN ET AL., 2005:340).

The Coase theorem implies that it is possible to arrive at an efficient solution without any government involvement through private bargaining and negotiation. However, an efficient outcome can only be achieved under several conditions. First, the initial property rights must be clear to all parties.[4] Second, people must be able to bargain at low transaction costs and finally, only a small number of parties can be involved in the negotiation process (CASE ET AL., 1999:372; PARKIN ET AL., 2005:335; PINDYCK AND RUBINFELD, 2003:895).[5]

Other possibilities of considering externalities include legal rules and procedures. In case of an external cost, the parties injured by the externality can obtain an injunction that forbids the damage-producing behaviour from continuing or a financial compensation for the damage caused. Rights can also be protected by liability rules. With such rules required compensation payments for the imposed damages are determined by law. This approach provides an incentive for decision makers to weigh all the consequences of their decisions (CASE ET AL., 1999:373f; PINDYCK AND RUBINFELD, 2003:896).

Instead of imposing taxes or charges it is possible to assign rights to deal with externalities. A famous example of this approach is the system of marketable emission certificates. In an emission trading system, a maximum of permitted emissions is primarily determined and allocated among potential polluters (PINDYCK AND RUBINFELD, 2003:883). The polluters have then the opportunity to buy and sell the assigned permits depending on their marginal cost of reducing pollution. Generally, this approach provides a stronger incentive to reduce pollution as compared to taxes or charges because the price of a permit increases as demand for permits rises (PARKIN ET AL., 2005:337).

---

[4] As an aside, the a priori allocation of property rights is irrelevant for the achievement of an efficient outcome.
[5] For detailed information on how efficient solutions can be achieved by private bargaining and negotiation please refer to COASE (1960).

The approaches presented above are methods of indirect regulation. Hence, taxes, subsidies, legal rules and marketable permits aim to induce decision makers to consider all – private and social – costs and benefits. However, for many externalities this indirect regulation appears to be an improper approach because they are too important. This is why externalities are often regulated directly by standards and statutory regulations (CASE ET AL., 1999:375).

Beside externalities, another source of market failure lies in public goods, also referred to as collective goods which bestow collective benefits on members of society. Generally, public goods exhibit two characteristics: non-rivalry in consumption and non-excludability.[6] First, a good is non-rival in consumption if its enjoyment by one person does not interfere with the consumption by another person. Thus, the benefits of public goods are collective meaning that they arise to the whole society. Second, non-excludability means that once a good has been produced it is impossible or prohibitively costly to prevent someone from enjoying its benefits[7] (CASE ET AL., 1999:375; PARKIN ET AL., 2005:348; PINDYCK AND RUBINFELD, 2003:902). A more precise and technical definition of public goods is provided by Samuelson (SAMUELSON, 1954 and 1955). According to Samuelson's definition, "a public good is a good that, once produced for some consumers, can be consumed by additional consumers at no additional cost" (HOLCOMBE, 1997:2). This concept refers to the jointness or non-rivalry in consumption explained above. Prior to Samuelson's definition, another understanding of public goods prevailed. Correspondingly, a public good was considered to be a good produced by government that is available for the benefits of the citizens. "Indeed, this more commonsense definition of public good was generally accepted by economists until Samuelson

---

[6] As an aside, some goods may be non-rival in consumption but excludable (PINDYCK AND RUBINFELD, 2003:902). Conversely, goods may generate collective benefits but still be rival in consumption (CASE ET AL., 1999:375). These cases cover a less restrictive definition of public goods.

[7] The characteristic of non-excludability imposes a problem called "free-riding". Because people cannot be excluded from enjoying the benefits of public goods they are likely to consume the good without paying, i.e. they will free ride. The free-rider problem causes market failure resulting in underproduction of the public good (HOLCOME, 1997:6; CASE ET AL., 1999:376).

made the definition more precise, and at the same time altered its meaning (HOLCOME, 1997:3).[8]

Generally, an economy will be efficient if it produces what people want. By the act of buying a private and excludable good at a certain price people reveal what the good is worth to them (CASE ET AL., 1999:377f). However, for public goods that are generally non-excludable and non-rival in consumption, this revealed preference mechanism does not work. For that reason, Samuelson argued that public goods will not be produced efficiently in the private sector. Instead, government or public production is required for an efficient allocation of resources. In Samuelson's view, the inefficiency of private sector production is resulting from the exclusion of potential consumers who attach any positive value to the good[9] (HOLCOMBE, 1997:4ff).

According to Samuelson's theory, there is an optimal provision of public goods. This optimal production of a public good depends on people's preferences and the cost of its production. Samuelson's theory presumes knowledge of people's preferences. More precisely, detailed knowledge on what people are willing to pay for the public good is presumed. In order to arrive at market demand, individual willingness to pay must be added up for each potential level of output of the public good. Hence, for public goods, individual demand curves are added vertically.[10] This is illustrated in Figure 3 for the simplified version of a two-person-economy.[11] The market demand curve reflects what society is willing to pay for the public

---

[8] Hence, public goods should not be confused with goods provided by the government. These goods are often excludable and rival in consumption. For instance, formal education is provided by the government because it creates positive externalities and not because education is a public good (PINDYCK AND RUBINFELD, 2003:903).
[9] Although public production is required from a theoretical point of view for the efficient provision of public goods, there is no anticipated guarantee that public production is more efficient than private production. This is because public sector production is usually tax-financed, and taxes create an excess burden on the economy. This excess burden must be weighted against the inefficiencies resulting from private-sector production (HOLCOMBE, 1997:6f).
[10] In the case of private goods, by contrast, the individual demand curves are added horizontally in order to arrive at market demand. Thus, the quantities that households consume at different prices are added up.
[11] The model of Samuelson represents a partial model that assumes for instance constant prices of other goods. Moreover, the simplified model does not consider the effect of taxes that are usually imposed to finance the provision of public goods.

good. This must be weighted against the marginal cost of producing the public good. The efficient or optimal level of provision for the public good is given at q* where society's total willingness to pay is equal to the marginal cost of production (CASE ET AL., 1999:377ff).

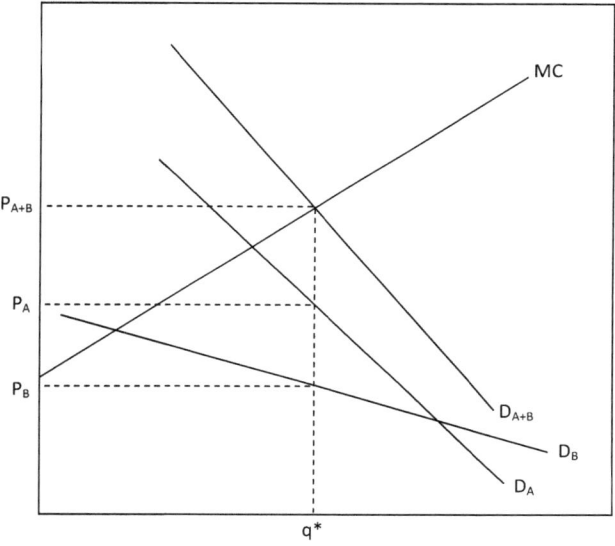

Figure 3: Optimal provision of a public good

Source: CASE ET AL. (1999:380); OWN REPRESENTATION

Yet, providing the optimal amount of a public good requires that governments have knowledge of people's preferences. This may be an improper assumption since public goods are non-excludable, and thus people have no incentive to reveal their preferences (CASE ET AL., 1999:380). For that reasons, political decision makers have to find ways to measure people's preferences.

As we have seen, externalities and public goods are major sources of market failure. "The presence of externalities is a significant phenomenon in modern life" (CASE ET AL., 1999:364). There are many examples for economic activities that create external costs or benefits. For instance, industrial production generates a negative externality through polluting the air. The air traffic is associated with negative external costs because of the noise pollution airplanes create. Or finally, smoking tobacco in a confined space harms the human health of the other people present. Positive externalities arise, for instance, by the inoculation against an infectious disease, the refurbishment of old buildings or formal education (PARKIN ET AL., 2005:330). Examples of public goods include national defence or public service broadcasting (PINDYCK AND RUBINFELD, 2003:902f).

The use of renewable energy sources belongs to the activities that create externalities. On the one hand, electricity production from renewable sources creates external benefits through reduced air emissions or employment creation. On the other hand, the environmental impacts associated with hydropower plants, wind parks or any other renewable power plant represent negative externalities, i.e. external costs to society (SUNDQVIST AND SÖDERHOLM, n.d.:1). Using a less restrictive definition of public goods under which the characteristics of non-excludability and non-rivalry are not necessarily applicable, renewable energy also exhibit public good characteristics. Accordingly, the use of renewable energy sources bestows collective benefits such as cleaner air from which nobody can be excluded.

The economic valuation of the external costs and benefits that are associated with the provision of public goods, as well as the study of externalities are major concerns of environmental economics. The issues have moved to the forefront of policy discussions (CARSON ET AL., 1998:314). Generally, there is an increasing tendency to legitimate political proposals such as energy policy decisions with monetary values. This includes the assignment of monetary values to externalities which is usually fraught with difficulties. Against this background, the development of methods to capture people's preferences gained increasing importance in the past (SCHLÄPFER AND ZWEIFEL, 2008:210; SLOMAN ET AL., 2012:343). This is the point where the concept of economic valuation starts.[12]

---

[12] A detailed description of methods and concepts to evaluate the positive and negative externalities that are associated with the expansion of renewable energy sources is given in chapter 4.

3 The significant role of renewable energy sources

3.1 Renewable energy investments

Investments in renewable energy sources are booming worldwide. A recently published report of the United Nations Environment Programme (UNEP, 2012) revealed that global investments in renewable energy (power and fuels) were € 185 billion in 2011[13], an 11.6 % increase compared to the preceding year. Since 2004 the investment volume has continuously gone up from € 31.8 billion to € 185.0 billion, marking a very dynamic development (see Figure 4).

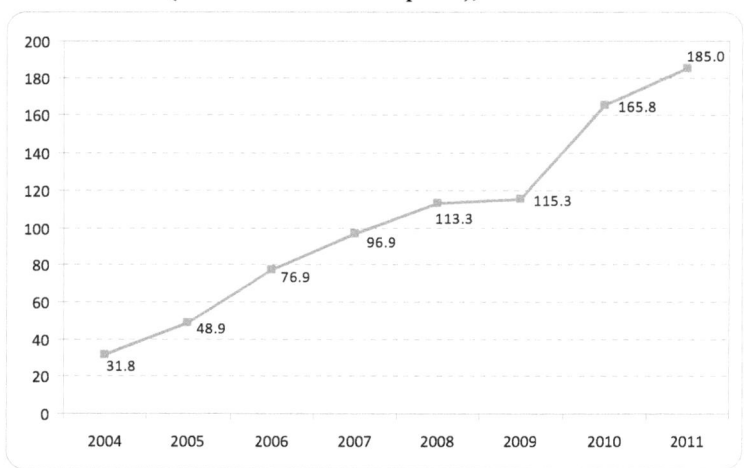

Figure 4: Global investments in renewable energy
(in billion €[14] at current prices), 2004-2011

Source: UNEP (2012:15); OWN CALCULATIONS AND REPRESENTATION

The biggest part of global investment in 2011 was recorded in the solar energy sector with € 105.9 billion; this corresponds to a share of 57.2 % of

---

[13] The main part of this amount (65 %) arises in developed economies, the rest (35 %) in developing countries.
[14] The values shown in Figure 4 have been converted from US-Dollars into Euros using the average yearly exchange rate of the regarding year 2004-2011 (EUROSTAT-DATABASE, 2012c, online). The same applies to Table 1.

the overall investment volume. Compared to the previous year, solar energy investments increased considerably by 44.9 %. Wind power investments account for 32.5 % of total investment, even though the investment volume declined by 16.4 % in 2011. The small hydropower sector was – similarly to the solar sector – also marked by a very dynamic development, though starting from a significantly lower investment level. Accordingly, global investments in small hydropower projects[15] increased by 53.4 % to € 4.2 billion in 2011. With this, small hydropower accounts for 2.3 % of total renewable energy investment. The remaining renewable energy sectors play only a minor role for global investment, especially due to the declining development (see Table 1). Hence, the considerable rise in overall investment was mainly based on increasing trends in the solar energy and hydropower sector.

Table 1: Global investments in renewable energy by sectors, 2011

Sector	Investments in billion €	in %	Growth 2010-11
Wind power	60.2	32.5 %	-16.4 %
Solar power	105.9	57.2 %	44.9 %
Biofuels	4.9	2.6 %	-23.8 %
Biomass & Waste to energy	7.6	4.1 %	-15.9 %
Small hydropower	4.2	2.3 %	53.4 %
Geothermal power	2.1	1.1 %	-10.9 %
Other	0.1	0.1 %	-36.5 %
In total	**185.0**	**100.0 %**	**11.6 %**

Source: UNEP (2012:15); OWN CALCULATIONS AND REPRESENTATION

Looking at worldwide added new generation capacity, the important role of renewable energy sources becomes further clear. In 2011, 44 % of the newly installed power generation capacity was raised by renewable sources.[16] By comparison, in 2004 this percentage was only 10 %. Furthermore, the total amount of power generated by renewables (again excluding large hydropower) was 6 % in 2011, a slight increase compared to the previous year (UNEP, 2012:11; DER SPIEGEL, 2012, online).

---

[15] In the UNEP (2012) report, small hydropower projects are defined as projects with a capacity of less or equal 50 megawatt (MW). Investments in large hydropower with a capacity of more than 50 MW are excluded from the analysis in the report.
[16] As before, this value excludes large hydropower with a power plant capacity of more than 50 MW.

**Figure 5: European investments in renewable energy
(in billion € at current prices), 2004-2011**

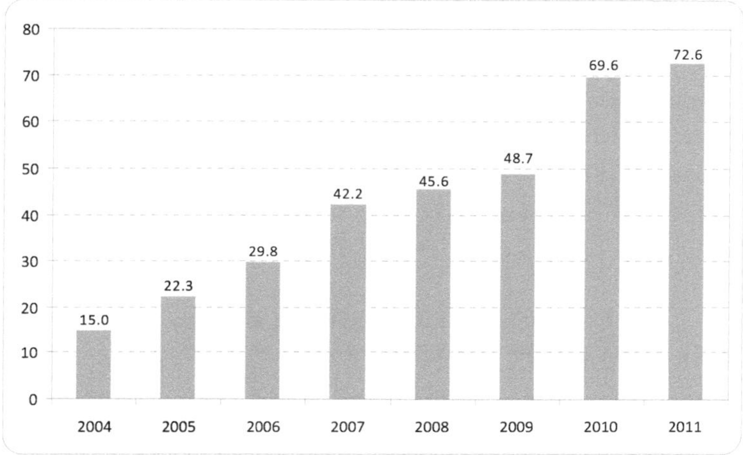

Year	Value
2004	15.0
2005	22.3
2006	29.8
2007	42.2
2008	45.6
2009	48.7
2010	69.6
2011	72.6

Source: UNEP (2012:15); OWN CALCULATIONS AND REPRESENTATION

"The European Union is often regarded as an international frontrunner concerning the development, promotion and implementation of renewable energy policy and technology" (ECORYS, 2010:15). Hence, Europe is the biggest investor in renewable energy with € 72.6 billion in 2011, followed by China (€ 37.5 billion) and the United States (€ 36.5 billion).[17] As can be seen from Figure 5, European investments in renewable energy followed a steady upward trend. Compared to the year 2004, the investment volume increased almost fivefold. A significant jump was recorded in 2010, with investments going up from € 48.7 billion to € 69.6 billion (+42.9 %), presumably due to the economic recovery after the financial and economic crisis in 2008/09. In 2011, renewable energy investments in Europe grew by 4.2 % which is significantly below global growth of 11.6 %.

In Austria, total investments in renewable power plants were € 353.9 million in 2010, a 26.1 % increase compared to 2008. 95.5 % or € 337.9 million of the overall investment volume were invested in hydropower

---

[17] Again, the values have been converted from US-Dollars into Euros using the average yearly exchange rate of the year 2011 (EUROSTAT-DATABASE, 2012c, online).

plants.[18] Wind turbines, photovoltaic installations and other renewable energy sources account for € 16.0 million (4.5 %) of total investment. Since 2008 hydropower investments have risen by 24.2 %. In contrast, wind power and solar energy (photovoltaics) were marked by a more dynamic development. Starting from a significantly lower investment level, the expenditures for wind parks have increased by 73.8 %, for photovoltaic installations by 187.0 % since 2008. Other renewable energy sources remained constant at an investment level of approximately € 2.0 million (see Figure 6).

Figure 6: Investments in renewable power plants in Austria (in million €), 2008 and 2010

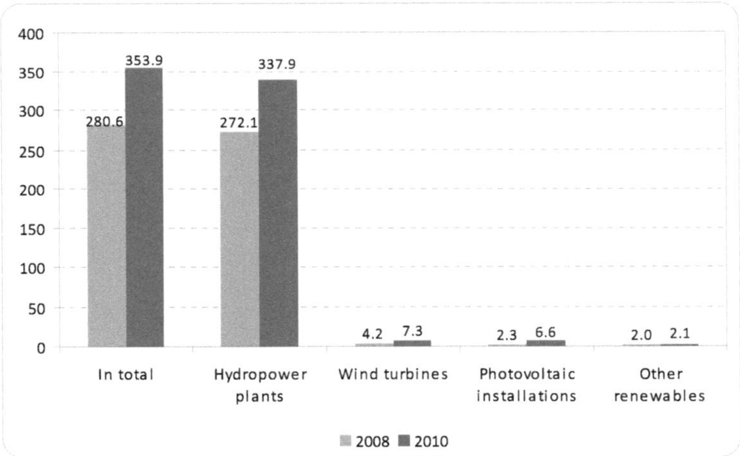

Category	2008	2010
In total	280.6	353.9
Hydropower plants	272.1	337.9
Wind turbines	4.2	7.3
Photovoltaic installations	2.3	6.6
Other renewables	2.0	2.1

Source: OESTERREICHS ENERGIE (2012:5); OWN REPRESENTATION

3.2 Global climate and energy goals

Investments in renewable energy sources are preferable due to security of supply issues as well as environmental (climate change) and dependency concerns (ECORYS, 2010:15; IEA, 2012:1). Hence, governments – especially in industrialised nations – are forced to further promote power generation by environmentally friendly technologies, particularly because of the

---

[18] This value includes investments in run-of-river and storage power plants, both large- and small-scale.

global environmental concerns over carbon dioxide (CO_2), or in the most general sense, greenhouse gas (GHG) emissions. The reduction of GHG emissions is one of the key goals of national and international climate and energy policies. The reduction targets are mainly determined by international agreements, as can be seen from the following remarks.

3.2.1 International agreements

Framework convention on climate change

In May 1992, the United Nations Framework Convention on Climate Change (UNFCCC) was adopted with the aim of limiting global temperature increases and the resulting climate change and to cope with the associated impacts. The treaty finally came into force in 1994 and involves 195 parties until now (UNFCCC, 2012, online). In particular, the aim of the UNFCCC is to stabilise GHG emissions on a level that prevents a harmful anthropogenic disturbance of the climate system. According to the Copenhagen Accord (UNFCCC, 2009, online), the result of the 2009 word climate conference in Copenhagen, a limitation of global temperature increases below 2°C is required in order to achieve the UNFCCC goals (UMWELTBUNDESAMT, 2012a:50). Within the scope of the world climate conference in Cancún (Mexico) in 2010, the stabilisation of GHG emissions so that global temperature rise is limited to 2°C has also been addressed and finally established in the Cancún Agreement. The agreement contains various measures describing ways to achieve this long-term climate goal (BUCHER, 2011:51; UMWELTBUNDESAMT, 2012a:50). At the UNFCCC conference in Durban (South Africa) in 2011 the 2°C target was acknowledged. In order to achieve this target, a reduction of GHG emissions by at least 80 % until 2050 (compared to the year 1990) is required in the industrialised countries (UMWELTBUNDESAMT, 2012b:14). For that reason, the European Commission started to promote long-term policy plans in the fields of energy, climate change and transport. Especially, the European roadmap for moving to a competitive low carbon economy in 2050 (EC, 2011a) was published in 2011 and set out the objective of reducing GHG emissions by 80-95 % compared to 1990 by 2050. The roadmap outlines possible measures for achieving this target (UMWELTBUNDESAMT, 2012b:14; EC, 2011a:3).

The Kyoto Protocol

During the 1990s countries realized that sufficient emission reductions are extremely important in view of the progressing climate change and therefore launched negotiations to strengthen the global response to this issue (UNFCCC, 2012, online). Thus, in December 1997 the Kyoto Protocol was adopted at the climate conference in Kyoto (Japan) and came into force eight years later in 2005. The treaty sets binding targets for 37 industrialized countries as well as the European Union (EU) to reduce the emissions over the five-year period 2008-2012 by at least 5 % compared to the base year 1990. The EU or more precisely, the EU-15 countries are committed to reduce its GHG emissions by 8 %. Within the European burden sharing, Austria is bound to reduce its emissions by 13 % (UNFCCC, 2008:13; UMWELTBUNDESAMT, 2012a:51).

The first Kyoto-period expires by the end of 2012. Yet, no fixed agreement over an international follow-up treaty has been reached. However, with the Cancún Agreement a continued existence of the climate change process of the United Nations is ensured (UMWELTBUNDESAMT, 2012a:51). Furthermore, at the UNFCCC conference in Durban by the end of 2011 the participating parties agreed over a second Kyoto commitment period that will start at the beginning of 2013. Regarding the end of the second commitment period, no general agreement could be reached. Thus, the second period will expire by the end of 2017 or 2020[19] (UMWELTBUNDESAMT, 2012b:62).

The EU climate and energy package
In addition to the ratification of the Kyoto protocol, climate and energy policy is strongly determined by EU initiatives. The year 2007 marked a turning point for the EU's climate and energy policy. "Europe showed itself ready to give global leadership: to tackle climate change, to face up to the challenge of secure, sustainable and competitive energy, and to make the European economy a model for sustainable development in the 21st century" (EC, 2008a:2). Thus, in December 2008 the EU member states reached

---

[19] The end of the second commitment period is planned to be set within the scope of the next UNFCCC conference by the end of 2012.

agreement over an extensive "Climate and Energy Package".[20] The package contains three major targets: the reduction of GHG emissions, the increase of renewable energy use, as well as an increase in energy efficiency (see Figure 7).

Figure 7: The major targets of the EU climate and energy package until 2020

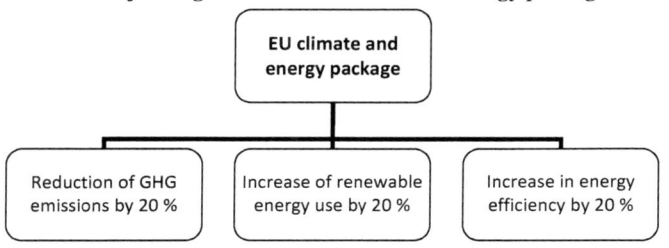

Source: EC (2008a:2ff); OWN REPRESENTATION

First, the EU package includes binding legislation to reduce overall GHG emissions by at least 20 % below 1990 levels until 2020.[21] This effort is divided between the EU Emissions Trading System (EU-ETS)[22] sector and the non-ETS sector. In total, about 40 % of total GHG emissions are covered in the EU-ETS sector; here a 21 % reduction compared to 2005 is required (EC, 2011c:15). The remaining 60 % of GHG emissions arise in areas like agriculture, transport, residential and service buildings, waste and small industrial installations.[23] These sectors fall under the threshold for inclusion in the EU-ETS and are therefore subject to a separate commitment. Hence, for GHG emissions beyond the emission trading system a 10 % reduction compared to 2005 is envisaged by 2020 (EC, 2008a:7;

---

[20] Prior to 2008, the European climate and energy policy was driven by a loose legislative framework, which set just non-binding targets (EC, 2011b:3).
[21] This is equivalent to a 14 % reduction compared to 2005.
[22] The EU-ETS is an important policy instrument which aims to reduce industrial GHG emissions cost-effectively. The system represents the first and biggest international scheme for the trading with GHG emission allowances. Currently, the EU-ETS covers about 11,000 power stations and industrial plants distributed across 30 countries (EC, 2010, online; CAPROS ET AL., 2008:1). The EU-ETS provides an attempt to internalise the external cost of air pollution (see chapter 2).
[23] In these fields of economic activity, the EU member states typically possess latitudes to implement policies and measures aiming at reducing the occurring GHG emissions (CAPROS ET AL., 2008:1).

BÖHRINGER ET AL., 2009:268).[24] For each member state individual targets are specified according to the effort sharing principle. Effort sharing means, that the individual member state reduction targets should be based on the principle of solidarity and the need for sustainable economic growth. Hence, relative GDP (Gross Domestic Product) per capita is taken into account for stipulating the member state goals. Accordingly, member states with a high GDP per capita are bound to reduce their GHG emissions. Conversely, relatively low levels of GDP per capita allow member states to increase their GHG emissions compared to 2005 (EC, 2008b:3). Following this, Austria is committed to achieve a 16 % reduction of its GHG emissions by 2020. The GHG emission limits of all member states are given in Table 2. As can be seen from this table, growing GHG emission limits are mainly allowed for the "new" member states of Eastern and South Eastern Europe.

Table 2: GHG emission reduction targets by 2020 for emissions beyond the EU-ETS

Member state	Target 2020	Member state	Target 2020
Austria	-16 %	Latvia	+17 %
Belgium	-15 %	Lithuania	+15 %
Bulgaria	+20 %	Luxembourg	-20 %
Czech Republic	+9 %	Malta	+5 %
Cyprus	-5 %	Netherlands	-16 %
Denmark	-20 %	Poland	+14 %
Germany	-14 %	Portugal	+1 %
Estonia	+11 %	Romania	+19 %
France	-14 %	Slovenia	+4 %
Finland	-16 %	Slovakia	+13 %
Greece	-4 %	Spain	-10 %
Hungary	+10 %	Sweden	-17 %
Ireland	-20 %	United Kingdom	-16 %
Italy	-13 %	**EU-27**	**-10 %**

Source: EC (2008b:15); OWN REPRESENTATION

The second pillar of the EU package refers to an increased use of renewable energy sources. "Renewable energy is crucial to any move towards a

---

[24] Generally, GHG emission reductions should take place between 2013 and 2020 (EC, 2008b:3). Prior to 2013 reduction targets are determined by the Kyoto commitment.

low carbon economy" (EC, 2011b:2). Due to the fact that 70 % to 80 % of total GHG emissions in developed countries stem from the energy system, i.e. from the combustion of fossil fuels, renewable energy can contribute substantially to the goal of reducing emissions (CAPROS ET AL., 2008:2; EC, 2008a:7). Hence, the climate and energy package contains binding target to increase the share of renewable energy sources in EU gross final energy consumption to 20 % by 2020.[25]

Table 3: Targets for the share of renewable energy sources in gross final consumption by 2020

Member state	Target 2020	Member state	Target 2020
Austria	34 %	Latvia	40 %
Belgium	13 %	Lithuania	23 %
Bulgaria	16 %	Luxembourg	11 %
Czech Republic	13 %	Malta	10 %
Cyprus	13 %	Netherlands	14 %
Denmark	30 %	Poland	15 %
Germany	18 %	Portugal	31 %
Estonia	25 %	Romania	24 %
France	23 %	Slovenia	25 %
Finland	38 %	Slovakia	14 %
Greece	18 %	Spain	20 %
Hungary	13 %	Sweden	49 %
Ireland	16 %	United Kingdom	15 %
Italy	17 %	**EU-27**	**20 %**

Source: EPC (2009:46); OWN REPRESENTATION

Since member states have different possibilities to exploit renewable energy, individual member state targets are stipulated, depending on different national starting points and potentials, the current energy mix and GDP per capita (EC, 2008a:7; E3G, 2009:4). Correspondingly, the mandatory targets of the EU member states range from 10 % to 49 %. As can be seen from

---

[25] Prior to the climate and energy package, the EU recognized the importance of an intensified use of renewable energy sources and issued Directive 2001/77/EC which aims to increase the generation of electricity from renewable energy sources in order to increase security of supply, diversify energy supply and prevent climate change. The basic EU aim is to reach a share of 21 % renewables in the electricity sector. Again national indicative targets for the contribution of electricity produced from renewable energy sources are stipulated. Accordingly, Austria's reference value is to achieve a 78.1 % contribution of renewables in electricity generation by 2010 (EPC, 2001:39).

Table 3, for Austria a share of 34 % is envisaged (EPC, 2009:46).[26] In particular, the renewable energy target includes the promotion of electricity from solar, wind, geothermal, hydro and biomass, as well as policies that support the use of renewable energy sources in the heating and cooling sector (E3G, 2009:4). Furthermore, the EU package stipulates a sub-target referring to an increase of renewable energy sources in the transport sector. More precisely, 10 % of overall fuel consumption should be raised by sustainable biofuels in 2020 (EC, 2008a:8).[27]

Finally, the EU climate and energy package is based on an increase in energy efficiency. The specific aim is to reduce energy consumption by 20 % until 2020 through increased energy efficiency. This is a crucial part of the "puzzle" since energy efficiency measures play a key role for the reduction of climate-damaging GHG emissions (EC, 2008a:8). However, up to now, the 20 % efficiency goal is not legally binding (UMWELTBUNDESAMT, 2012b:13).[28]

3.2.2 European progress towards the climate and energy goals

The 2011 published "Renewable Energy Progress Report" (EC, 2011b) highlights the European progress towards the ambitious 20/20/20 targets described above. In 2010, the share of renewable energy sources (RES) in gross final consumption was 12.5 % in the EU-27 countries. Compared to 2006, the proportion of RES increased by 3.5 percentage points (see Table 4). According to the progress report, the 20 % target will be achieved or even exceeded in 2020, given that all production forecasts are fulfilled (EC, 2011b:4). However, the progress differs significantly between countries. About half of the member states like Austria, Germany, Spain or France are on a good way to reach their national renewable energy targets; they will even generate more energy from renewable sources than required by the EU energy package. In contrast, other member states as for instance Italy or

---

[26] The promotion of the use of energy from renewable sources is regulated within Directive 2009/28/EC, also known as "Renewable Energy Directive" (see EPC, 2009).
[27] This target is regulated within Directive 2003/30/EC (EPC, 2003).
[28] By the end of 2012, the European Union issued a Directive on energy efficiency (see EPC, 2012).

Luxembourg are expected to lag behind their renewable energy targets (ENDS, 2010:4; EC, 2011b:4).

Table 4: Progress in the field of renewable energy in the EU-27 countries, 2006-2010

Year	Share of RES in gross final consumption	Share of RES in the transport sector
2006	9.0 %	1.9 %
2007	9.9 %	2.7 %
2008	10.5 %	3.5 %
2009	11.7 %	4.2 %
2010	12.5 %	4.7 %
Target 2020	**20.0 %**	**10.0 %**

Source: EUROSTAT-DATABASE (2012a and 2012b, online); OWN REPRESENTATION

The European share of RES in the transport sector increased from 1.9 % in 2006 to 4.7 % in 2010 (see Table 4). Until 2020, the predominant renewable energy source in transport will be first-generation biofuels.[29] By contrast, second-generation biofuels[30] and electric vehicles are expected to make up only a small part of RES in the transport sector by 2020. In order to achieve the 10 % target, continued investment in research for advanced renewable energy technologies and a reduction in costs for electric cars and second-generation biofuels are required (EC, 2011b:6f).

Regarding the past development of overall GHG emissions, Figure 8 shows a downward trend over the last 20 years. In 2010, total GHG emissions[31] amounted to 3,798 million tonnes (Mt) CO_2 equivalents in the EU-15. Compared to 1990, GHG emissions decreased by 10.6 %, even though emissions recently increased by 2.1 % in 2010. As can be further seen from

---

[29] The so-called first-generation biofuels are fuels that have been derived from cereal crops (e.g. wheat, maize), oil crops (e.g. rape oil) or sugar crops (e.g. sugar cane). The most popular types of first-generation biofuels are biodiesel, bioethanol or biogas (EUROPEAN BIOFUELS TECHNOLOGY PLATFORM, 2012, online).

[30] Second-generation biofuels are produced from cellulosic materials. In contrast to first-generation biofuels, the raw materials used to derive second-generation biofuels are considered more sustainable und do not compete directly with food. Second-generation biofuels include for instance hydrocarbons, biohydrogen or bioelectricity (EUROPEAN BIOFUELS TECHNOLOGY PLATFORM, 2012, online).

[31] In the following, total GHG emissions exclude emissions and removals from Land Use, Land Use Change and Forestry (LULUCF).

Figure 8, total GHG emissions have been below the EU-15 Kyoto target since 2009 (EEA, 2012:7). Hence, the EU-15 is well on track to achieve or even exceed its Kyoto commitment (EC, 2011c:3).

Figure 8: Total GHG emissions in the EU-15 (in Mt CO$_2$ equivalents), 1990-2010

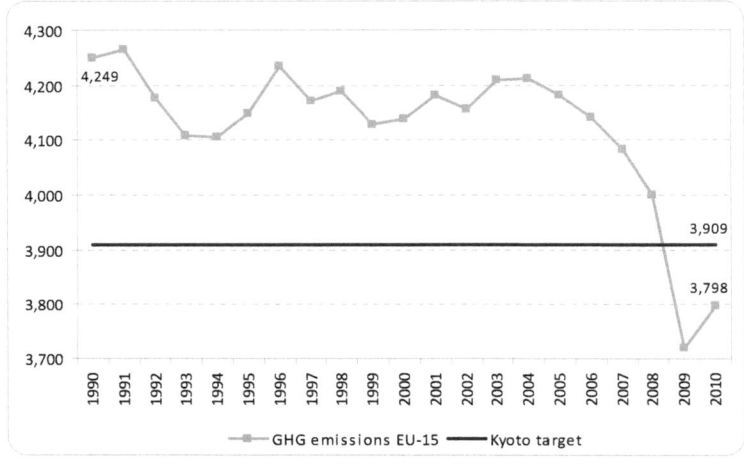

Source: EUROSTAT-DATABASE (2012d, online);
OWN CALCULATIONS AND REPRESENTATION

For the EU-27 the Kyoto objective is not applicable since the common commitment under the Kyoto Protocol applies only to the EU-15 countries. However, it is possible to examine the European progress towards the 2020 emission reduction goal. Generally, the development of GHG emissions in the EU-27 is very similar to the emission trend in the EU-15. Between 1990 and 2010 overall GHG emissions fell from 5,583 Mt to 4,721 Mt CO$_2$ equivalents. This corresponds to a decrease of 15.4 %. Despite this downward trend, emissions increased by 2.4 % between 2009 and 2010. This was partly due to the economic recovery from the 2009 recession in many European countries (EEA, 2012:9).[32] The EU objective is to reduce overall GHG emissions by 20 % compared to 1990 from 2013 to 2020. As can be seen from Figure 9, emission levels approach to the EU's 2020 target but are still above the target value. According to the progress report on GHG

---

[32] In contrast, the substantial decrease of GHG emissions between 2008 and 2009, both in the EU-15 (-7.0 %) as well as the EU-27 countries (-7.3 %), was mainly caused by the financial and economic crisis itself.

emissions, the existing and planned measures in the field of climate mitigation are not yet sufficient to reach the 2020 target, although there is a positive trend towards an achievement of the Kyoto Protocol commitment. Hence, more effort and additional policies are necessary (EC, 2011c:4ff).

Figure 9: Total GHG emissions in the EU-27 (in Mt CO_2 equivalents), 1990-2010

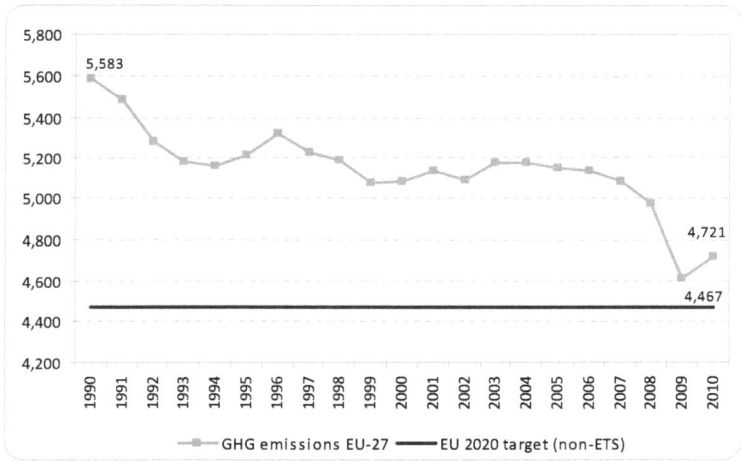

Source: EUROSTAT-DATABASE (2012d, online);
OWN CALCULATIONS AND REPRESENTATION

To conclude, in order to reach the 20/20/20 targets a huge mobilisation of renewable energy investments is required in the coming decade (DE JAGER ET AL., 2011:9). Generally, the member states can achieve their targets by adopting two major measures. First, the adjustment of the overall energy mix by increasing the use of renewable energy sources in the electricity, the heating and cooling sector and second, the reduction of total energy use as a result of increasing energy efficiency (ENDS, 2010:4).

3.2.3 Austrian progress towards the stipulated climate and energy goals

Although the overall European development towards the stipulated climate and renewable energy objectives looks at least partly promising, past progress and future prospects differ considerably between the individual member states. This raises the question of whether Austria will be able to reach its renewable energy and GHG emission reduction targets.

First, Austrian progress in the field of renewable energy is analysed. In 2010, the share of energy from renewable sources in gross final consumption was 30.1 %. Between 2006 and 2010 this share increased by 3.5 percentage points. Within the EU, Austria belongs to the countries that are planning to exceed their own targets and be able to provide surpluses for other member states (EC, 2011b:4). Hence, until 2020 the share of RES in gross final consumption is projected to achieve a level of 34.2 % which is slightly above the target value of 34.0 % (see Figure 10).

Figure 10: Austrian share of renewable energy sources in gross final consumption, 2006-2020

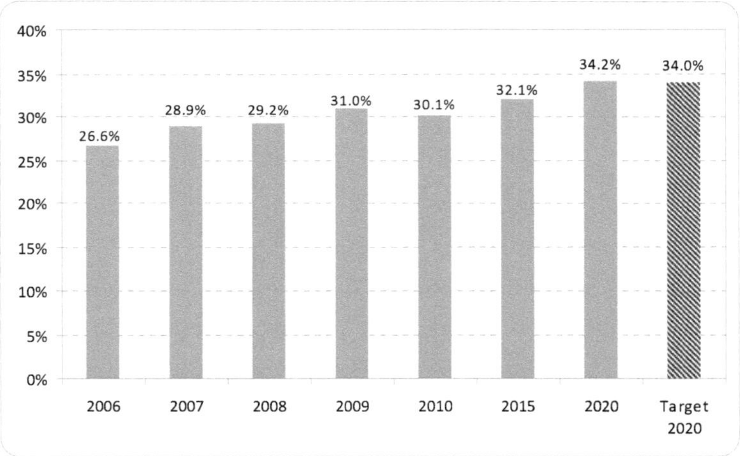

Year	Share
2006	26.6%
2007	28.9%
2008	29.2%
2009	31.0%
2010	30.1%
2015	32.1%
2020	34.2%
Target 2020	34.0%

Source: EUROSTAT-DATABASE (2012a, online); BMWFJ (2010a:9); OWN REPRESENTATION

Figure 11: Austrian share of renewable energy sources in the transport sector, 2005-2020

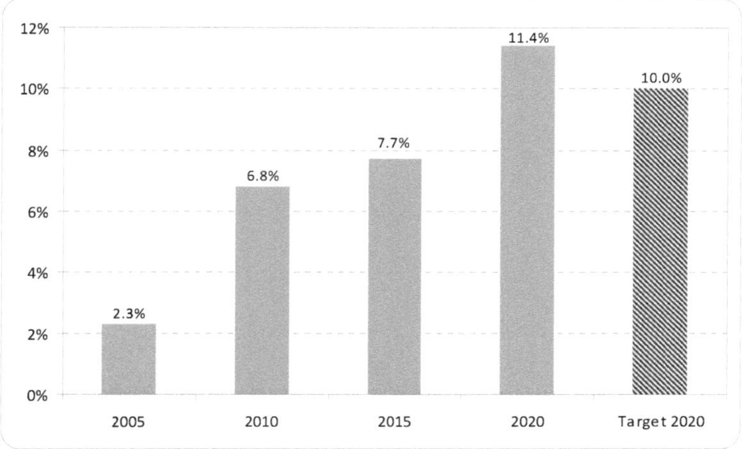

Year	Share
2005	2.3%
2010	6.8%
2015	7.7%
2020	11.4%
Target 2020	10.0%

Source: BMWFJ (2010a:9); OWN REPRESENTATION

A similarly positive performance shows up in the transport sector. In 2010, the proportion of RES (including electricity) in the transport sector was 6.8 %. Since 2005 the share has tripled. The future prospects appear to be very auspicious. In 2015, the share of RES in transport is predicted to be 7.7 %; until 2020 the share will even rise to 11.4 % exceeding the renewable energy sub-target of 10.0 % in transport (see Figure 11).

The development of Austrian GHG emissions shows a somewhat different picture. As can be seen from Figure 12, total GHG emissions amounted to 84.6 Mt CO_2 equivalents in 2010. Emissions in 2010 were therefore 15.8 Mt above the annual mean value of the Kyoto target stipulated for 2008-2012, which are 68.8 Mt CO_2 equivalents. Compared to the emission level of 1990 (the base year of the Kyoto commitment), emissions increased by 8.2 %. The main reason for this rise was the growing fossil fuel use and the associated rise of CO_2 emissions. Since 2005, a declining trend in Austrian GHG emissions has been observed. Between 2008 and 2009, emissions declined substantially by 8.4 %. This decrease is attributable to the damped economic activity in consequence of the financial and economic crisis. In 2010, the economy recovered and led in turn to a 6.1 % increase of GHG emissions.[33] In particular, growing emissions in 2010 were caused by an increased transport of goods, boosted electricity consumption, the growing industrial production of energy-intensive products like steel, as well as the cold weather conditions (UMWELTBUNDESAMT, 2012a:52).

Figure 12: GHG emissions in Austria (in Mt CO_2 equivalents), 1990-2010

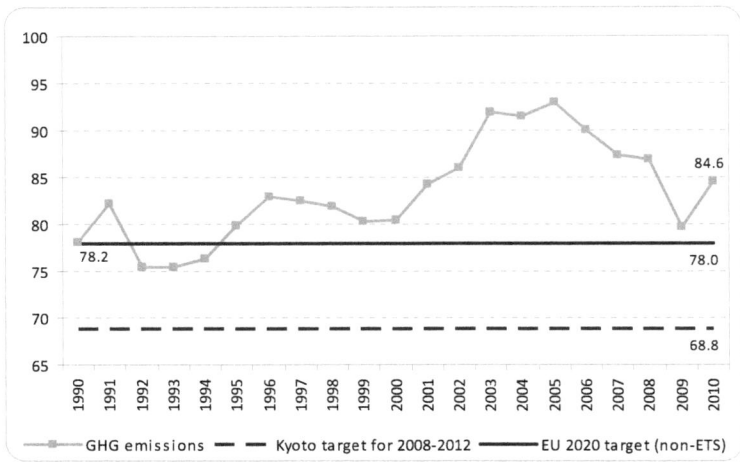

Source: UMWELTBUNDESAMT (2012a:106); OWN CALCULATIONS AND REPRESENTATION

With this development Austria is – beside Italy and Luxembourg – among the European countries that are likely to face difficulties with achieving their Kyoto target (EC, 2011c:4). Austrian compliance with the Kyoto commitment can only be achieved by purchasing further emission reduction units, since sufficient domestic emission reductions are rather unrealistic within the remaining months of the commitment period (UMWELTBUNDESAMT, 2012b:20). Regarding achievement of the EU 2020 target, which is to reduce GHG emissions in the non-ETS sector by 16 % until 2020 compared to 2005, Figure 12 shows a general convergence to the target value. However, projections based on the existing measures in the field of climate and energy policy indicate that Austria is likely to fail the target of -16 %. More precisely, the future scenario predicts a decrease of GHG emissions by merely 9.5 %. By contrast, in a scenario with additional poli-

---
[33] Despite this increase, GHG emissions in 2010 remained below the level of 2008 so that the declining trend has continued since 2005.

cy measures the required emission reduction of at least 16 % can be achieved (UMWELTBUNDESAMT, 2012b:60).

The major part of overall GHG emissions in Austria arises in the industry sector (31.1 %). Additionally, about 26.6 % of GHG emissions are caused by transport, 17.5 % by energy supply and 13.5 % by small-scale consumption. The agricultural sector accounts for 8.8 % of total GHG emissions. Other sectors are responsible for 2.5 % of overall emissions (see Figure 13).

Figure 13: GHG emissions in Austria by sectors (in %), 2010

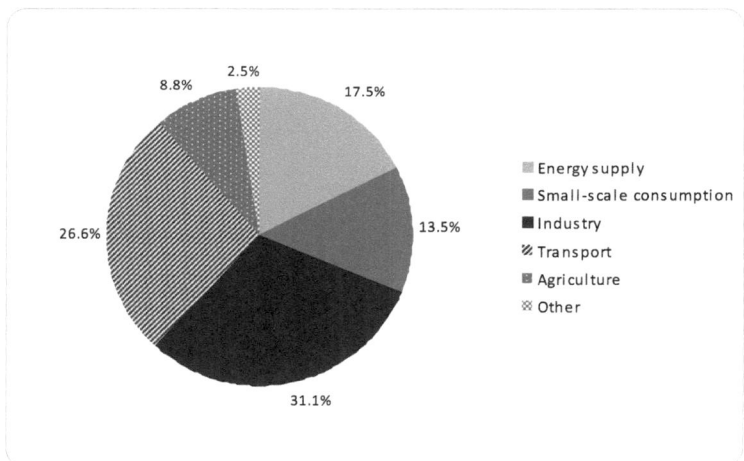

Source: UMWELTBUNDESAMT (2012a:106); OWN CALCULATIONS AND REPRESENTATION

Compared to the year 1990, the transport sector recorded the highest emission growth rate (+59.9 %), followed by industry (+14.9 %). In the energy supply sector emissions have increased by 4.6 % since 1990. On the contrary, in the small-scale consumption sector (-20.8 %), agriculture (-13.0 %) and other sectors (-48.0 %) considerable emission reductions have been achieved (see Figure 14).

The substantial rise of GHG emissions in the industry sector was mainly due to production increases in the iron and steel industry, the chemical industry and other manufacturing branches. Furthermore, the increasing use of low-carbon fuels like gas and renewable energy sources, as well as improvements in energy efficiency led to a partial decoupling of output and emissions. The sharp decrease of GHG emissions in the industry sector between 2008 and 2009 is attributable to a drop of the production of energy-intensive products goods like iron or steel as a result of the economic crisis. Conversely, in 2010 the economic recovery caused a significant increase of GHG emissions (UMWELTBUNDESAMT, 2012a:55).

**Figure 14: GHG emissions in Austria by sectors
(in Mt CO_2 equivalents), 1990-2010**

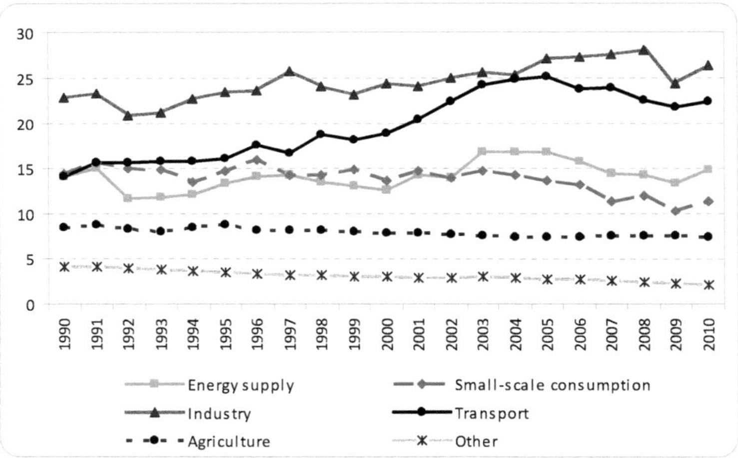

Source: UMWELTBUNDESAMT (2012a:106); OWN REPRESENTATION

The sharp increase of GHG emissions in the transport sector compared to 1990 was mainly due to the rising traffic volume on Austrian streets and fuel exports. Emission reductions in the energy supply sector have been caused by the intensified use of gas, biomass and renewable energy sources (especially hydropower), as well as raising efficiency. Emissions in the small-scale consumption sector strongly depend on weather conditions and the associated heating demand. A general trend towards renewable sources, the intensified use of district heating and the improved thermal condition of buildings due to refurbishment measures have induced a general reduction of emissions in this sector (UMWELTBUNDESAMT, 2012a:54f).

3.3 Austrian climate and energy policy

In order to achieve the goals of international agreements like the Kyoto Protocol or the EU climate and energy package (see section 3.2) a series of strategies has been launched in Austria (see Figure 15).

Figure 15: The main initiatives of the Austrian climate and energy policy

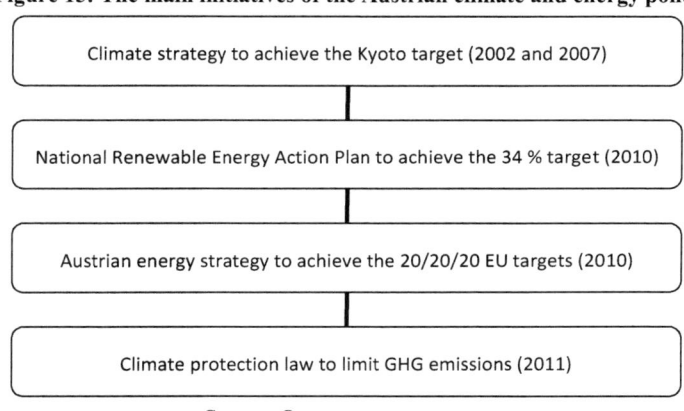

Source: OWN REPRESENTATION

First, in 2002 the Federal Government adopted an Austrian climate strategy to achieve the Kyoto target (UMWELTBUNDESAMT, 2012a:51).[34] However, the 2005 published progress report on the climate strategy showed that Austria was lagging behind its targets, although various climate protection measures have been implemented. For that reason, the climate strategy has been revised from 2005 to 2007 setting a new strategic focus (BMLFUW, 2007a). Accordingly, the adjusted climate strategy relies on the intensified use of available technologies in the fields of energy efficiency and renewables, as well as the promotion of new technologies. The major aims of the revised climate strategy are to boost efficiency increases in energy consumption and to promote the further use of renewable energy sources through an improvement of the green electricity funding.[35] Furthermore, the adapted climate strategy includes measures like an increase of renewable energies in electricity generation and the further promotion of biofuels.

---

[34] For more information see
http://www.lebensministerium.at/umwelt/klimaschutz/klimapolitik_national/klimastrategie.html.
[35] More on green electricity funding in Austria can be found in section 3.5 of this work.

Finally, the strategy relates to organisational issues like the creation of fair conditions for emission trading or the strengthening of the collaboration of local authorities (BMLFUW, 2007a:7ff).

In order to achieve the national targets on renewable energy, each EU member state was obliged to submit its renewable energy action plan by June 2010. The National Renewable Energy Action Plan (BMWFJ, 2010a) shows that Austria will be able to achieve its national target of 34 % renewables in gross final consumption. By adopting additional measures, the target value can even be exceeded. The policies and measures to promote the use of energy from renewable sources divide into regulatory and financial measures, as well as strategies and information campaigns. The major measures of the National Renewable Energy Action Plan (NREAP) include the introduction of a CO_2-tax in the non-ETS sector, the funding of investments in the field of solar thermal energy and biomass, a new law on green electricity, an extension plan for biogas and finally, the promotion of thermal building refurbishment. Additionally, the Austrian renewable energy action plan includes measures in the field of public relations, public awareness and enlightenment (ÖSTERREICHISCHER BIOMASSE-VERBAND, 2010:1).

Figure 16: The three pillars of the Austrian energy strategy

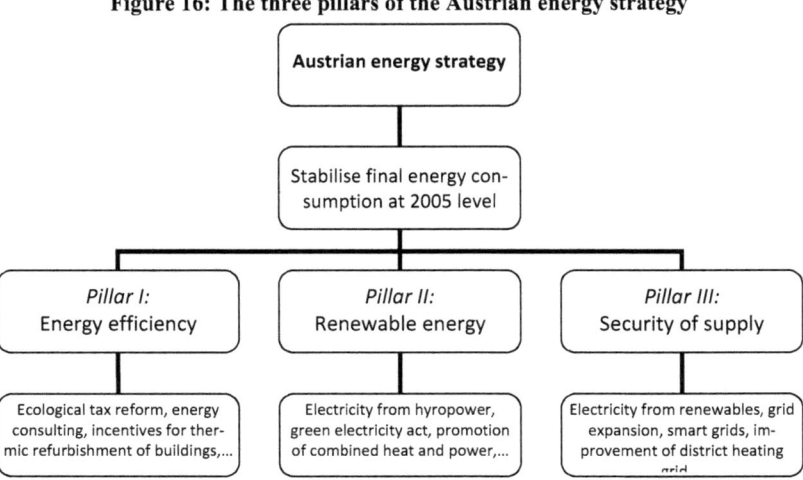

Source: BMLFUW (2010:7f and 123f); OWN REPRESENTATION

One of the overall strategies launched within the scope of the NREAP was the Austrian energy strategy. The energy strategy was presented in 2010 by the Federal Ministry of Agriculture, Forestry, Environment and Water Management (BMLFUW) and the Federal Ministry of Economy, Family and Youth (BMWFJ) and represents the groundwork for achieving the 20/20/20 EU targets (BMLFUW, 2010a). Within the scope of a comprehensive stakeholder process, 180 experts developed nearly 400 recommendations for measures which were incorporated and clustered in the final energy strategy (BMWFJ, 2010b:1f). The overall goal of the energy strategy is to stabilise final energy consumption at the level of 2005 which is 1,100 Petajoule (PJ). Hence, the strategy aims to combat the trend of steadily increasing energy consumption[36] in order to meet the EU objectives presented above (BMLFUW, 2010a:9; BMWFJ, 2010b:3). In addition to that, the Austrian energy strategy is based on three pillars (see Figure 16). First, the strategy aims to increase energy efficiency in the main sectors energy consumption of households and firms, mobility, buildings and the use of primary energy. The second pillar of the Austrian energy strategy refers to renewable energy. Especially, the aim is to increase the use of renewable energy sources in electricity generation (utilisation and exploitation of the potentials in the field of hydropower, wind power, biomass and photovoltaics), the heating sector (promotion of district heating and heating on the basis of solar thermal energy and biomass) and the transport sector (substitution of conventional fuels by biofuels and electric mobility). The long-term security of energy supply represents the third pillar of the Austrian energy strategy and aims to increase security of supply cost-efficiently through district heating and cooling, new transmission networks or the diversification of supply sources and routes (BMLFUW, 2010a:7f; BMWFJ, 2010:3). Exemplary measures belonging to the three pillars of the energy strategy are given in Figure 16.[37] The implementation of the proposed measures will contribute to an achievement of the 34 % renewable target by 2020. According to the

---

[36] According to the Monitoring Report of Energie-Control Austria, final energy consumption is predicted to rise on average by 1.24 % (or 761 GWh) per year until 2020 (ENERGIE-CONTROL AUSTRIA, 2011a:5). This development is opposed to the EU target of reducing energy demand through efficiency increases.

[37] As an aside, the stated measures may belong to more than one pillar of the energy strategy. For instance, an ecological tax reform may contribute to an increase of the share of renewables, an increase of energy efficiency and a reduction of GHG emissions.

Austrian energy strategy, the share of renewable energies in gross final consumption is predicted to be 35.5 % in 2020 (BMLFUW, 2010a:11; BLIEM ET AL., 2011:34ff).

One of the proposed measures of the Austrian energy strategy was the implementation of a climate protection law (BMLFUW, 2010a:123) with the aim of reducing GHG emissions, increasing energy efficiency and raising the share of renewables in final consumption. Hence, the Austrian climate protection law was passed in November 2011 (BGBL, 2011a). The climate protection law set maximum quantities for GHG emissions for the Kyoto commitment period 2008-2012 in the heat sector, electricity generation, waste management, transport, manufacturing industry, agriculture, fluorinated gases and other emissions. Another aim of the climate protection law is the coordinated implementation of effective measures for climate protection, especially in the fields of energy efficiency, renewable energies, mobility management and waste avoidance. The law further stipulates that the adopted measures must be associated with a measurable and verifiable reduction of GHG emissions (UMWELTBUNDESAMT, 2012b:57f).

3.4 The structure of the Austrian electricity sector

As shown before, renewable energy sources play a key role for the reduction of GHG emissions, the increase of the share of renewables in final consumption and the increase of energy efficiency. The electricity sector is one of the key elements for moving towards a low carbon economy (EC, 2011a:7). The major part (70-80 %) of total GHG emissions in developed countries stems from fossil fuel combustion such as coal, oil or gas (CAPROS ET AL., 2008:1). Hence, the increase of renewable energy sources in the electricity sector may contribute significantly to the reduction of climate-damaging GHG emissions. Thus, the following section aims to elucidate the role of renewable energy sources in the Austrian electricity sector.

Figure 17: Gross domestic electricity production (in %), 2011

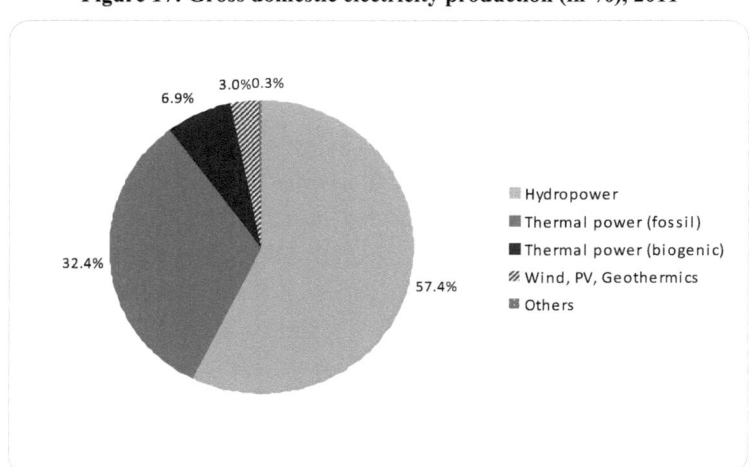

Source: ENERGIE-CONTROL AUSTRIA (2012a, online);
OWN CALCULATIONS AND REPRESENTATION

Due to the natural and environmental conditions, the Austrian electricity generation is largely based on hydropower. As can be seen from Figure 17, currently (2011) more than half (57.4 %) of total electricity produced in Austria comes from hydroelectric power stations; this corresponds to an amount of annually 37,701 gigawatt hours (GWh). Another 21,277 GWh or 32.4 % are produced in thermal power plants fueled by fossil energy sources like coal, gas or oil. Biogenic fuel combustion in thermal power plants accounts for 6.9 % (or 4,555 GWh) of total gross domestic electricity production.[38] The share of wind power, photovoltaics (PV) and geothermal energy amounts to 3.0 % which are 1,985 GWh in total. The major part of this amount (1,934 GWh) is raised by wind power; photovoltaics and geothermal energy, by contrast, play only a minor role.[39] To sum up, the electricity sector in Austria is largely based on renewable energy sources which account for 67.4 % of gross domestic generation, including hydropower, biomass, wind power, photovoltaics and geothermal energy. With this, Austria is among the European countries with the highest proportions

---

[38] As an aside, about 32.1 % of the total electricity production in thermal power plants is generated in combined heat and power facilities.
[39] The remaining 0.3 % or 170 GWh belong to the category "other sources" which are not classifiable.

of renewable energy sources in electricity generation. Beside Austria, only Norway (90.0 %) and Switzerland (54.8 %) exhibit a renewable energy share in the electricity sector of more than 50 % (EUROSTAT-DATABASE, 2012e, online; GRUBER, 2011:45).

Until the late 1990s, electricity generation from hydropower plants steadily increased. By contrast, during the last ten years annual hydropower production levelled out at approximately 40,000 GWh, depending on weather conditions and the associated amount of water. Since 2009, however, electricity generation from hydropower has continuously declined (see Figure 18). Electricity generation from thermal power plants (including biogenic fuel firing) is marked by a continuous upward trend and has nearly tripled since 1970, although starting from a significantly lower level compared to hydropower. Beside that, electric power generation from wind, photovoltaics and geothermal energy increased substantially from 67 GWh in 2000 to 1,985 GWh in 2011. This dynamic development is mainly due to the system of green electricity funding in Austria (for details see section 3.5).

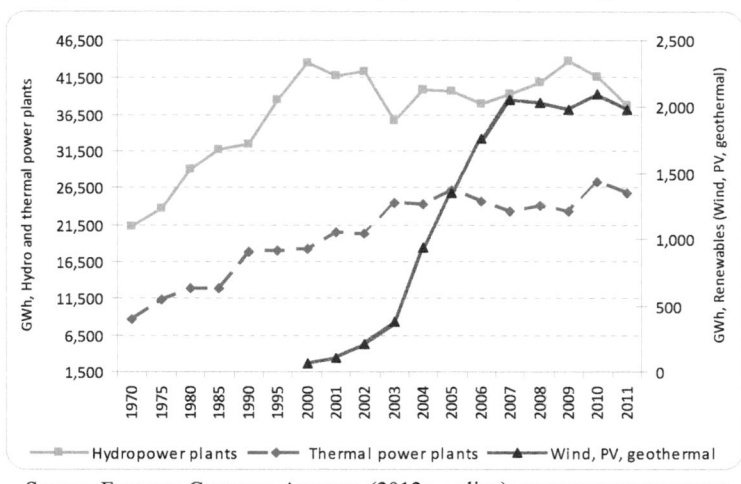

Figure 18: Gross domestic electricity production (in GWh), 1970-2011

Source: ENERGIE-CONTROL AUSTRIA (2012a, online); OWN REPRESENTATION

The share of hydropower in relation to total domestic electricity generation has decreased over time from more than 70 % in 1970 to fewer than 60 %.

Due to the decelerated expansion of hydropower and the simultaneous increase of electricity consumption in the last years, other forms of electricity generation gained increasing importance (BMLFUW, 2007a:33). Hence, the share of thermal power plant production (including biogenic fuel firing) increased considerably from 29.3 % in 1970 to 39.3 % in 2011. The share is, however, still below the proportion of hydropower (see Table 5).

Table 5: Shares of hydropower and thermal power plant production (in %), 1970 and 2011

Year	Hydropower plants	Thermal power plants
1970	70.7 %	29.3 %
2011	57.4 %	39.3 %

Source: ENERGIE-CONTROL AUSTRIA (2012a, online); OWN CALCULATIONS AND REPRESENTATION

The Austrian electricity generation system consists of various types of power plants, as shown in Table 6. The total number of hydropower plants in Austria is 2,671 with an entire installed capacity of 13,200 megawatt (MW). In addition, there exist 583 thermal power plants with a total capacity of 8,249 MW. This number includes thermal power plants that are fired with biogenic fuels. Finally, there is a huge number of renewable energy plants (10,573) accounting for a capacity of only 1,179 MW.

Table 6: Power plants and installed capacity (in MW) in Austria, 2011

Type of power plant	Number of plants	Installed capacity in MW
Hydropower plants	2,671	13,200
Thermal power plants	583	8,249
Renewable energy plants	10,573	1,179
In total	**13,827**	**22,628**

Source: ENERGIE-CONTROL AUSTRIA (2012b, online); OWN REPRESENTATION

Figure 19: Hydropower plants in Austria by technology, 2011

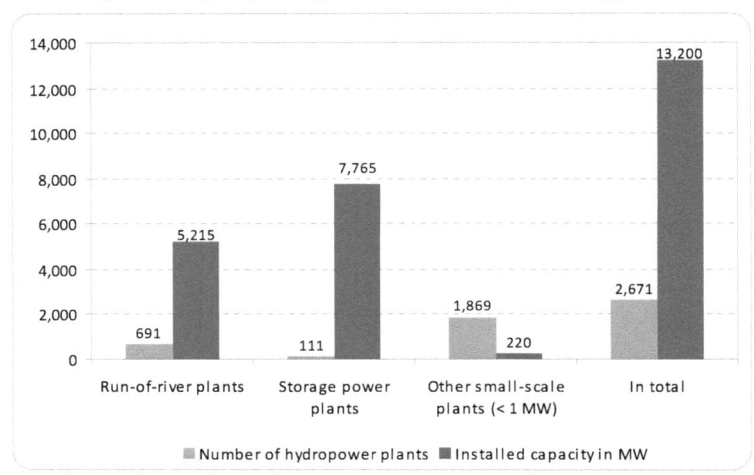

Source: ENERGIE-CONTROL AUSTRIA (2012b, online); OWN REPRESENTATION

Hydropower is usually classified by size (generating capacity) and the type of the scheme (IRENA, 2012:11). Hence, hydro plants divide into small-scale and large-scale plants, and according to technology, into run-of-river and storage power plants.[40] Following this, 691 out of the total number of hydropower plants are run-of-river and 111 storage power plants (including pumped storage technologies). While the existing storage power plants exhibit a total capacity of 7,765 MW, the installed power plant capacity of the run-of-river plants is significantly lower and amounts to 5,215 MW. Most run-of-river plants operate as base load, while storage power plants generally run in the case of high demand, i.e. provide peak load power (HOR-

---

[40] Run-of-river plants transform the power of the water flowing in a river into electric energy, delivering mainly base-load power. The amount of producible electricity depends on the penstock and the natural water flow. Run-of-river schemes have little or no storage possibilities. The most important run-of-river plants are located in the eastern part of Austria along the major rivers like Danube, Inn, Enns, Mur and Drave. Storage schemes, by contrast, have water reservoirs to store the water. Hence, the generation of electricity can be decoupled from the timing of rainfall or glacial and snow melt. The capacity of a storage power plant depends on the altitude difference between the water reservoir and the power house. In case of high power demand the plants can immediately start running, i.e. electricity can exactly be generated when it is required. Hence, storage power plants are usually running at peak times (peak load). A special form of storage plants are pumped storage hydropower technologies. Here, off-peak electricity is used to pump water from a river or lower reservoir up to a higher reservoir to allow its release during peak times. In Austria the large storage power plants are exclusively located in the alpine areas of western Austria (KLEINWASSERKRAFT ÖSTERREICH, 2012a, online; IRENA, 2012:9ff).

LACHER, 2007:95). Lastly, there exists a large number of very small-scale (so-called "mini") hydropower plants (1,869) with a capacity lower than one MW, accounting for only 220 MW of the total installed capacity[41] (see Figure 19).

Figure 20: Hydropower plants in Austria by size, 2011

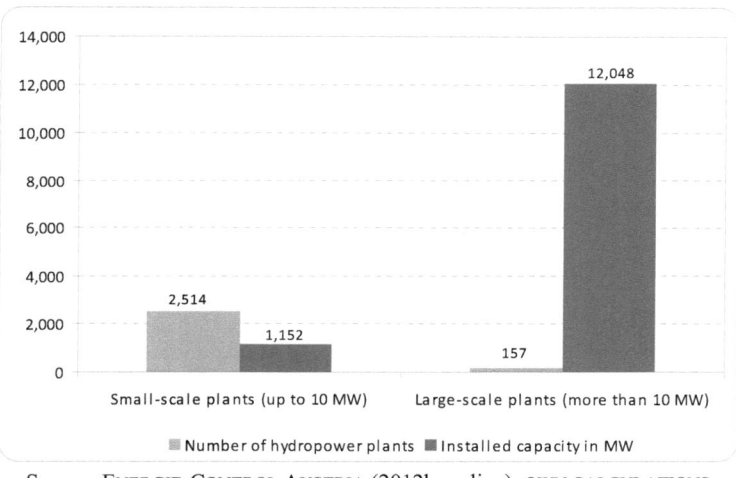

Source: ENERGIE-CONTROL AUSTRIA (2012b, online); OWN CALCULATIONS AND REPRESENTATION

Regarding the number of facilities, the Austrian hydropower sector is largely based on small-scale hydropower.[42] 94.1 % (2,514) of the total number of hydropower stations are small-scaled with a capacity of less than or equal to 10 MW. These facilities have a total capacity of 1,152 MW. In contrast, there exist only 157 large-scale hydropower plants (> 10 MW) with a total installed capacity of 12,048 MW which is way above the ca-

---

[41] For these small-scale hydropower plants an attribution to a power plant technology (run-of-river or storage power plant) is not possible.
[42] There is no global agreement upon the classification of "small" and "large" hydropower. What constitutes "small" varies from country to country. In Austria, small-scale hydropower facilities are defined as plants with an installed capacity of up to 10 MW. Hydropower plants with a capacity of more than 10 MW, by contrast, can be classified as large-scale. The definition of what is subsumed under small hydropower systems is especially important because it can determine which schemes are covered by support policies for small-scale hydropower and which are covered by those for large hydro (IRENA, 2012:11; KLEINWASSERKRAFT ÖSTERREICH, 2012b, online).

pacity of small-scale facilities (see Figure 20).[43] Most of the small-scale hydropower facilities, namely those with a capacity of less than 1 MW, cannot be assigned to a specific technology. Usually, small hydropower plants are more likely to be run-of-river facilities than large hydropower schemes (IRENA, 2012:12). Correspondingly, for hydropower plants with a capacity between 1 and 10 MW, there is a strong tendency towards run-of-river technologies. Thus, 93.2 % of these hydropower stations are run-of-river plants. Large-scale hydropower plants are equally allocated among run-of-river and storage technologies.[44]

As far as the producible electricity is concerned, small-scale hydropower stations generate only 4,681 GWh per year; these are 12.4 % of total hydropower generation. The remaining 33,020 GWh or 87.6 % are generated by the few large-scale facilities. This is shown in Table 7. Additionally, the table highlights the important role of run-of-river plants in gross electric power generation. The main part of hydropower generated electricity in Austria – 25,276 GWh or 67.0 % – is produced by run-of-river schemes, while only 12,425 GWh (33.0 %) are generated by storage power plants (see Table 7).

Table 7: Electricity generation by size and technology (in GWh), 2011

Power plant size	in GWh	in %
Up to 10 MW	4,681	12.4 %
More than 10 MW	33,020	87.6 %
Power plant technology	*in GWh*	*in %*
Run-of-river plants	25,276	67.0 %
Storage power plants	12,425	33.0 %

Source: ENERGIE-CONTROL AUSTRIA (2012b, online);
OWN CALCULATIONS AND REPRESENTATION

---

[43] Compared to the previous year, the total number of hydropower plants increased by 73. Hence, in 2011 73 new hydropower stations have been put into operation. The major part of these new hydropower stations (54) are small-scale facilities with a capacity of less than 1 MW. In addition, 19 small-scale run-of-river power plants (capacity < 10 MW) have been taken into operation in the last year. Finally, no large-scale run-of-river schemes (capacity > 10 MW) or storage power plants have been completed in the past year (ENERGIE-CONTROL AUSTRIA, 2011c, online).

[44] More precisely, 42.7 % of the large-scale hydropower facilities are attributable to storage, the remaining 57.3 % to run-of-river plants.

As mentioned above, the Austrian electricity sector is, beside hydropower, also based on thermal power and renewables. The total number of fossil fuel fired power plants is 96 with an installed capacity of 7,158 MW. In comparison, the total capacity of the existing 107 power plants fired with biogenic fuels, is only 507 MW. In addition, there are 10 mixed fired plants (497 MW capacity) and 370 other thermal power plants with a capacity of 87 MW that cannot be classified with respect to fuel type. The number of wind power plants, PV and geothermal facilities amounts to 198. The installed capacity of these plants is 1,170 MW. Finally, there exist – similar to the hydropower sector – a huge number of 10,375 small-scale renewable facilities which cannot be classified by technology. With regard to installed capacity (72 MW), these facilities play only a minor role (see Figure 21).

Figure 21: Thermal power plants and renewable energy facilities in Austria, 2011

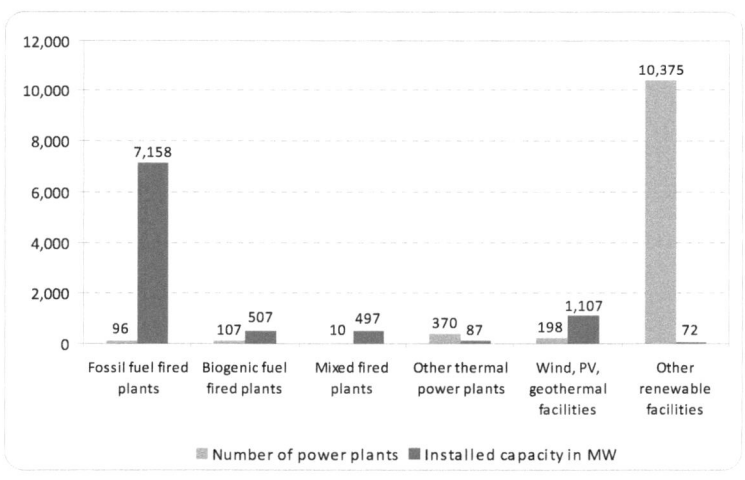

Source: ENERGIE-CONTROL AUSTRIA (2012b, online);
OWN CALCULATIONS AND REPRESENTATION

3.5 Green electricity funding in Austria

In Austria, the generation of green electricity from sources like biomass, hydropower, geothermal energy, photovoltaics and wind power is promoted financially.[45] The legal basis for this funding is the green electricity act in its latest version from 2012 ("Ökostromgesetz 2012", BGBL, 2011b). In view of climate and environment protection, as well as security of supply issues, the green electricity act 2012 aims to promote the generation of green electricity, to increase the share of green electricity[46], to ensure the energy-efficient generation of green electricity, to use the financial resources for the promotion of renewable energy sources efficiently, to guarantee investments for existing and future facilities, to completely reduce the dependency from electricity imports based on nuclear power, and finally, to set technology political priorities in order to achieve market maturity for green technologies (§ 1).

The basic system of green electricity funding in Austria is shown in Figure 22. This is a simplified representation of the Austrian system to promote green electricity. Generally, the operators of funded green electricity facilities sell their generated electricity to the central settlement agent, known as OeMAG ("Abwicklungsstelle für Ökostrom Österreich"), and receive in exchange the prescribed feed-in tariff. The feed-in tariff depends on the type of technology and the installed capacity of the facility as shown in Table 8 (ENERGIE-CONTROL AUSTRIA, 2012c, online and 2011b:18f).[47] The highest feed-in tariffs are granted for photovoltaic installations, followed by biogas and solid biomass like wood chips or straw. Additionally, for liquid biomass and biogas facilities there exists a supplement of 2 cent/kWh for the efficient generation in combined heat and power (CHP) cycles (BGBL, 2012b, §§ 9-10).

---

[45] With this financial promotion it is taken account of the general requirement to subsidise economic activities that create external benefits. For details on the theoretical assessment of externalities please refer to chapter 2.

[46] The latest version of the green electricity act contains concrete renewable energy expansion targets for 2020 (ENERGIE-CONTROL AUSTRIA, 2011b:26). A detailed description of these expansion targets is given in section 3.6.

[47] The levels of the feed-in tariffs are issued yearly in the "Ökostromverordnung" (ENERGIE-CONTROL AUSTRIA, 2011b:17).

Figure 22: The basic system of green electricity funding in Austria

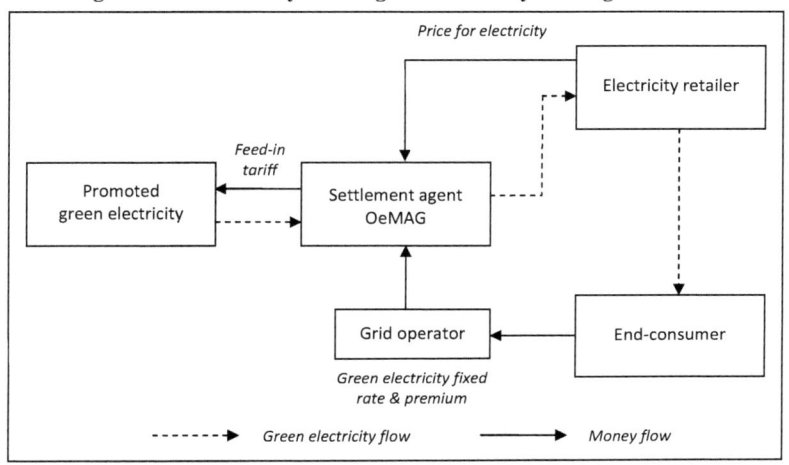

Source: ENERGIE-CONTROL AUSTRIA (2012c, online); OWN REPRESENTATION

The electricity retailer, who obtains the green electricity from the OeMAG at a certain transfer price[48], is then committed to distribute the received electricity via its grid to the end-consumers. The amount of green electricity the retailer receives from the OeMAG depends on the amount of the electricity which is usually distributed to end-consumers. For instance, an electricity retailer with a market share of 5 % is allocated with a proportion of 5 % of the total amount of green power hold by the OeMAG. The end-consumers, in turn, have to make a contribution to the funding of green electricity. First, final electricity consumers have to pay a fixed rate for green electricity ("Ökostrompauschale") which amounts to € 11 per household and year.[49] Additionally, end-consumers are subject to a green electricity premium ("Ökostromförderbeitrag") on top the grid use and network loss charges (ENERGIE-CONTROL AUSTRIA, 2012c and 2012d, online). This premium is currently 15.4 % (BGBL, 2012a). Both, the green electricity fixed rate and the premium are redistributed via the grid operators to the OeMAG which in turn finances green electricity via feed-in tariffs (see Figure 22; ENERGIE-CONTROL AUSTRIA, 2012c, online and 2011b:18ff).

---

[48] The transfer prices charged by the OeMAG for the allocated green electricity to electricity retailers are set annually (ENERGIE-CONTROL AUSTRIA, 2011b:17).
[49] The green electricity fixed rate is staggered according to the grid level. The household fixed rate of € 11 corresponds to grid level 7.

Table 8: Feed-in tariffs for new green power facilities (in cent/kWh), 2012

Technology	Feed-in tariff (cent/kWh)
Photovoltaics	19.0 – 27.6
Wind power	9.5
Geothermal power	7.5
Solid biomass	10.0 – 14.98
Liquid biomass	5.8
Biogas	13.0 – 18.5
Landfill and sewage gas	5.0 – 6.0

Source: BGBL (2012b, §§ 5-11); OWN REPRESENTATION

As can be seen from Table 8, no feed-in tariffs are indicated for hydropower. This is due to the fact that hydropower is usually promoted by investment grants. Accordingly, the construction, as well as the revitalisation of small-scale hydropower plants can be subsidized.[50] For hydropower facilities with an installed capacity of 0.5 MW, the investment grant is at most 30 % of the required investment volume. Hydropower facilities with a capacity of 2 MW can be funded with a subsidy of 20 % at the utmost. For power plants with a capacity of 10 MW the investment grant amounts to 10 %. Furthermore, for medium-sized hydropower stations with a capacity of up to 20 MW, investment grants in the amount of maximum 10 % can be allowed (see Table 9; BGBL, 2011b, §§ 26-27). Prospectively, small-scale hydropower (capacity ≤ 2 MW) will be promoted by feed-in tariffs as well. This represents an important improvement of the latest green electricity law amendment (ENERGIE-CONTROL AUSTRIA, 2011b:27; BGBL, 2011b, § 12).

Table 9: Investment grants for small- and medium-scale hydropower plants

Hydropower capacity	Investment grant (in % of required investment volume)
Capacity 0.5 MW	30 %
Capacity 2 MW	20 %
Capacity 10 MW	10 %
Capacity up to 20 MW	10 %

Source: BGBL (2011b, §§ 26-27); OWN REPRESENTATION

---

[50] The funding of revitalisation measures will only be permitted if the investment leads to an increase of the standard operating capacity by at least 15 %.

The total installed capacity of Austrian green electricity facilities that are in contractual relationship with the OeMAG and thus already in operation is 1,784 MW. Compared to 2006, installed green capacity increased by 8.8 %. As shown in Table 10, wind power, solid biomass and small-scale hydropower (capacity < 10 MW) play the most important role. Together, these technologies make up 91.0 % of total installed green power capacity. Beside that, biogas and solar energy (photovoltaics) are also rather important, while the remaining technologies play only a minor role. Although photovoltaic energy makes up only 3.1 % (54.7 MW) of total capacity, the technology has experienced the most dynamic development since 2006. In particular, solar energy capacity went up by 257.5 %, hence more than tripled. By contrast, wind power capacity grew by only 10.7 % and small-scale hydropower capacity even decreased (-24.5 %). Moreover, solid biomass gained increasing importance in the past with an installed capacity rising from 257.9 MW in 2006 to 325.4 MW in 2011 (+26.2 %).

Table 10: Installed capacity of Austrian green electricity facilities (in MW), 2006 and 2011

Green technology	Capacity 2006	Capacity 2011	Growth 2006-11
Biogas	62.5	79.8	27.7 %
Solid biomass	257.9	325.4	26.2 %
Liquid biomass	14.7	9.4	-36.1 %
Landfill and sewage gas	13.7	16.0	16.8 %
Geothermics	0.9	0.9	0.0 %
Photovoltaics	15.3	54.7	257.5 %
Wind power	953.5	1,055.8	10.7 %
Small-scale hydropower	320.9	242.2	-24.5 %
In total	**1,639.4**	**1,784.2**	**8.8 %**

Source: ENERGIE-CONTROL AUSTRIA (2012e, online);
OWN CALCULATIONS AND REPRESENTATION

The total number of green electricity facilities in Austria (excluding small-scale hydropower) was 6,900 in 2011. These are facilities that are currently in operation and in contractual relationship with the OeMAG. [51] Most of the green electricity facilities, namely 6,253, are solar energy installations in-

---

[51] In contrast, there are 31,297 authorized green electricity facilities in Austria. However, most of these facilities have not been constructed yet and are thus not in operation. Instead, they have just been approved from the federal government.

cluding building integrated systems, as well as open space photovoltaics. In addition, there exist 288 operating biogas plants, 147 wind power stations and 121 power plants fired with solid biomass. As mentioned above, liquid biomass facilities (45), landfill and sewage gas (44), as well as geothermics (2) play a less important role (ENERGIE-CONTROL AUSTRIA, 2012f, online).

3.6 Renewable energy potentials

Although more than two thirds of the total electricity produced in Austria already comes from renewable energy sources, there is still substantial development potential, especially in the field of hydropower. According to the hydropower potential study of PÖYRY ENERGY (2008), the techno-economic potential which is worth being explored is estimated at 56,100 GWh. This estimation is based on the classification of Austrian rivers according to their degree of utilisation. A large part of the Austrian techno-economic potential has already been opened up (38,200 GWh), but 17,900 GWh are not used yet. This value corresponds to the estimate of residual techno-economic potential. 16,500 GWh of the residual techno-economic potential can be explored by new hydropower plants, the remaining part (1,400 GWh) by the optimisation of existing facilities (see Figure 23). However, these estimates do not consider possible reductions of the residual techno-economic potential due to environmental and socio-economic restrictions. An estimate of reduced techno-economic potential therefore excludes potentials located in regions with a high degree of sensibility such as national parks and world heritages. This leaves a value of 12,700 GWh which is effectively exploitable. However, the estimation of reduced techno-economic potential does not consider reductions due to the possible restrictions imposed by the European Water Framework Directive (WFD). The implementation of the WFD may imply a further reduction of the expandable hydropower potential in Austria[52] (PÖYRY ENERGY, 2008:64).

---

[52] The consequences of the WFD for hydropower use in Austria are elucidated in greater detail in section 3.7.

Figure 23: Hydropower potential in Austria (in GWh)

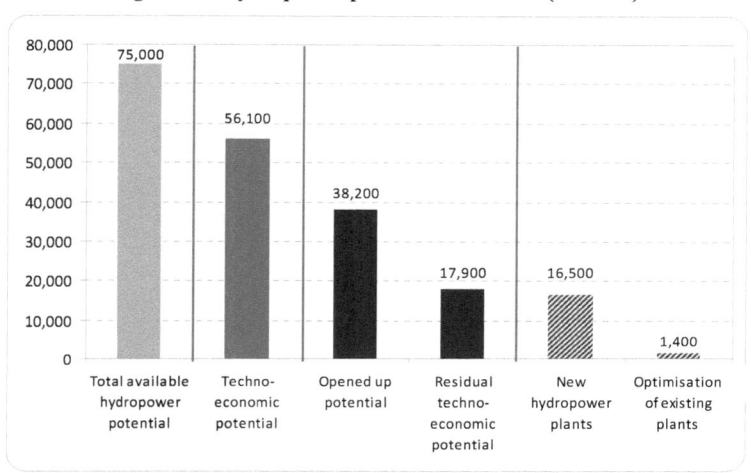

Source: PÖYRY ENERGY (2008:63); OWN REPRESENTATION

As can be seen from Figure 24, the highest realisable hydropower potential is located in Tyrol with a total reduced techno-economic potential of 5,300 GWh. Beside that, Salzburg and Styria exhibit considerable hydropower potentials which are theoretically exploitable. These potentials amount to 2,100 GWh (Salzburg) and 1,600 GWh (Styria) respectively. A considerable reduced techno-economic potential is also given in Vorarlberg (1,200 GWh) and Carinthia (1,100 GWh). In Upper Austria, Lower Austria and Vienna, by contrast, the available hydropower potentials have already been widely exhausted leaving only small expandable potentials. Another reason for the low development potential in these states lies in the fact that the theoretically available hydropower potentials are mainly located in protected areas (like national parks), and thus cannot be explored (PÖYRY ENERGY, 2008:59; STANZER ET AL., 2010:109).

Figure 24: Hydropower potentials across Austrian federal states (in GWh)

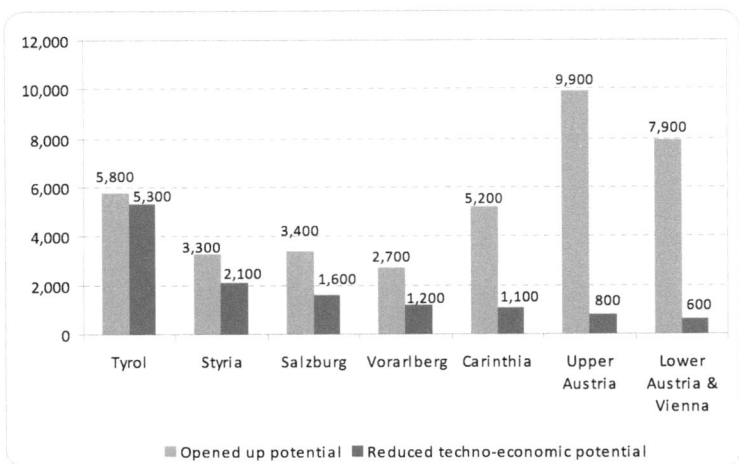

Source: PÖYRY ENERGY (2008:59); OWN REPRESENTATION

The entire available hydropower potential presented above cannot be exploited, especially due to nature and water protection. Nevertheless, it is important to open up at least a significant part of the realisable potential. In order to promote the intensified use of hydropower in the future, several initiatives have been launched in Austria. First, in 2008 the master plan for the expansion of hydropower utilisation was presented. It aims at increasing the use of hydropower by 7,000 GWh until 2020, especially in order to reduce climate-damaging GHG emissions and to increase the share of renewable energy sources in gross final consumption (VEÖ, 2008:10ff). In the Austrian energy strategy, presented in 2010, the intensified use of renewable energy sources represents the core element of a sustainable and future-oriented energy policy too (see also section 3.3). In the energy strategy, a realisable hydropower expansion of 3,500 GWh is stipulated. This expansion target considers both, ecological requirements, as well as economic aspects. Beside that, the Austrian energy strategy focuses on the further development of wind power capacities, photovoltaics, and in the field of heat generation the use of biomass (BMLFUW, 2010a:79ff).[53]

---

[53] For wind power, an extension target of approximately 2,800 GWh until 2020 is stipulated (BMLFUW, 2010:83).

According to the simulations of "REGIO Energy" (see STANZER ET AL., 2010), hydroelectric power can be extended by 4,170 GWh in a future scenario for 2020. For wind power, an even higher development potential is stipulated, namely 5,922 GWh. Finally, solar thermal energy and photovoltaics are expected to rise by 2,174 GWh in the 2020 future scenario (see Table 11).[54]

Table 11: Additional renewable energy capacities
in a future scenario for 2020 (in GWh)

Technology	Additional capacity by 2020 (in GWh)
Wind power	5,922
Hydropower	4,170
Photovoltaics & solar thermal energy	2,174

Source: STANZER ET AL. (2010:190ff); OWN CALCULATIONS AND REPRESENTATION

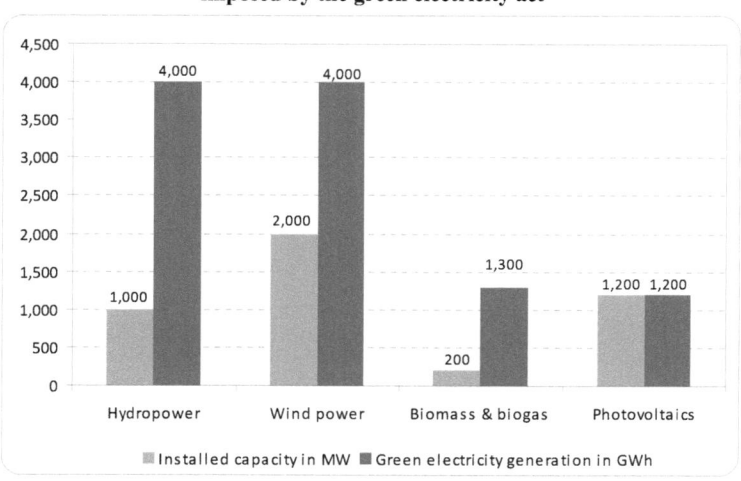

Figure 25: Expansion targets for renewable energy
imposed by the green electricity act

Source: BGBL (2011b, § 4); OWN REPRESENTATION

---

[54] The indicated values refer to the "Midi" scenario which is based on a linear extrapolation of the best renewable energy growth rates and substantial renewable energy incentives in the field of economics and politics. The additional renewable energy capacities indicated above result from the comparison of the total available capacity in 2020 and the starting value of the simulation model (STANZER ET AL., 2010:17ff).

In addition, the intensified use of renewable energy sources was established by law in 2011 (see Figure 25). Accordingly, the green electricity act aims to increase hydropower generation by approximately 4,000 GWh in the period 2010 to 2020.[55] This value corresponds to an additional installed capacity of 1,000 MW. For wind power an expansion target of 4,000 GWh (= 2,000 MW installed capacity) is striven for. The goals for hydro and wind power are under the premise that appropriate locations for new facilities are available. Provided that the required raw materials are available, green electricity generation from biomass and biogas is stipulated to rise by roughly 1,300 GWh which is equivalent to an installed capacity of 200 MW. Finally, photovoltaic generation should be increased by 1,200 GWh corresponding to an additional capacity of 1,200 MW (BGBL, 2011b, § 4).

Until 2020, the Austrian electricity industry plans to invest about € 8.1 billion into the power generation system. The major part of these power plant investments will arise in the renewable energy sector. Hence, € 5.6 billion or 69.1 % will be invested in the construction of new hydropower plants. Additionally, 6.2 % (€ 0.5 billion) of overall power plant investments are proposed for wind power, photovoltaics and biomass installations. Further € 2.0 billion (24.7 %) will flow into the extension of the thermal power plant sector. With this, the Austrian electricity industry aims to strengthen the flexible Austrian electricity generation mix which is largely based on renewable energy sources (OESTERREICHS ENERGIE, 2012:8).[56]

Table 12: Proposed power plant investments in Austria until 2020 (in billion €)

Type of power plant	*Investments in billion €*	*in %*
Thermal power plants	2.0	24.7 %
Hydropower plants	5.6	69.1 %
Wind, PV and biomass	0.5	6.2 %
In total	**8.1**	**100.0 %**

Source: OESTERREICHS ENERGIE (2012:8); OWN REPRESENTATION

---

[55] This target value includes the effects of revitalisation measures and the extension of existing facilities.
[56] In addition, approximately € 8.2 billion are planned to be invested in the extension, maintenance and modernisation of the Austrian power grid structure.

Figure 26: Hydropower plants in the process of construction among Austrian federal states

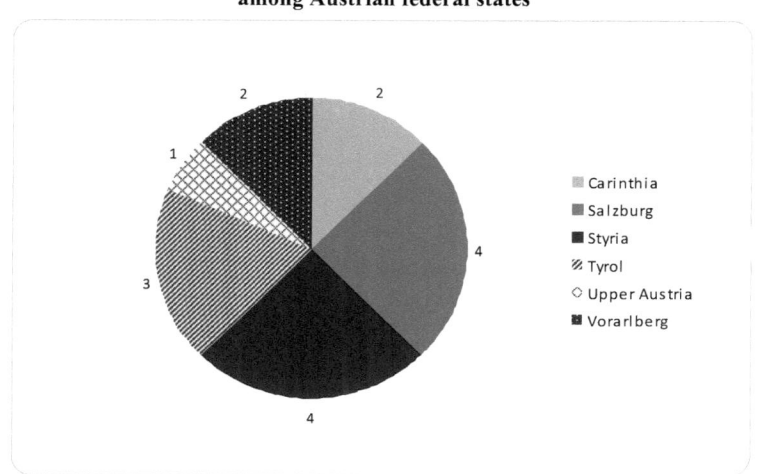

Source: OESTERREICHS ENERGIE (2012:16); OWN CALCULATIONS AND REPRESENTATION

Currently, 16 hydropower projects with an additional yearly electricity generation of approximately 290 GWh are nationwide in construction process.[57] Most of these plants are built in Salzburg (4), Styria (4) and Tyrol (3). Two power stations each are established in Carinthia and Vorarlberg and one new hydropower installation is built in Upper Austria (see Figure 26). Regarding the time horizon, the power stations will be finished between 2013 and 2018.

Nine out of the 16 hydropower plants in construction are small-scaled with a capacity less than 10 MW. The remaining 7 power stations exhibit a capacity of more than 10 MW and can therefore be classified as large-scaled. Regarding the power plant technology, most of the plants in construction are run-of-river power plants, but also 6 (pumped) storage power plants with capacities between 360 and 480 MW are built (see Table 13).[58]

---

[57] This value includes the revitalisation of the hydropower plant Pernegg in Styria and the efficiency improvement of the pumped storage power plant Zillertal in Tyrol.
[58] Usually, (pumped) storage power plants are large-scaled with capacities far beyond the 10 MW threshold value while the run-of-river plants under construction are mostly small-scaled. Moreover, it is important to point out that run-of-river and (pumped) storage power plants perform different roles in the electricity system. Run-of-river plants are designed to meet base load demand, while storage power plants only run at peak times.

Table 13: Hydropower plants in the process of construction by size and technology

Power plant size	Number of plants	in %
Small-scaled (< 10 MW)	9	56.2 %
Large-scaled (> 10 MW)	7	43.8 %
Power plant technology	Number of plants	in %
Run-of-river plants	10	62.5 %
(Pumped) storage power plants	6	37.5 %

Source: OESTERREICHS ENERGIE (2012:16); OWN CALCULATIONS AND REPRESENTATION

In addition to the hydropower stations, four concrete renewable energy projects are currently under construction, one waste material heat power station in Upper Austria and three wind parks in Lower Austria. The total installed capacity of these projects will be 63 MW, able to produce 230 GWh of electricity. Beside these specific projects various wind power plants, not particularly defined yet, with an electricity generation amount of 805 GWh are to be built (OESTERREICHS ENERGIE, 2012:18).

In the hydropower sector, 30 concrete facilities are currently in the planning stage, most of them in Salzburg (7), Tyrol (7) and Vorarlberg (5). Another four hydropower stations are planned to be built in Styria, three plants each in Lower Austria and Upper Austria and only one hydropower station in Carinthia (see Figure 27).[59] With this, an additional electricity amount of approximately 3,400 GWh can be generated. 40 % of the planned hydropower stations (12) have a capacity of less than 10 MW (small hydro). The remaining 18 (60.0 %) are large-scaled exhibiting a capacity of more than 10 MW. As far as the power plant technology is concerned, the focus is on run-of-river technologies accounting for more than three quarters of the total number of planned hydropower stations. However, a number of large-scale pumped storage power stations is also planned to be built, especially to balance the fluctuating electricity supply from wind power and photovoltaics (see Table 14). Generally, pumped storage technologies provide large-scale energy storage and can thus be a useful tool to improve energy system security, i.e. to provide grid stability and to integrate renewables, such as wind or solar in the electricity mix (IRENA, 2012:9; JRC, 2011:29).

---

[59] The indicated values include the extension of two existing facilities in Tyrol, as well as measures to increase the efficiency of three hydropower facilities in Lower Austria, Salzburg and Styria.

In addition to the concrete hydropower projects, measures to increase the efficiency of existing facilities are planned. These efficiency increases yield an installed capacity of 875 MW and a corresponding electricity generation of 375 GWh. Moreover, several undefined pumped storage power plants are quoted in the current list of planned hydropower projects with an added capacity of 800 MW. Finally, various small-scale hydropower stations, not particularly defined yet, with a total capacity of 80 MW and an electricity generation amount of 321 GWh are scheduled (OESTERREICHS ENERGIE, 2012:16f).

Figure 27: Hydropower plants in the planning stage among Austrian federal states

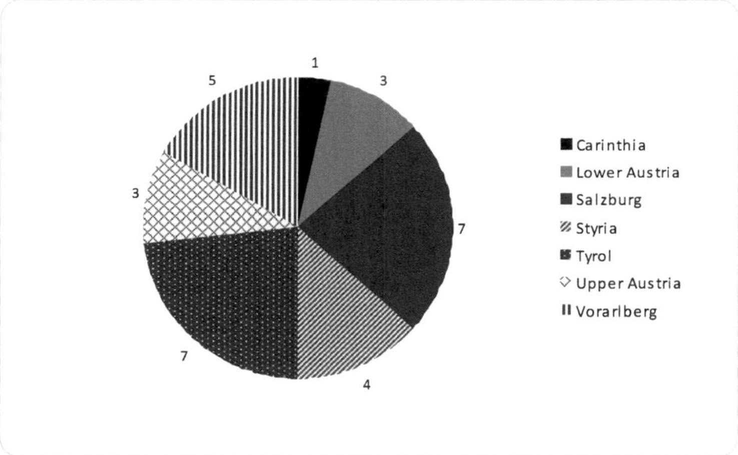

Source: OESTERREICHS ENERGIE (2012:16f); OWN CALCULATIONS AND REPRESENTATION

Table 14: Hydropower plants in the planning stage by size and technology

Power plant size	Number of plants	in %
Small-scaled (< 10 MW)	12	40.0 %
Large-scaled (> 10 MW)	18	60.0 %
Power plant technology	Number of plants	in %
Run-of-river plants	23	76.6 %
(Pumped) storage power plants	7	23.3 %

Source: OESTERREICHS ENERGIE (2012:16f); OWN CALCULATIONS AND REPRESENTATION

Future renewable energy projects focus, beside hydropower, also on other technologies. Accordingly, one combined heat and power plant fired with biomass is planned to be built in Upper Austria generating 80.0 GWh of electricity. Furthermore, two concrete wind power plants, one in Upper and one in Lower Austria, are scheduled. The installed capacity of these projects amounts to 18.0 MW and a corresponding electricity generation of 41.0 GWh. Finally, five very small photovoltaic installations in Carinthia, Upper Austria and Vienna are currently in the stage of planning producing only 2.2 GWh of electricity. In total, the additional yearly electricity generation of the planned renewable energy projects is 123.2 GWh, a value far beyond the generation capacity of the future hydropower plants (see Table 15).[60]

Table 15: Renewable energy power plants in the planning stage

Type of power plant	Number of plants	Installed capacity	Electricity generation
Biomass	1	10.0 MW	80.0 GWh
Wind power plants	2	18.0 MW	41.0 GWh
Photovoltaic installations	5	2.7 MW	2.2 GWh
In total	**8**	**30.7 MW**	**123.2 GWh**

Source: OESTERREICHS ENERGIE (2012:18); OWN CALCULATIONS AND REPRESENTATION

To summarise, the renewable energy extension plans of the Austrian electricity industry are mainly based on hydropower. Here, the available potentials can be exhausted optimally. Together, the hydropower plants that are currently under construction, as well as the planned hydropower projects will be able to generate an additional electricity amount of approximately 4,400 GWh. This value coincides with the mandatory expansion target stipulated in the latest version of the green electricity act. Regarding the expansion of other renewable energy sources, the concrete plans of the Austrian electricity industry lag behind the stated expansion targets. Altogether, only about 1,160 GWh of electricity can be generated by the biomass, photovoltaic, wind power and waste material installations that are currently under construction or planned for the future. So, the theoretically available potentials for these technologies have not been exhausted yet. Further initiatives are required.

---

[60] Moreover, several undefined wind power projects in Lower Austria and Vienna with an additional installed capacity of 400 MW are planned.

3.7 Assets and drawbacks of renewable energy expansion

Investments in hydropower or in the most general sense renewable energy are associated with both, advantages as well as disadvantages. As shown in the previous section, hydropower plays the most important role within the plans of the Austrian electricity industry to expand generation capacities from renewable sources.

Hydropower is the most mature renewable power generation technology available (IRENA, 2012:4). The long lasting approved technology is, furthermore, highly reliable and extremely energy efficient. Correspondingly, hydropower plants exhibit the highest level of efficiency (90 %) compared to other power generation facilities (HORLACHER, 2007:96; TRUFFER ET AL., 2001:21; SETCOM, n.d.:13). Moreover, hydropower is the only technology available today that enables the storage of energy to a large extent and in a cost-efficient way (IRENA, 2012:4; BARD, 2006:56). The stored energy is instantaneously available at the moment of demand (TRUFFER ET AL., 2001:21). Generally, hydropower schemes possess significant flexibility in their design. They can be designed to meet base-load demands on the one side or, on the other side, much larger shares of peak demand (IRENA, 2012:4; JRC, 2011:29). Another advantage of the hydropower technology is the long lifetime of the plants which is usually between 30 and 80 years and can even go up to 100 years. Although the construction of a hydropower plant requires high capital expenditures, the operating and maintenance costs are very low (HORLACHER, 2007:96; IRENA, 2012:7; JRC, 2011:30; SETCOM, n.d.:7). This gives rise to one of the most important advantages of the technology. Hydropower is relatively cheap, hence a cost-effective possibility to generate electricity from a renewable source (IRENA, 2012:7; JRC, 2011:29; STERNBERG, 2008:1591).

Figure 28: Electricity production costs of renewable energy sources in Austria (in cent/kWh)

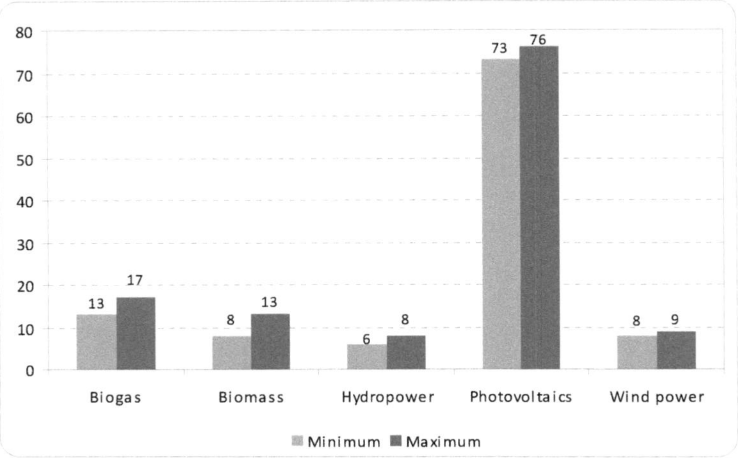

[Bar chart showing Minimum and Maximum electricity production costs:
- Biogas: 13, 17
- Biomass: 8, 13
- Hydropower: 6, 8
- Photovoltaics: 73, 76
- Wind power: 8, 9]

Source: FANINGER (2008:213); OWN REPRESENTATION

The electricity production costs are a key criterion for the use of renewable energy sources. They consider capital and operating costs and vary significantly across renewable technologies (EDER, 2009:3). As can be seen from Figure 28, hydropower exhibits the lowest electricity production costs ranging from 6 to 8 cents per kWh generated electricity. Beside that, wind power generation is relatively cheap with production costs between 8 and 9 cents per kWh. The electricity production costs from biomass and biogas range between 8 and 17 cents per kWh. The by far most expensive renewable power generation technology is photovoltaics. The generation of solar power is associated with costs between 73 and 76 cents per kWh (FANINGER, 2008:213; KOHL, n.d.: 11).[61] Although renewable energies, except for hydropower, are currently more expensive than non sustainable technologies like coal or gas[62], there is a downward tendency of electricity gen-

---

[61] Similar cost dimensions are given by OWEN (2006). For hydropower, current cost of one kWh delivered energy is between 2 and 9 cents. A likewise magnitude is indicated for land-based wind power (4-9 cents/kWh) and biomass (2-12 cents/kWh). Again photovoltaics is the most expensive renewable energy source with electricity production costs ranging from € 0.45 to € 1.34 per kWh (all values converted from US-Dollars into Euros using the average yearly exchange rate from EUROSTAT-DATABASE, 2012c, online; OWEN, 2006:212).

[62] As shown by OWEN (2006), electricity production costs for coal and gas lie between 3 and 4 cents per kWh which is by tendency below the cost of generating electricity from renewable sources (OWEN, 2006:212).

eration costs. Hence, technology costs for renewable sources were recently falling, especially for photovoltaic modules and onshore wind turbines. These cost reductions make renewable energy sources more competitive with fossil-fuel alternatives (UNEP, 2012:11; OWEN, 2006:212).[63] However, from the level of electricity production costs no conclusion regarding the economic efficiency of a project can be drawn. For this purpose, not only the costs but also the revenues have to be taken into consideration. Due to the system of green electricity funding in Austria (feed-inn tariffs, see section 3.5), facilities with comparatively high production costs can even be profitable (NEUBARTH, n.d.:10).

Figure 29: CO_2 emissions from different technologies (in g CO_2 equivalent per kWh)

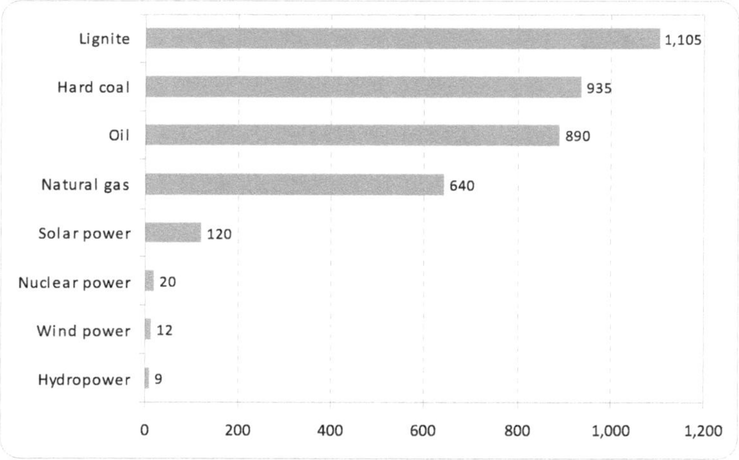

Technology	g CO_2 eq./kWh
Lignite	1,105
Hard coal	935
Oil	890
Natural gas	640
Solar power	120
Nuclear power	20
Wind power	12
Hydropower	9

Source: EURELECTRIC (2012, online); OWN CALCULATIONS AND REPRESENTATION

One of the major advantages of renewable energies is that the technologies create by far less CO_2 emissions than coal and oil-based technologies. They are generally CO_2-free in operation, but there are GHG emissions from the construction of renewable energy power plants, as for instance hydropower schemes or wind turbines (IRENA, 2012:5; OWEN, 2006:212). Hydropower

---

[63] If the externalities of producing energy from fossil fuels like for instance air pollution are furthermore internalised into the price of electricity output, renewable technologies become indeed financially competitive with fossil-fuel fired generation (MENEGAKI, 2008:2423; OWEN, 2006:213).

is environmentally next to wind energy the cleanest energy source and therefore ranks among the ecologically most preferable energy systems (STERNBERG, 2008:1591; TRUFFER AND BRATRICH, 1998:2). As apparent from Figure 29, the carbon footprint of hydropower is in the range of 9 g CO_2 equivalent per kWh which is the lowest value in comparison with other electricity generation technologies. A wind power plant creates on average 12 g CO_2 equivalent per kWh of generated electricity. CO_2 emissions from solar power plants are significantly higher amounting to 120 g CO_2 equivalent on average. Nevertheless, this value is still considerably below the carbon footprints of non-sustainable technologies. Accordingly, lignite (1,105 g) and hard coal (935 g) create the highest levels of CO_2 emissions, followed by oil and natural gas with 890 g respectively 640 g CO_2 equivalent per kWh (EURELECTRIC, 2012, online).[64]

Based on the values presented above, it can be concluded that investments in renewable energy sources, especially hydropower, can contribute substantially to the reduction of CO_2 emissions and in further consequence the prevention of global climate change (HORLACHER, 2007:95; MEYERHOFF AND PETSCHOW, 1998:26; OTT ET AL., 2008:35).[65]

Beside the positive effect on CO_2 emissions, policies that support renewable energy sources boost the economy. A 2009 published report analysed the effects of renewable energy investments (RAGWITZ ET AL., 2009) on economic growth and employment in the EU. The analysis showed that achievement of the 2020 targets leads to total gross value added in the renewable energy sector of € 129 billion in the EU-27; this is 1.1 % of total Gross Domestic Product (GDP). Compared to a hypothetical baseline scenario in which all renewable energy support policies are abandoned, measures being taken to achieve the 2020 targets induce a net GDP growth between 0.23 % and 0.25 % in the EU-27. In addition, investments in renewable energies lead to job creation. In 2020 about 2.3 million people in the EU-27 will be employed in the renewable energy sector. This outcome

---

[64] The indicated values correspond to the midpoint estimates of the stated CO_2 emission ranges by EURELECTRIC (2012, online).
[65] For instance, 3.1 Mt CO_2 can be saved until 2020 by the expansion of hydropower capacities to the extent of 7,000 GWh (VEÖ, 2008:10).

is based on the premise that the adopted renewable energy support policies lead to an achievement of the 2020 goals. In comparison to the hypothetical baseline scenario, approximately 410,000 net additional jobs can be generated in the renewable energy sector. These employment effects are mainly due to the higher labour intensity of renewable energy sources and increased expenditure on electricity. Furthermore, the agricultural sector will especially benefit from promoting renewable energies due to the increased demand for biomass (RAGWITZ ET AL., 2009:5ff; MENEGAKI, 2008:2423).

Similar to the effects at EU level, the expansion of renewable energy sources in Austria is equally accompanied by positive impacts for the local economy (BMLFUW, 2010a:7). According to BODENHÖFER ET AL. (2004:90ff), the construction, as well as the operation of green electricity facilities induce positive net effects on value added and employment. Furthermore, the implementation of the Austrian green electricity act is associated with yearly investments in the electricity industry up to € 1 billion. These investments will create about 9,500 domestically available jobs (GRUBER, 2011:50). The expansion of hydropower utilisation creates value added and employment effects too, especially in peripheral regions (OTT ET AL., 2008:35). Accordingly, the Austrian master plan for promoting hydropower states that 6,000 jobs can be created over a period of ten years by increasing hydropower utilisation in the amount of 7,000 GWh (VEÖ, 2008:15). Moreover, TICHLER AND KOLLMANN (2005) examined the economic effects associated with the exploitation of small-scale hydropower in Austria. According to their simulations, about 4,160 additional jobs can be created by the promotion of small-scale hydropower until 2020.[66] This value is based on the assumption that a 10 % contribution of small-scale hydropower to the entire electricity consumption is achieved by 2020 (TICHLER AND KOLLMANN, 2005:12ff).[67]

---

[66] Generally, the presented figures represent gross effects instead of marginal or net effects. Regarding employment effects, however, one should keep in mind how many jobs would have been created if the investment had been undertaken elsewhere (e.g. in another sector). So, from the perspective of job, hydropower is probably less positive (smaller "job intensity") than other renewable technologies.

[67] An issue that is discussed controversially in the literature, is the substitution effect associated with renewable energy investments. Usually, renewable energies substitute fossil energy sources so that investments into conventional technologies go down, leading to a loss of jobs in the traditional energy sector. Additionally, due to the fact that some renewable energy sources

Further economic benefits associated with the promotion of renewable energy sources involve the reduced dependency on imported fossil fuels or nuclear energy and the resulting increased security of supply (RAGWITZ ET AL., 2009:4; RWI, 2009:8), as well as the new market opportunities "green technologies" open up for firms (HANNA ET AL., 2011:147). The vast importance of renewable energy expansion for the strengthening of energy self-sufficiency and the associated security of supply, as well as the economic competitiveness is further emphasised in the Austrian energy strategy (BMLFUW, 2010a:7). Especially, hydropower expansion is assumed to contribute essentially to the autonomy of the Austrian electricity supply (VEÖ, 2008:7).

Another advantage of the hydropower technology is the multi-functionality of hydropower projects (STERNBERG, 2008:1591). Beside energy generation, hydropower plants provide further important water supply services as for instance flood control or irrigation (HORLACHER, 2007:95; IRENA, 2012:5). In addition, hydropower projects may facilitate the attractiveness of a region for recreational purposes, i.e. create recreational areas for the adjacent population (OTT ET AL., 2008:36; VEÖ, 2008:7).[68]

Although the need for renewable energy investments is undisputed due to security of supply issues as well as climate change and dependency con-

---

are still not price-competitive compared to non-sustainable technologies people have to make a contribution to the funding of green electricity (see also section 3.5). The corresponding money amount is subsequently not available for the purpose of other consumption which in turn affects employment negatively. This is formally known as the income effect (KRATZAT AND LEHR, 2006:144f). Both, the substitution and the income effect need to be taken into account in order to make a valid statement about the long-run employment effects associated with renewable energy investments. This has at least been partly done in the studies presented above. Generally, the consideration of negative employment effects in other sectors of the economy often lead to the conclusion that in the long-run, renewable energy investments do not lead to significant job creation. Hence, employment creation should only be considered as a side effect of renewable energy investments. The main reason for the promotion of renewable energy sources should be rather founded in the contribution of the technologies to prevent climate change, i.e. reduce GHG emissions (HÄDER AND SCHULZ, 2005:475).

[68] However, there are ambiguous empirical results regarding the impact of hydro-electric power plants on recreational opportunities. A study from GETZNER (2012) in the River Basin Mur in the Austrian province of Styria revealed that recreational activities like walking, cycling, swimming or boating are primarily done along the free-flowing sections of the river Mur. Furthermore, within the scope of a travel cost analysis it was found that people value recreational activities along free-flowing rivers higher than recreational use along dammed river sections (GETZNER, 2012:1).

cerns (ECORYS, 2010:15), green technologies such as biomass, hydropower, photovoltaics or wind power are subject to some disadvantages. First, building renewable energy facilities (e.g. hydropower plants or wind parks) raises an issue of social acceptance. Depending on the type of technology, renewable energy plants are often seen as a blot on the landscape and a threat for the ecosystem (HANNA ET AL., 2011:146). Accordingly, the rapid expansion of renewable energy technologies entailed increasing opposition in parts of the affected local population due to growing amenity impacts (KREWITT, 2002:844). Wind turbines, for instance, are associated with aesthetic impacts on landscape, noise disturbances and a threat for birdlife (STRUTZMANN, 2011:19; SETCOM, n.d.:12). Hydropower installations are affiliated to social disfavour as well (STERNBERG, 2008:1589). Although electricity from hydropower creates no air pollution and thus contributes to the prevention of climate change, hydro-developments on rivers are subject to serious environmental impacts. Hence, the expansion of hydropower utilisation is between the priorities of climate protection, energy generation and the preservation of nature ("trade-off"; KEITE, 2004:21; RIPL, 2004:53; TRUFFER AND BRATRICH, 1998:2; TRUFFER ET AL., 2001:21).

Generally, hydropower plants intervene in the natural water flow of a river. The larger this intervention is, the greater the environmental impacts (JRC, 2011:33). The exploitation of rivers for electricity generation purposes affects the amount of undisturbed nature, wildlife, fish stocks, outdoor recreation and water supply (CARLSEN ET AL., 1993:201). More precisely, ecological impacts are caused by the damming of watercourses, the construction of transverse structures like dams or weirs, as well as diverted reaches (see Figure 30). First, the damming of watercourses leads to an alteration of flow conditions causing erosion and sedimentation. Moreover, an increase of the water temperature and associated oxygen-deficiency problems may be induced by damming up a river. These impacts seriously affect fish and other water-dependent wildlife and may result in a loss of aquatic habitats and an alteration of biodiversity (KRUCK AND ELTROP, n.d.:28; TRUFFER ET AL., 2001:21). Second, one of the most significant environmental impacts associated with the utilisation of hydropower is the disruption of the water body continuity (RIPL, 2004:55). More specifically, transverse structures such as dams are migration barriers for fish and other aquatic organisms. This may affect the composition of aquatic biodiversity substantially. Finally, the operation of diversion plants[69] leads to a reduction of the water flow in the main river channel which in turn alters flow conditions, causes oxygen-deficiency and correspondingly affects fish life (KRUCK AND ELTROP, n.d.:28; STRUTZMANN, 2011:18).[70] Although the construction, as well as the operation of a hydropower plant undoubtedly cause a significant intrusion on the natural environment, the ecological impacts can be reduced by planning hydropower schemes in an environmentally friendly way (HORLACHER, 2007:95; KNÖDLER ET AL., 2007:10). "One of the greatest challenges with the development of hydropower is ensuring that the design and construction of hydropower projects is truly sustainable" (IRENA, 2012:5). Hence, negative impacts on ecosystems and biodiversity need to be mitigated in the project plan. An ecological method of construction may contribute substantially to the preservation of natural habitats (VEÖ, 2008:7). For that purpose, specific criteria for the evaluation of a sustainable hydropower utilisation has been elaborated taking into account ecological, water management-related and energy efficiency considerations ("Kriterienkatalog Wasserkraft"; BMLFUW, 2012:3).

Figure 30: The main environmental impacts associated with the use of hydropower

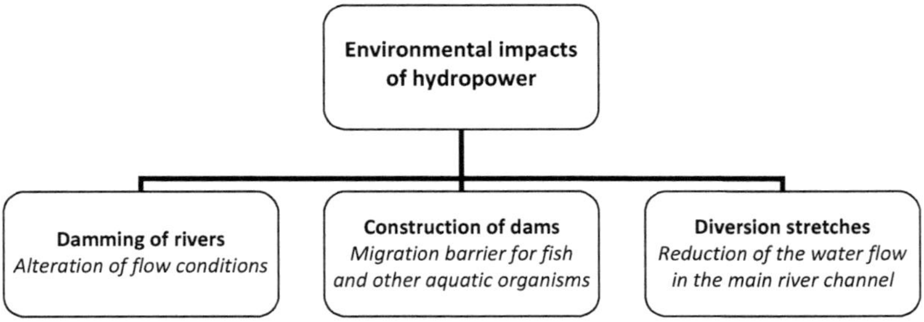

```
                    ┌─────────────────────┐
                    │ Environmental impacts│
                    │    of hydropower     │
                    └──────────┬──────────┘
          ┌────────────────────┼────────────────────┐
┌─────────┴─────────┐ ┌────────┴────────┐ ┌─────────┴─────────┐
│ Damming of rivers │ │Construction of dams│ │Diversion stretches│
│Alteration of flow │ │Migration barrier for│ │Reduction of the water│
│    conditions     │ │fish and other aquatic│ │flow in the main river│
│                   │ │     organisms      │ │      channel      │
└───────────────────┘ └─────────────────┘ └───────────────────┘
```

Source: KRUCK AND ELTROP (n.d.:28); OWN REPRESENTATION

---

[69] Diversion power plants represent a particular type of the run-of-river power plant. Here, the water is dammed by a weir and redirected through a separate intake canal to drive the turbines and generate electricity (KEITE, 2004:21).
[70] There is a lot of literature addressing the ecological impacts of hydropower utilisation. For more information see BUNGE ET AL. (2001), KNÖDLER ET AL. (2007), MEYERHOFF AND PETSCHOW (1997) or WURZEL AND PETERMANN (2006).

To conclude, hydropower facilities interfere deeply in the river ecosystem causing a general conflict of interest with the objectives of nature and water protection as for instance the European Water Framework Directive (WFD; KEITE, 2004:21; MACLEOD ET AL., 2006:2048). Principally, the WFD represents a legal framework for the protection of water bodies throughout the European Union (EPC, 2000). It became European law in December 2000 and was transferred into national law three years later by the amendment of the Austrian Water Law Act, in December 2003. The main objective of the WFD is to achieve a good ecological and chemical status for ground and surface water bodies until the year 2015 (BMLFUW, 2007b:6f; PABBRUWEE, 2006:21).[71] The problem, however, is that numerous water bodies have been altered in the past by anthropogenic action like flood-protection measures or the utilisation of hydropower. These water bodies can be designated as "Heavily Modified Water Bodies" (HMWB).[72] For the category of HMWB the aim of a good ecological potential is applied representing a compromise between the alteration of rivers resulting from hydropower use or hydraulic engineering measures, and water protection (BMLFUW, 2007b:25; BUNGE ET AL., 2001:16; PABBRUWEE, 2006:25). In addition, the WFD represents a legal framework to ensure that human agency does not lead to a deterioration of the water body status ("Verschlechterungsverbot"; BMLFUW, 2006:9; EPC, 2000:9).

Basically, the good ecological status is given if the structure of the surface waters and their symbiotic communities are only marginally affected by human development activities (BMLFUW, 2006:7; EPC, 2000:38). Hence, the good status generally allows the exploitation of waters for activities like power generation or irrigation, but only if the ecological functions of the water body are negligibly affected (BUNGE ET AL., 2001:15). For the deter-

---

[71] Due to the fact that surface water bodies or more precisely, rivers are mainly affected by the utilisation of hydropower, the following remarks simply refer to these water bodies.

[72] More precisely, a surface water body can be designated as heavily modified, if the hydro morphological changes that would be necessary to achieve a good ecological status have significant adverse effects on activities for the purposes of drinking water supply, power generation or irrigation, water regulation, flood protection, navigation, recreation or other equally important sustainable human development activities (PABBRUWEE, 2006:21). In Austria, 328 water bodies with a total length of 4,998 km are designated as heavily modified. This corresponds to a share of 44 % of the total water network (BMLFUW, 2005:8).

mination of the water body status/potential, biological, hydro morphological and physico-chemical quality elements are considered (see Table 16).

Table 16: Quality elements for the determination of the water body status in rivers

Biological quality elements	– Phytoplankton – Macrophytes and phytobenthos – Benthic invertebrate fauna – Fish fauna
Hydro morphological quality elements	– Hydrological regime – River continuity – Morphological conditions
Physico-chemical quality elements	– Specific synthetic pollutants – Specific non-synthetic pollutants

Source: EPC (2000:39ff); OWN REPRESENTATION

For hydropower use, the hydro morphological quality elements are of particular relevance. Generally, the construction of a hydropower facility requires a modification of the river. This affects the main quality elements which are usually taken into account for the determination of the good ecological status/potential. More precisely, the utilisation of hydropower affects the hydro morphological condition of a water body significantly since hydro plants disrupt the river continuity or lead to an alteration of flow conditions. These effects influence the water body classification negatively (BUNGE ET AL., 2001:15; PABBRUWEE, 2006:26). Hence, the implementation of the WFD affects the Austrian plans to expand hydropower utilisation considerably.[73] In particular, the WFD sets high standards for the approval and the operation of new hydropower schemes. According to the principle of no further deterioration, new hydropower facilities are prohibited to cause a decrease in ecological quality. This may be guaranteed by the preservation of the river continuity and sufficient residual water flow. Thus, the installation of fish ladders and a minimum amount of residual water flow are obligatory when new hydropower stations are built (KNÖDLER ET AL., 2007:11; PABBRUWEE, 2006:26).[74]

---

[73] The WFD represents an example for a situation in which negative externalities arising from an economic activity are governed by legal regulations.
[74] Generally, the implementation of the WFD does not prevent the construction of new hydropower facilities. However, new plants must fulfil the ecological requirements of the directive

In 2005, a comprehensive analysis of Austrian rivers was carried out aiming at assessing the Austrian potential to achieve the good ecological status/potential in 2015. Altogether, 940 water bodies comprising a water network with a total length of 11,488 km have been analysed. Regarding physico-chemical quality elements, 770 water bodies with a total length of 8,900 km (78 % of the water network) are assumed to match the target set by the WFD. As far as chemical pollutants are concerned, even 90 % of the analysed rivers are predicted to achieve the good status. The situation of Austrian rivers with respect to hydro morphological parameters is less convenient. Particularly, 450 water bodies with a total length of 6,387 km (56 % of the analysed water network) are projected to fail the attainment of the good status. The main reasons for this failure are migration barriers, damming and insufficient residual water flows. These hydro morphological pressures are primarily caused by hydropower use and flood-protection measures (BMLFUW, 2005:6ff).

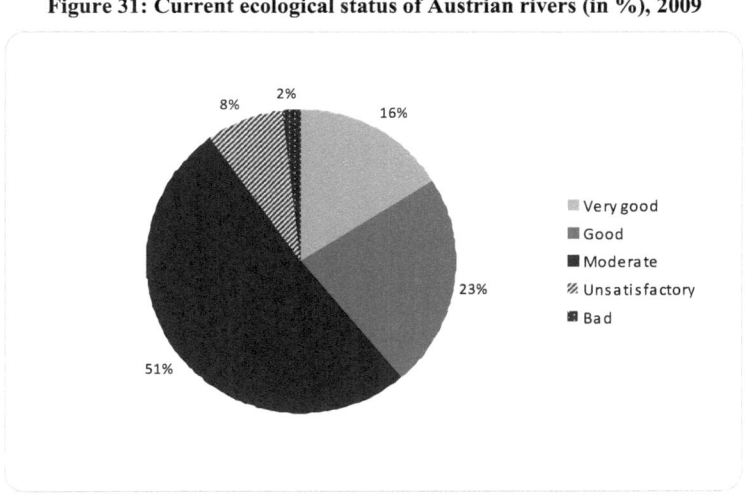

Figure 31: Current ecological status of Austrian rivers (in %), 2009

Source: BMLFUW (2010b:71); OWN REPRESENTATION

Similar results are given by the National River Basin Management Plan (RBMP) 2009. The RBMP is the main instrument for the realisation of the

---

(STIGLER ET AL., 2005:12). These environmental standards may act as a limiting factor for the approval of new projects or the quantity of exploitable resource (JRC, 2011:33).

objectives of the WFD and has to be issued every six years. Currently, 39 % of the Austrian rivers have a very good or good ecological status. Slightly more than half the rivers exhibit a moderate ecological status. The proportion of rivers with an unsatisfactory or bad performance is 10 % in total (see Figure 31). Until 2015 the share of rivers with a very good or good water body status is predicted to increase to 42 %; in 2027 all Austrian rivers will be in very good or good ecological condition (BMLFUW, 2010b:71ff). Furthermore, the RBMP shows that the main part of the rivers designated as heavily modified (87 %) does not correspond to the aim of good ecological potential. Conversely, only 13 % of the HMWB exhibit the required good ecological potential. In 2015, the share of HWMB with very good or good ecological potential is expected to equal 23 % (BMLFUW, 2010b:81f).

Additionally, the RBMP shows that Austrian rivers exhibit significant hydro morphological pressures mainly caused by the use of hydropower and flood prevention measures. More specifically, 10 % of the Austrian rivers are significantly affected by insufficient residual water flows, 2.6 % by hydropeaking and 3.4 % by damming. Additionally, about 28,800 migration barriers in 4,125 water bodies have been localized. 10 % of these migration barriers are attributable to hydropower use. All in all, 52 % of the analysed water courses have a high risk to fail the aim of the good ecological status (BMLFUW, 2010b:34ff). Accordingly, improving hydromorphology and establishing the longitudinal connectivity of the river are the main tasks for the implementation of the WFD. However, these measures are associated with substantial costs. Currently, the main part (90 %) of the existing small-scale hydropower facilities (capacity < 10 MW) is not passable for fish and other aquatic organisms. The implementation of measures to establish the passability of these hydropower plants would cost about € 90 million. For large-scale hydropower, the costs for the establishment of river connectivity, the implementation of restructuring measures, as well as the connection of subsidiary water bodies amount to € 144 Mio. Another cost factor caused by the realisation of the WFD are production losses due to the mandatory residual water flow. These production losses amount to 10 % to 32 % in the case of small-scale hydropower, and 3 % to 10 % for hydropower plants with a capacity of more than 10 MW (STIGLER ET AL., 2005:1f).

3.8 Research question

In light of the preceding explanations, it was shown that the use of hydropower is associated with a considerable conflict potential. On the one hand there are the targets of climate and energy policy like the reduction of GHG emissions or the intensified use of renewable energy sources, and on the other hand the objectives of nature and water protection as for instance the European Water Framework Directive (WFD). Positive effects arising from the use of hydropower especially involve the emission-free generation of electricity and the associated CO_2 avoidance. In addition, hydropower investments can have positive impacts on the local economy (especially employment effects) and contribute to domestic energy security. Important environmental concerns related to the operation of hydropower plants are the visual impact of a power plant on landscape, as well as erosion, sedimentation and oxygen-deficiency problems due to the alteration of the water flow in the river, and correspondingly, the impacts of these changes on fish and other water-dependent wildlife (see Figure 32).

A similar argumentation applies for the expansion of other renewable energy sources. While the intensified use of biomass, wind power or photovoltaics contributes considerably to the reduction of GHG emissions, the economic development, and security of supply, the technologies are subject to negative social and environmental impacts. With the development of wind power, for example, alterations of the natural landscape, noise disturbances and threats for birdlife are associated (STRUTZMANN, 2011:19). Solar parks that take up several hectares in size negatively affect landscape and thus raise an issue of social acceptance as well. Moreover, solar parks may affect the micro-climate because the solar panels cover large areas of landscape. Finally, biomass plants represent a blot on the landscape too. Furthermore, the required large-scale cultivation of biomass may imply negative effects on biodiversity through pesticide pollution (KOHL, n.d.:9ff).

Figure 32: Conflict of interests and ecological trade-offs of hydropower use

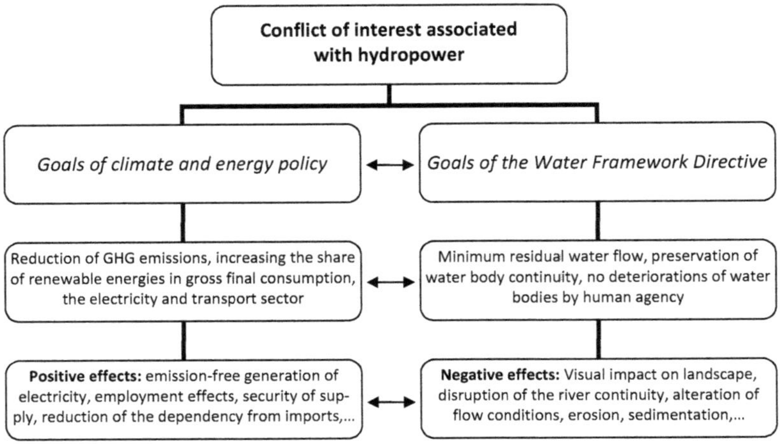

```
            ┌─────────────────────────────┐
            │  Conflict of interest       │
            │  associated with hydropower │
            └──────────────┬──────────────┘
                ┌──────────┴──────────┐
    ┌───────────────────────┐   ┌────────────────────────────┐
    │ Goals of climate and  │◄─►│ Goals of the Water         │
    │ energy policy         │   │ Framework Directive        │
    └───────────┬───────────┘   └─────────────┬──────────────┘
                │                              │
 ┌──────────────────────────────┐  ┌──────────────────────────────────┐
 │ Reduction of GHG emissions,  │  │ Minimum residual water flow,     │
 │ increasing the share of      │◄►│ preservation of water body       │
 │ renewable energies in gross  │  │ continuity, no deteriorations    │
 │ final consumption, the       │  │ of water bodies by human agency  │
 │ electricity and transport    │  │                                  │
 │ sector                       │  │                                  │
 └──────────────┬───────────────┘  └──────────────┬───────────────────┘
 ┌──────────────────────────────┐  ┌──────────────────────────────────┐
 │ Positive effects: emission-  │  │ Negative effects: Visual impact  │
 │ free generation of           │◄►│ on landscape, disruption of the  │
 │ electricity, employment      │  │ river continuity, alteration of  │
 │ effects, security of supply, │  │ flow conditions, erosion,        │
 │ reduction of the dependency  │  │ sedimentation,...                │
 │ from imports,...             │  │                                  │
 └──────────────────────────────┘  └──────────────────────────────────┘
```

Source: OWN REPRESENTATION

Generally, "the conflict between the preservation of natural environmental assets, and their development have been one of the longest-standing concerns in environmental economics" (HYNES AND HANLEY, 2006:170). "If an environmental policy decision involves a trade-off in the choice between providing one ecosystem service (such as a particular habitat or ecological service) and providing another good or service (such as agricultural output), then information about the relative values of these alternative goods and services (through nonmarket valuation) can lead to better-informed and more defensible choices" (STEPHENSON AND SHABMAN, 2008:17f). This is why a policy that promotes the intensified use of renewable energy in general or hydropower in particular, should take all the social and economic benefits, as well as environmental impacts into account. It is very important to gain knowledge on how consumers of renewable energy weigh the contribution of an intensified use to the economic development, the environment or the security of supply. Generally, there exists considerable demand for the conservation of nature and landscape throughout the population (MEYERHOFF AND PETSCHOW, 1998:27). In principle, consumers appreciate products that possess properties they value more highly than those in a standard product. Although hydropower or renewable energy sources create negative environmental side effects, they are prime candidates for delivering electricity to environmentally conscious consumers (TRUFFER ET AL., 2001:20).

Figure 33: Basic research questions of the dissertation

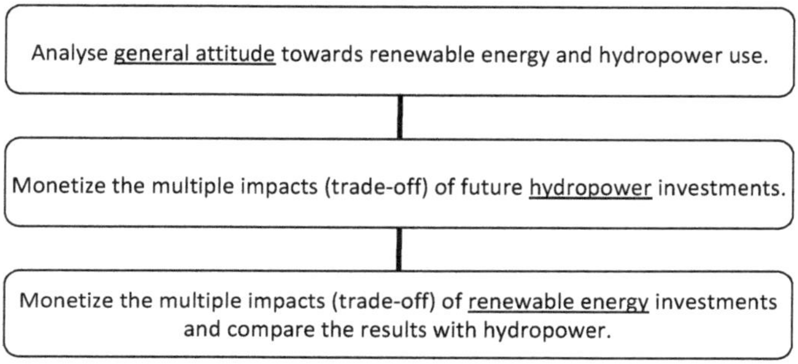

```
┌─────────────────────────────────────────────────────────────────┐
│  Analyse general attitude towards renewable energy and hydropower use. │
└─────────────────────────────────────────────────────────────────┘
                                 │
┌─────────────────────────────────────────────────────────────────┐
│  Monetize the multiple impacts (trade-off) of future hydropower investments. │
└─────────────────────────────────────────────────────────────────┘
                                 │
┌─────────────────────────────────────────────────────────────────┐
│  Monetize the multiple impacts (trade-off) of renewable energy investments │
│  and compare the results with hydropower.                       │
└─────────────────────────────────────────────────────────────────┘
```

Source: OWN REPRESENTATION

The general aim of this work is to examine public perception and preferences for an expansion of hydroelectric power and other renewable energy sources in Austria. First, the dissertation aims to assess people's general attitude towards renewable energy and hydropower use. Next, this work refers to the monetary valuation of the multiple impacts associated with future hydropower investments. The last part of the dissertation deals with the quantification of public preferences for the multiple impacts of different renewable technologies (see Figure 33). This is especially important since not only hydropower but also other renewable energy sources such as biomass, photovoltaics or wind power play a role for future renewable energy expansion.

However, a general issue that occurs when thinking about the multiple impacts of hydropower or renewable energy in general is the problem of non-market goods and services. Correspondingly, some of the effects associated with future renewable energy or hydropower investments, as for instance environmental damages, landscape impacts or reduced air pollution are not traded in commercial markets, i.e. have no market prices. This is why we have to draw on non-market valuation techniques which enable to assign a monetary value on something that is not traded in conventional markets. This is important because "failure to monetize preferences for environmental services will result in ignoring these preferences when decisions are made (STEPHENSON AND SHABMAN, 2008:18). The various concepts of non-market valuation are elucidated explicitly in the following chapter.

4 Economic valuation with Stated Preference techniques

4.1 General remarks

"Economic values are needed for a variety of purposes and underpin a range of different decisions" (PEARCE ET AL., 2002:18). Accordingly, the results of economic valuation may represent important inputs for cost benefit analysis (CBA) in order to conclude whether a project or policy is acceptable or not. Furthermore, economic values may simply be used to demonstrate the importance of an issue or to prioritise political decisions (PEARCE ET AL., 2002:18f). More specifically, measures of economic value may be required to analyse policies that affect the environment and to evaluate environmental damages resulting from human action (KAHNEMAN AND KNETSCH, 1992:57). Generally speaking, the aim of economic valuation is to uncover the total economic value (TEV) of the good or service in question. "TEV identifies all changes in human wellbeing that accrue from a change in the provision of the good. Those values may accrue to users – persons who make direct or indirect use of the good – and to non-users – persons who are willing to pay for the change in the provision of the good but who make no direct use of the good" (PEARCE ET AL., 2002:17). According to that, the TEV of a good or service typically consists of the sum of use and non-use values (PEARCE AND SECCOMBE-HETT, 2000:1420). That is:

$$TEV = USEVAL + NONUSEVAL \qquad (4.1)$$

Use values are generally categorised into direct and indirect use values. The basic assumption is that use values are associated with the direct use of a good or more precisely an environmental amenity (ADAMOWICZ ET AL., 1998a:4; MITCHELL AND CARSON, 1990:62). In the case of renewable energy sources as the environmental good to be valued, the use value is derived for instance from the fact that people purchase "green electricity" to get their homes electrified (MENEGAKI, 2008:2424). However, during the last years economists paid increasing attention to non-use values (CHEE, 2004:553). The concept of non-use values, which are categorised into option, bequest and existence values, refers to the value that people derive independent of any (present or future) use (HAUSMAN, 1993:6; LIEBE AND MEYERHOFF, 2005:4). First, the bequest value arises from the desire of individuals to preserve goods for the use of future generations (altruistic motives). For instance, individuals benefit from bequeathing a clearer environment to the next generation as a result of reduced emissions. Second, the existence value reflects benefits from simply knowing that a good or service exists, even though an individual makes no direct use of it. For example, people may benefit from the preservation of a natural landscape, although they do not intend to make any kind of recreational use of it (MENEGAKI, 2008:2424; LIEBE AND MEYERHOFF, 2005:5f). Finally, people may be willing to pay for an environmental amenity to conserve the option of future use. This intention constitutes an option value (see Table 17; PEARCE ET AL., 2002:23).[75]

Table 17: **The categorisation of TEV into use and non-use values**

Use values	Non-use values
– Direct use value	– Bequest value
– Indirect use value	– Existence value
	– Option value

Source: LIEBE AND MEYERHOFF (2005:5); OWN REPRESENTATION

The essence of economic valuation is the assignment of monetary values to either changes in environmental services and functions or to an environmental amenity itself, considering both use values as well as non-use values (PEARCE AND SECCOMBE-HETT, 2000:1419). Usually, the price system of conventional markets is doing the elicitation work, meaning that the price paid for a good indicates what it is worth to people (HAUSMAN, 1993:6; HANEMANN, 1994:20). However, most environmental goods (e.g. natural resource sites) are not traded in markets. Consequently, it is very difficult to assign a monetary value on something that is not traded and

---

[75] The components of TEV are classified differently by different authors. For instance, the option value is often regarded as a use value since the option is related to an individual's own future use of the good or service (MENEGAKI, 2008:2424).

does not affect individual actions in the normal manner (HAUSMAN, 1993:4; CARSON, 1999:1). Basically, there are two ways of estimating values attached to non-market goods and services (and bads): *revealed* and *stated preference techniques* (see Figure 34).

Figure 34: Economic valuation techniques in relation to the concept of total economic value

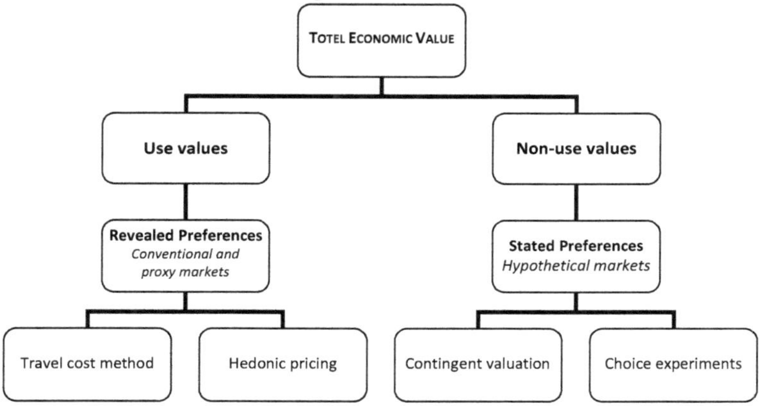

```
                        TOTEL ECONOMIC VALUE
                       /                    \
              Use values                  Non-use values
                  |                             |
         Revealed Preferences           Stated Preferences
         Conventional and               Hypothetical markets
         proxy markets
         /              \                 /              \
  Travel cost    Hedonic pricing    Contingent      Choice
    method                          valuation       experiments
```

Source: PEARCE AND SECCOMBE-HETT (2000:1420); OWN REPRESENTATION

Environmental goods or generally speaking non-market goods that create benefits through direct or indirect use can be valued by means of revealed preference methods. Revealed preference approaches quantify the value of a non-market good by studying actual (revealed) behaviour on a closely related market, the proxy market. The two most popular revealed preference methods are the travel cost method and the hedonic pricing method (ALPIZAR ET AL., 2001:2; ADAMOWICZ ET AL., 1994:271; CARSON AND FLORES, 2000:5). The basic idea of the travel cost approach is to value environmental or recreational assets (e.g. national parks etc.) via the expenditures on travelling to the site (travel costs, entrance fees etc.). Hedonic pricing is based on the idea that market prices reflect the prices of the individual attributes of a good. The method therefore refers to the measurement of effects, which show up in real markets like labour or property markets (PEARCE ET AL., 2002:31; CHEE, 2004:555).[76]

---

[76] For instance, property markets may be used to capture the relationship between house prices and wind turbines located in the neighbouring area (LADENBURG AND DUBGAARD, 2007:4060).

The problem is that people may value an environmental good at least partly for reasons unrelated to their consumption (non-use value). If this is the case, revealed preference methods capture just part of people's value, namely only the use value component (HANEMANN, 1994:20). Hence, the most notable disadvantage of revealed preference techniques is the impossibility of measuring non-use values associated with an environmental good or service. Another weakness of revealed preference techniques is the impossibility to value future environmental changes or policies. For example, if a recreational site or a policy for promoting renewable energy sources has not yet been implemented there are no revealed preferences to study. By contrast, with stated preference methods future developments can easily be captured (ALRIKSSON AND ÖBERG, 2008:244; LIEBE AND MEYERHOFF, 2005:8). For those reasons, research in the area of non-market valuation has increasingly focussed on stated preference methods over the last decades (ALPIZAR ET AL., 2001:2; VENKATACHALAM, 2004:90).

4.2 Contingent valuation

Stated preference techniques are commonly based on constructed (or hypothetical) markets, meaning that people are induced to state their preferences for an environmental good within the scope of a survey (PEARCE ET AL., 2002:31; HANLEY ET AL., 1998a:2; LIEBE AND MEYERHOFF, 2005:7). The most common stated preference technique in economic analysis is the contingent valuation (CV) method (ADAMOWICZ ET AL., 1994:272; ADAMOWICZ ET AL., 1998a:2). CV has been used since the early 1970s to value environmental goods that are not traded in conventional markets, for instance recreational sites, wetlands, air or water quality (MITCHELL AND CARSON, 1990:2; MACMILLAN ET AL., 2006:299). The basic idea of the CV method is to elicit people's preferences for a specific environmental good by asking directly what they would be willing to pay at the maximum for a change of the regarded good or service.[77] This approach circumvents the absence of conventional markets by creating a hypothetical market in which people have the opportunity to buy the good in question (MITCHELL AND CARSON, 1990:2f; PEARCE ET AL., 2002:16; CARSON, 1999:1). More

---

[77] One can also ask for the minimum compensation requirement for a loss of the environmental good. This refers to the concept of willingness to accept (WTA).

precisely, CV focuses on a precise scenario (e.g. increased use of renewable energy sources) and aims to gather information on people's willingness to pay (WTP) regarding this precise scenario (ADAMOWICZ ET AL., 1998b:65). In doing so, the WTP question can have various elicitation formats. The elicitation technique is an important component of any CV application and can be divided into four major interrogative forms that are shown in Table 18 (VENKATACHALAM, 2004:105).

Table 18: Elicitation techniques for contingent valuation studies

Elicitation technique	Example
Open-ended question	How much are you at maximum willing to pay in addition to your monthly electricity bill for the increased use of renewable energy sources?
Bidding game	Are you willing to pay € 10 in addition to your monthly electricity bill for the increased use of renewable energy sources? 1. case "yes": The interviewer increases the WTP amount until the respondent refuses to pay. 2. case "no": The interviewer decreases the WTP amount until the respondent accepts or zero WTP is reached.
Payment card	Which of the following money amounts describes best your maximum willingness to pay for an increased use of renewable energy sources? € 0 € 1 € 2 … € 10 … > € 100
Single-bounded dichotomous choice	Are you willing to pay € 10 in addition to your monthly electricity bill for the increased use of renewable energy sources? *(Stated money amounts vary within the sample)*
Double-bounded dichotomous choice	Are you willing to pay € 10 in addition to your monthly electricity bill for the increased use of renewable energy sources? *(Stated money amounts vary within the sample)* 1. case "yes": Are you willing to pay € 20? 2. case "no": Are you willing to pay € 5?

Source: LIEBE AND MEYERHOFF (2005:14); OWN REPRESENTATION

First, the most convenient way of elevating WTP for a public good or policy is to use an open-ended question. With this elicitation technique, information on the maximum WTP is gathered through direct questions like "What are you willing to pay for an expansion of renewable energy sources?" (VENKATACHALAM, 2004:106f; WHITEHEAD, 2000:9f). Thus, respondents have to find a money amount that reflects their preferences for the regarding environmental good on their own (LIEBE AND MEYERHOFF, 2005:13). However, with such an interrogative form people are sometimes overextended. Respondents may be unfamiliar with stating maximum WTP values, because "real-life" market transactions usually involve only the decision whether or not to buy a good at a certain price (PEARCE ET AL., 2002:50; HANEMANN, 1994:23). Moreover, the open-ended approach is often criticized due to a large number of non-responses, zero bids and outliers, i.e. unrealistically large bids (LIEBE AND MEYERHOFF, 2005:13), as well as the attraction of strategic behaviour, like free-riding, over- or underbidding (WHITEHEAD, 2000:9). "Zero bids may come from respondents who genuinely believe that what is being valued is not worth anything, but can also arise from a host of other reasons" (CHEE, 2004:556). Accordingly, zero bids may represent protest zeros. In this case respondents state a zero WTP although they have a positive WTP for the good to be valued (CARSON, 1999:10; PEARCE ET AL., 2002:59). This may have various reasons which are usually unknown to the researcher without digging deeper. Untrue zero WTP may be an expression of protest against the proposed payment vehicle or people might believe that paying for the good or service in question is the responsibility of others (CHEE, 2004:556). Furthermore, non-responses may also be a form of protest (PEARCE ET AL., 2002:59). Free-riding means that people consciously indicate an amount far below their true WTP, only because they know that other people might pay for the environmental good in question (VENKATACHALAM, 2004:112). Underbidding, which is also a form of free riding, means that respondents may understate their "true" WTP if they believe that their stated WTP will actually be collected. On the contrary, stated WTP bids are often overstated due to the hypothetical character of CV studies, i.e. due to the fact that people know that no money will directly change hands. These forms of strategic behaviour are fraught with problems because they are extremely difficult to detect in CV applications (CARSON AND FLORES, 2000:25f; CHEE, 2004:556; PEARCE ET AL., 2002:59).

Out of this criticism other elicitation techniques evolved. The bidding game question represents an attempt to gradually converge to the maximum WTP. In particular, people are for instance asked "Are you willing to pay € 3 for an expansion of renewable energy sources?" (PEARCE ET AL., 2002:51). Respondents would then have to answer "yes" or "no" to that particular bid. In case the answer is "yes", the bid will be increased until the respondent refuses to pay the stated amount, i.e. the highest positive response is recorded. Conversely, if the answer is "no" the interviewer will state a lower bid until the respondent accepts the stated WTP amount. The main advantage of the bidding game is that this elicitation format represents a "market-like" situation to the respondents and therefore provides relatively better results. However, the implementation of a bidding game requires the presence of interviewers during the interview resulting in comparatively higher survey costs.[78] Another problem of bidding games is the so called starting point bias, which implies that the starting bid might influence the final value of the stated WTP (VENKATACHALAM, 2004:105f; WHITEHEAD, 2000:9). More precisely, in the presence of starting point bias WTP is anchored on the initially stated value in the bidding game. For instance, if the starting point is, say € 5, average willingness to pay will end up lower than if the starting point is € 25 (PEARCE ET AL., 2002:59; WHITEHEAD, 2000:9). Finally, the bidding game technique might accelerate the problem of socially desirable answers. This corresponds to the problem of "yea-saying" meaning that the respondent accepts a stated bid simply to please the interviewer although his or her true WTP is below the suggested bid (CARSON, 1999:10; LIEBE AND MEYERHOFF, 2005:15).

Another technique to elicit WTP is the payment card approach. With the payment card people are confronted with a range of WTP values from which they have to choose their maximum WTP value that corresponds best with their preferences for the good under question. Problems that may arise using the payment card approach are distortions of stated WTP due to the chosen range of WTP amounts ("range bias") and the selected endpoint (LIEBE AND MEYERHOFF, 2005:15; VENKATACHALAM, 2004:106). This

---

[78] Thus, the bidding game is usually impracticable in mail surveys (VENKATACHALAM, 2004:106).

means that average WTP is likely to rise if another higher response category is included in the payment card or a higher endpoint is chosen. This is because survey respondents are usually very open to suggestions when answering unfamiliar questions (WHITEHEAD, 2000:9).

The latest elicitation approach, overcoming some of the problems associated with the other techniques, is the dichotomous choice (referendum method). The dichotomous question can be single-bounded or double-bounded. The single bounded dichotomous choice corresponds to a take-it-or-leave-it situation. Respondents are asked to state only "yes" or "no" to a single predetermined bid that potentially reflects the maximum WTP of the respondent; this predefined bid varies within the sample (PEARCE ET AL., 2002:52; VENKATACHALAM, 2004:107). An extension of the take-it-or-leave-it approach is the double-bounded dichotomous choice. This approach involves a follow-up question after the initial single bound question, whose direction depends on the respondent's answer ("yes" or "no") to the initial bid (for a detailed example see Table 18). This procedure enables the researcher to bound WTP between two threshold values (VENKATACHALAM, 2004:107; WHITEHEAD, 2000:10). The main advantage of dichotomous choice questions is that respondents are confronted with a simple yes/no decision similar to "real-life" market transactions where people usually face discrete choices (e.g. buying a supermarket good at a certain price). Furthermore, this technique minimises non-responses and avoids outliers (LIEBE AND MEYERHOFF, 2005:13; PEARE ET AL., 2002:52). Disadvantages associated with dichotomous choice questions (single or double-bounded) refer to yea-saying problems and starting point bias (see before). Additionally, referendum methods provide less information for each respondent, since the researcher only knows whether WTP is above or below a certain threshold value. With double-bounded questions WTP can at least be localized between two threshold levels. Nevertheless, with dichotomous techniques larger samples and more sophisticated econometric methods are required to gain information on average WTP (PEARCE ET AL., 2002:52; WHITEHEAD, 2000:10).

To sum up, there is no general agreement upon which elicitation technique is the best to inquire WTP for a non-market (environmental) good. The selection of the elicitation approach may in detail depend on the survey method (mail survey or interviews), as well as the budget available for the implementation of a CV study. However, one must keep in mind that the elicitation technique determines which kind of WTP measure can be calculated. Accordingly, WTP estimates can differ substantially dependent on the way people were asked to state their WTP. For instance, open-ended questions usually yield lower WTP values than closed-ended formats like dichotomous choices (CARSON, 1999:9; HANLEY ET AL., 1998a:2; PEARCE ET AL., 2002:50).

"The standard interpretation of CVM results is that the WTP for a good is a measure of the economic value associated with that good, which is fully comparable to values derived from market exchanges and on the basis of which allocative efficiency judgements can be made" (KAHNEMAN AND KNETSCH, 1992:58). However, embedding effects and the hypothesis of "warm glow" raise doubt on this interpretation. Embedding, also known as part-whole effect, means that WTP for the same good varies depending on whether the good is valued on its own or as a part of a more inclusive bundle (VENKATACHALAM, 2004:96). People are usually willing to pay somewhat more for more of a good provided that individuals are not satiated (ARROW ET AL., 1993:11). However, in case of embedding this relationship does not hold. Hence, "the same good is assigned a lower value if WTP for it is inferred from WTP for a more inclusive good rather than if the particular good is evaluated on its OWN" (KAHNEMAN AND KNETSCH, 1992:58). In other words, embedding occurs "when the WTP for one good is found to be insignificantly different from the WTP for a more inclusive good" (HARRISON, 1992:248). For instance, if people's WTP for the expansion of renewable energy sources is only slightly higher than WTP for hydropower expansion which is only part of renewable energy expansion, then embedding will occur. The occurrence of embedding may reflect standard economic theory, i.e. the phenomenon of decreasing marginal utility. Thus, marginal utility of an individual declines with any additional bundle of the good they consume (VENKATACHALAM, 2004:99). Other reasons for embedding driven by the researcher may be a poor design of the survey instrument, an ineligible implementation of the survey and the sampling procedure, or the inability of respondents to understand the survey questions (VENKATACHALAM, 2004:102).

Another problem that may arise in CV studies are "warm glow" effects indicating that people express a willingness to acquire a sense of moral satisfaction by contributing to the provision of a good (KAHNEMAN AND KNETSCH, 1992:64). Hence, people derive utility from the act of giving, i.e. they feel the "warm glow" from donating to worthy causes (ARROW ET AL., 1993:17; CARSON, 1999:8). The degree of moral satisfaction people derive from the act of paying differs depending on the type of the good to be valued. For instance, "saving the panda may well be more satisfying for most people than saving an endangered insect" (KAHNEMAN AND KNETSCH, 1992:64). If warm glow effects occur, CV responses may represent important indicates of approval for the environmental program or policy in question, rather than reliable estimates of true WTP (ARROW ET AL., 1993:17).[79]

4.3 Choice modelling

In contrast to the contingent valuation method, choice experiments (CE) represent a relatively innovative valuation technique, which has been deployed to assess the monetary value of non-market environmental goods and policies for several years. The CE approach originates from the field of marketing (conjoint analysis) and transport economics (LIEBE AND MEYERHOFF, 2005:15; ALRIKSSON AND ÖBERG, 2008:245; BOXALL ET AL., 1996:252; HANLEY ET AL., 1998a:3). It is based on the assumption that consumers derive utility from the properties or characteristics of a good and not from the good per se. This is formally known as the "Characteristics theory of value" first presented by LANCASTER (1966:133), and implies that the value of a good, service or policy (e.g. a hydropower expansion strategy) can be expressed by its characteristics or attributes (RYAN ET AL., 2001:55; LOUVIERE ET AL., 2000:2). These attributes have in turn different

---

[79] Latest discussion on the method of contingent valuation is given in CARSON (2012) and HAUSMAN (2012).

levels. By varying attribute levels, representing the experimental design of a CE, "packages" or "bundles" of attributes that reflect different states of the good in question are created. Individuals are then asked to choose their preferred alternative from a selection of two or more different "packages", which are described in terms of their attributes and levels (BOXALL ET AL., 1996:244; HANLEY ET AL., 1998b:44).[80] Such a selection of "packages" is known as the "choice set" or "choice card" (BOXALL ET AL., 1996:244). Typically, one of the attributes used to describe the good in question is a price or cost factor (see Figure 35). Furthermore, respondents are usually asked to make a sequence of choices (HANLEY ET AL., 1998b:414).[81]

Figure 35: Basic idea of implementing a choice experiment

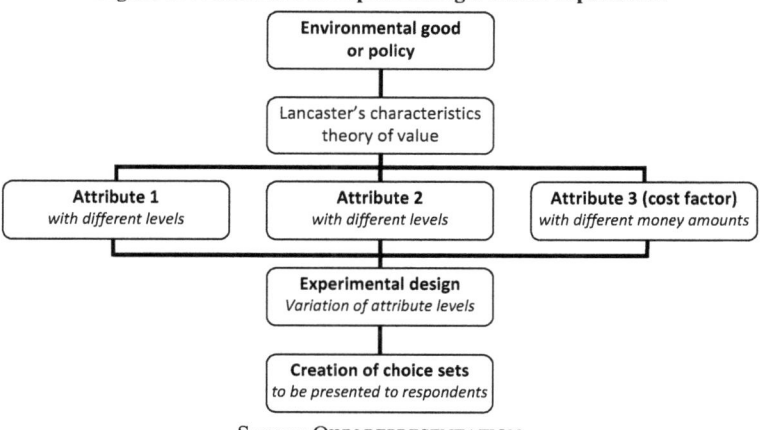

Source: OWN REPRESENTATION

The sequence of choice outcomes enables the analyst to gain four major pieces of information. First, it can be shown which attributes significantly influence respondent's choice. Second, it is possible to gain information on the implied ranking of the attributes used in the CE. Third, the inclusion of a monetary attribute enables to elicit marginal WTP for a one unit change in any significant attribute, and finally, attribute-based stated choice methods allow the researcher to value situational changes, i.e. to estimate WTP

---

[80] For further information see also ALRIKSSON AND ÖBERG (2008:245f), BENNETT AND BLAMEY (2001:6), HANLEY ET AL. (1998a:2) or LIEBE AND MEYERHOFF (2005:15f).
[81] See also BENNETT AND BLAMEY (2001:6), BOXALL ET AL. (1996:244) or HANLEY ET AL. (1998a:2).

for a policy which changes more than one attribute simultaneously (BOX-ALL ET AL., 1996:244f; ADAMOWICZ ET AL., 1998b:65; HANLEY ET AL., 1998a:2f; LANCSAR AND SAVAGE, 2004:1; LIEBE AND MEYERHOFF, 2005:15ff). The theoretical and econometric framework behind the choice experiment approach is elaborated in greater detail in section 9.1 of this work. This chapter aims to simply describe the basic idea of choice experiments.

Other forms of choice modelling include contingent ranking and contingent rating methods. With contingent ranking respondents are asked to rank the alternatives presented in the choice set according to their preferences instead of choosing one alternative. Within the scope of a contingent rating experiment, respondents are asked to independently assign predetermined scores to all of the alternatives (ALRIKSSON AND ÖBERG, 2008:246f; PEARCE ET AL., 2002:56).

4.4 Comparison of the two methods and general limitations

In contrast to the contingent valuation approach, choice experiments involve a broader attribute based perspective (ADAMOWICZ ET AL., 1998a:29). While CV focuses on the valuation of a precise scenario which presents for instance a potential environmental quality change, the CE approach aims to assess people's preferences over the attributes of the scenario rather than the specific scenario at hand (ADAMOWICZ ET AL., 1998b:65; ROLFE ET AL., 2000:290). Hence, the CE method enables to elicit trade-offs between the attributes used to describe the environmental good to be valued[82.] (POWE ET AL., 2005:514; ADAMOWICZ ET AL., 1998b:65; BOXALL ET AL., 1996:244). The CV method, by contrast, primarily focuses on evaluating the price of one specific scenario and therefore represents only a single trade-off approach (ADAMOWICZ ET AL., 1998a:29; ROLFE ET AL., 2000:290).

Generally, choice experiments overcome some of the problems associated with CV. Thus, the CE approach possesses several advantages when compared to the CV method. First, choice modelling considers the multidimen-

---

[82] For example, people may be willing to accept the impact of wind turbines on landscape in exchange for a clearer atmosphere through reduced air emissions.

sionality of an environmental good (PEARCE ET AL., 2002:32; HANLEY ET AL., 1998b:425). With the attribute based perspective of CEs, it is easier to disaggregate values for environmental goods into the values of the attributes that characterise this good (HANLEY ET AL., 1998a:4 and 1998b:416). Consequently, the CE approach is very flexible with respect to modelling the complex trade-offs between these attributes (ADAMOWICZ ET AL., 1998b:65). This is important since the marginal value of changing specific attributes might produce more interest to managers and policy makers than the value of the environmental good as a whole (HANLEY ET AL., 1998a:4). Management or policy decisions are usually not concerned with losing or gaining the environmental good as a whole[83], but with rather changing attribute levels (HANLEY ET AL., 1998b:416). Furthermore, CEs avoid the so-called "yea-saying" problem which is often dominant in CV studies using dichotomous question formats (HANLEY ET AL., 1998a:4; PEARCE ET AL., 2002:33). In a CE respondents are not faced with an "all or nothing" choice. Instead, they have to choose between two or more "bundles" of attributes in repeated choice situations. Hence, respondents have repeated opportunities to express their individual preferences (HANLEY ET AL., 1998b:416). The multidimensionality of the CE approach additionally reduces strategic behaviour, when compared to CV surveys, since CE approaches do not ask directly for the price (WTP)[84]. Instead, CE methodscreate multiple choice outcomes between "bundles" of attributes versus an "all or nothing" choice in CV (ALPIZAR ET AL., 2001:24; BAUMGART, 2005:55). In addition, the CE method avoids embedding or part-whole bias problems which are often present in CV studies (HANLEY ET AL., 1998a:4). Embedding generally means that the sequence of questions or the nesting of the good into a broader framework affects the results of the environmental valuation. In CEs the individual attributes represent the parts of the good in question and these parts are embedded into the whole good. Thus, respondents are constrained to weigh between the attributes (parts) of the good avoiding implausible high valuations of individual attributes (BAUMGART, 2005:55). Another advantage of CE over CV is the possibility of

---

[83] This is the situation usually able to be addressed by contingent valuation.
[84] As mentioned above, in a CE monetary values (marginal WTP) are determined indirectly through the inclusion of a price factor as one of the attributes.

benefits transfer[85] provided that the measurable attributes of an environmental good can be quantified into monetary values and that socio-demographic characteristics are included in the choice model. Finally, the repeated choices made in CE studies enable the analyst to test for internal consistency since models can be fitted on subsets of the collected data (HANLEY ET AL., 1998b:416; BAUMGART, 2005:55).

Even though stated preference approaches, especially choice experiments, provide valuable advantages, the method suffers from several weaknesses and difficulties. First of all, stated preference techniques, both CE and CV, are generally criticised due to the hypothetical character of the questions and the fact that they are not based on actual behaviour (ADAMOWICZ ET AL., 1994:271). The hypothetical situations presented to respondents are constructed by the researcher making the whole exercise expert-driven. This applies to both, the CE and the CV method. Within the scope of CEs there is no guarantee that the attributes and levels chosen by the researcher to describe the good in question reflect the true preferences of the respondents. Another general problem in stated preference studies is the hypothetical bias meaning that respondents may state high WTP values knowing that there will not be any real consequences from their indicated answers or choices. Other challenges stated preference techniques are confronted with involve the problem of different estimations of economic value resulting from different elicitation formats, self-selection bias[86] or providing the right amount of information to respondents (ALRIKSSON AND ÖBERG, 2008:250). Finally, there is lots of discussion whether the results of a stated preference study actually reflect people's preferences or rather attitudinal intentions towards the good in question (ALRIKSSON AND ÖBERG, 2008:250; LIEBE AND MEYERHOFF, 2005:27).

---

[85] Benefits transfer means that economic values from one context/site estimated within the scope of existing empirical work can be "transferred" or applied to other contexts or sites of interest (HANLEY ET AL., 1998a:11; PEARCE ET AL., 2002:35).

[86] Self-selection bias means that people with a strong opinion on the subject rather participate in the survey than people without any affiliation to the subject (ALRIKSSON AND ÖBERG, 2008:250). For example, people who are very affine to environmental protection and pollution control are more likely to respond to a survey on renewable energy than others.

5 Previous research

Although stated preference techniques face general limitations, they are the only viable method to assign a monetary value to non-use components held by most environmental goods. Contingent valuation methods have a long tradition in environmental economics. In contrast, choice experiments represent a more recent approach first used in marketing, transport and health economics. Based on the valuable advantages choice experiments possess, the technique gained increasing importance in the field of environmental valuation and represents a valuable alternative and/or complement to CV methods (ADAMOWICZ ET AL., 1998b). Thus, stated preference techniques are widely used to assess the value of renewable energy sources. In the following, CE and CV applications carried out in the past ten years in Europe, the USA and Asia to estimate WTP for renewable energy are reviewed. Beside the focus on renewable energy in general, two particular energy sources, namely hydro and wind power, are considered.

5.1 Renewable energy studies

An overview of the reviewed stated preference applications on the subject of renewable energy is given in Table 19. Beside the study year, the methodological approach, the survey method, the sample size and the econometric model used to quantify preferences, the last column indicates converted WTP values of the CE attributes and the CV-based WTP estimates. In order to make the estimated WTP values from different study sites (with different currencies) comparable, the indicated figures have been transformed into US Dollar (USD) using Purchasing Power Parities (PPP) of the year in which the particular survey was implemented (OECD-DATABASE, 2012, online).[87] These values have then been converted from USD into Euros using the average yearly exchange rate of the regarding year (EUROSTAT-DATABASE, 2012c, online).

---

[87] Purchasing power parities (PPP) are currency converstion rates that eliminate the differences in price levels between countries within the scope of the conversion process (OECD, 2012, online).

The first study covered in this review refers to BERGMANN ET AL. (2004) which tried to value the external costs and benefits associated with renewable energy investments in Scotland using a choice experiment. The renewable technologies considered within the scope of the study are hydropower, on-shore and off-shore wind power and biomass. The attributes used to describe the renewable energy sources have been determined within the scope of several focus group sessions. These are *landscape impact* (with the levels none, low, moderate or high), *wildlife impact* (slight improvement, no impact, slight harm), *air pollution* (none, slight increase), *local jobs* excluding temporary employment during the construction phase of renewable energy projects (1-3, 8-12, 20-25 jobs), and finally, *price* defined as an annual increase in household's electricity bill (£ 0, 7, 16, 29, 45). The sample contains 211 randomly selected Scottish individuals; the response rate of the mail survey was 43 %. A standard Multinomial Logit (MNL) model including CE attributes and socio-demographic characteristics was used to estimate respondents' preferences and to calculate marginal WTP for the individual attributes.[88] First, households possess a positive WTP for a decrease from a high landscape impact to no impact. Wildlife improvements are, compared to a slight increase in harm, also associated with a positive WTP. Furthermore, renewable energy projects that do not cause any increase in air pollution are valued positively. Employment effects did not show up to be a statistically significant determinant of choice in the econometric model. Finally, welfare analysis showed that, compared to a base case[89], Scottish households place the greatest value on off-shore wind farms, mainly due to the absence of landscape impacts. Biomass projects are second best valued especially because of the high employment effects associated with this type of technology.

BERGMANN ET AL. (2008) extended the investigation of BERGMANN ET AL. (2004) in order to account for heterogeneous preferences and to differentiate between urban and rural preferences. First, differences in individual's taste, i.e. preference heterogeneity, were captured using a Random Parame-

---

[88] Detailed estimates of marginal WTP are indicated in the last column of Table 19.
[89] The base case is defined as follows: fossil fuel expansion, low landscape impact, no wildlife impact but an increase in air pollution.

ter Logit (RPL) model instead of a standard MNL model. Significance and signs of the estimated coefficients remained unchanged as compared to the model in BERGMANN ET AL. (2004). Second, distinctions between urban and rural preferences were addressed by dividing the whole sample into an urban and a rural subset. For each subsample separate models were estimated. The results indeed differ between the two groups, especially regarding employment creation. While the creation of permanent jobs from renewable energy investments is no statistically significant determinant of choice in the urban-sample model, rural residents exhibit a positive WTP for each additional full time job created by renewable projects. Moreover, wildlife benefits and reductions in air pollution are valued more highly by rural people than by urban ones. Finally, alternative renewable energy development scenarios create different welfare implications to urban and rural residents. Hence, in the urban subsample only large off-shore wind farm projects are associated with a significant positive WTP. By contrast, in the rural sample all renewable energy project scenarios create a positive welfare, with biomass projects ranked first followed by large off-shore wind farms.

Another study from LONGO ET AL. (2008) aimed to investigate people's preferences in Bath (England) for a hypothetical programme that promotes the use of renewable energy sources. Within the scope of focus groups and one-on-one interviews, the following attributes for describing the hypothetical renewable energy policies were specified: *Reduction of greenhouse gas (GHG) emissions* (with the levels -1 %, -2 % or -3 % per year), *length of electricity shortages* (30, 60 or 120 minutes per year), *employment in the electricity sector* (+1000, -1000 or 0 jobs) and finally, *increase in quarterly electricity bill* (£ 6.5, 16, 25 or 38). The survey was carried out by professional interviewers asking 300 respondents in central areas of Bath (e.g. shopping centres, public parks). The results of a standard MNL model[90] show that GHG emissions, electricity blackouts, jobs and price are statistically significant determinants of choice with the expected signs. The subsequent calculation of WTP reveals that respondents exhibit a positive WTP for a 1 % reduction of GHG emissions, a decrease of electricity

---

[90] Due to the fact that random parameter standard deviations did not show up to be statistically significant, the authors sticked to the classical MNL model.

shortages by 1 minute a year and for every additional permanent job in the energy sector. Compared with the results of previous studies (e.g. BERGMANN ET AL., 2004), renewable energy policies that increase employment are valued significantly positive. Finally, the investigation concluded that people consider energy security as an important externality justifying public efforts to make improvements in this area.

An analysis from FIMERELI ET AL. (2008) attempted to investigate public preferences in South-East England for the use of low-carbon energy technologies in electricity production. The energy sources considered within the scope of this study are on-shore wind power, biomass and nuclear power. These technologies are described by five attributes, namely *distance to respondent's home* (with the levels 0.4, 1, 6 or 10 km), *local biodiversity impact* (no change, more, less biodiversity), CO_2 *emission reduction* associated with 20 % of electricity generation (-50 %, -90 %, -95 %, -97 %, -99 %), *total land use* required to generate 20 % of electricity by 2020 (568 ha, 5,832 ha, 816,000 ha) and *cost* defined as an increase in household's annual electricity bill (£ 20, 40, 67, 90 or 143).[91] The technologies themselves represent the labels (names) of the CE alternatives. Identification and refinement of the attributes and their levels was carried out in the course of expert discussions, focus groups, pilot interviews and a small-scale pilot survey. The main survey was distributed among residents in three randomly selected towns (with more than 100,000 inhabitants) in South-East England. In total, 1,200 questionnaires were sent out; the response rate was 31 % resulting in a sample size of 376 respondents. A standard MNL model was applied to quantify people's preferences over the selected low-carbon technologies. Based on the statistically best fit model it was found that respondents prefer energy options that are further away from their home and lead to an increase in local biodiversity. In addition, choice is positively affected by higher CO_2 emission reductions and negatively affected by higher electricity costs. Finally, the type of technology plays a significant role for choice. Respondents have a strong preference for on-shore wind power and an aversion to nuclear power. WTP analysis

---

[91] The levels of the attributes *local biodiversity* and CO_2 *emission reduction* are classed with technology, i.e. each technology has its own attribute levels. *Total land use* is fixed to technology; the 5,832 ha are for wind power, 816,000 ha for biomass and 568 ha for nuclear power.

shows that respondents are willing to incur extra costs for the development of on-shore wind power and biomass projects. Furthermore, people exhibit a positive WTP for every mile a power station is located further away from their homes, an increase in biodiversity as well as every percent reduction in CO_2 emissions. Another interesting result of the study is that previous knowledge of and experience with the regarded energy technologies significantly influenced respondent's answers in the choice exercise.

Northern European countries heavily rely on the development of renewable energy sources like wind or hydropower instead of continuing to invest in fossil fuels. Therefore NAVRUD AND BRATEN (2007) analysed people's preferences for different renewable energy sources and their characteristics in Norway. They developed a rather simple CE considering the *technologies* wind power, hydropower and natural gas (label). Environmental impacts associated with the development of these technologies were described to the respondents rather than using them as a separate attribute. The attributes, by contrast, are the *size of power plants* (few large power plants, more medium sized power plants or many small power plants) and *price* of the renewable energy option defined as an annual fee on the electricity bill (0, 300, 700, 1200, 1800 or 3500 NOK). In total, 189 respondents were personally interviewed, 91 of them in a rural area named Smøla island and 98 in the urban area in and around the capital of Norway (Oslo/Akershus).[92] A standard MNL model revealed that people strongly prefer wind power over sustained imports from Danish coal fired power plants (which corresponds to the current situation or "status quo"). This is shown by the positive WTP calculated for wind power technology. Hence, people accept the negative visual impacts of wind farms on landscape in exchange for a reduction of global and regional air pollution. Hydropower schemes and natural gas fired power plants are, by contrast, less preferred over imported coal-based electricity. This is reflected by the negative WTP. In addition, people prefer few large power plants compared to many small or more medium sized ones. This is due to the concentration of the visual impacts at selected sites large power plants are associated with, leaving the possibility to preserve

---

[92] In Smøla island, the largest wind farm in Norway is located. In the urban region around Oslo/Akershus, by contrast, there are no renewable energy projects.

the landscape elsewhere. A final result worth to be emphasized refers to the "Not in my backyard" (NIMBY) effect the authors were able to detect. By estimating two separate models for the rural and urban sample it was shown that people in the rural area value a replacement of imported fossil fuels by local production of renewable energy significantly lower than people from the urban area. This provides confirmation of the NIMBY effect because renewable energy plants are more likely to be established in rural regions. This means that rural respondents have a greater risk of placing renewable energy plants near their residence, simply due to the fact that there is more space.

Finally, two non-European CE applications, one from Asia and one from the United States, are considered within this literature review. Generally, renewable energy investments play an important role not only in Europe but also in other parts of the world. Thus, KU AND YOO (2010) addressed public preferences for the benefits of renewable energy investments in Korea using a CE. In Korea, renewable energy sources account for only 2 % of the total energy supply. Thus, the country puts a lot of efforts to achieve an 11 % contribution of renewable energy sources to total primary energy supply by 2030. In order to measure the various economic values of renewable energies the following four attributes were identified: *landscape improvement* (0 %, 25 %, 50 % improvement), *wildlife improvement* (0 %, 25 %, 50 % more biodiversity in the vicinity of the renewable energy plant), *decrease in air pollution* (0 %, 70 %, 100 % reduction) and *local long-term employment* (0, 10, 30 jobs), all compared to a fossil-fuel power plant (which was defined as the "status quo"). The monetary attribute was specified as a monthly increase in household's electricity bill (0, 1000, 2000, 4000, 7000 Korean won). In total, 774 applicable face-to-face interviews were conducted in the metropolitan areas of Korea (Seoul, Incheon and Gyeonggi) by professional interviewers of an opinion research institute. The authors used a Multinomial Probit (MNP) model to quantify respondent's preferences over the impacts of renewable energies. This model specification considers unobservable heterogeneous preferences in the population which cannot be captured by a standard MNL model. The statistical results show that wildlife improvements, additional employment and emission reductions positively affect people's choice. For those attributes respondents exhibit a positive WTP. The landscape attribute, however, is statistically not significant indicating that people do not derive utility from the improvement of landscape.

BORCHERS ET AL. (2007) tried to estimate public preferences for voluntary participation in green electricity programmes in New Castle County (Delaware, USA). In doing so, the authors distinguished between a generic green energy source and specific renewable technologies including wind power, solar power, farm methane and biomass. These technologies represent the levels of the first attribute. The second attribute used to describe the green energy programme was the percentage of respondent's monthly electricity consumption that would come from the regarding green energy source (levels: 0 %, 10 %, and 25 %). The payment vehicle was specified as an increase in respondent's monthly electricity bill (with the levels US-$ 0, 50, 10, 15, 20 or 30). Sampling was done by making interviews at the Departments of Motor Vehicle locations in New Castle County since all Delaware drivers have to renew their licenses periodically at these locations. This procedure thus allows for a nearly random sample. In total, 128 interviews were conducted. The authors used a Nested Logit (NL) model to quantify people's preferences over green energy programmes. More specifically, the NL model allows for a relaxation of the independence of irrelevant alternatives (IIA)[93] assumption, which is an underlying part of the classic MNL model. In the NL model, choice is partitioned into two parts. First, respondents decide between the "status quo" (no green energy) and participation in a green energy program. In the second step, choices between different green energy programmes are analysed including those respondents who have opted for green energy in the first step. The results of the NL model revealed that first, higher quantities of green energy increase, and second, higher monthly electricity costs decrease the probability of choosing a green programme. However, the most important results refer to the perception of different renewable energy sources. The basic message obtained from the model is that people do not perceive green energy sources

---

[93] IIA means that the relative probabilities of two options being selected are unaffected by the introduction or removal of additional alternatives (LOUVIERE ET AL., 2000:44).

as equivalent. Instead, specific renewable technologies are more preferred compared to others. Thus, solar power is the most preferred energy source compared to wind power and generic green energy.[94] In contrast, biomass and farm methane represent the least preferred technologies. This preference ordering is also reflected when looking at WTP for green energy programs with 25 % of the electricity coming from the specific source (for details see Table 19).

Beside the application of the CE approach to value the external costs and benefits of renewable energy sources, research has intensely focussed on the use of Contingent Valuation (CV) methods. The recently published study of ZORIC AND HROVATIN (2012) analysed households' WTP for renewable energy based electricity in Slovenia. The authors relied on a two-stage procedure asking first whether people would participate in a green electricity programme when it is offered to them by their current electricity supplier (yes/no decision). Subsequently, respondents who were generally willing to participate (yes-answers) have been asked to state how much more they would be willing to pay in addition to their monthly electricity bill. The elicitation technique used for the WTP question was an open-ended format. The whole survey was conducted in terms of an online-based format in cooperation with a professional market research institute. In addition, a field survey was implemented in order to capture respondents older than 65 years; these people are usually not familiar with internet surveys. The final sample used for analysis contained 450 respondents in total. On average, people are willing to pay € 4.1 on top of their monthly electricity bill for green electricity; this corresponds to a yearly value of approximately € 49.2.[95] The main determinants of stated WTP are age, household size, household income and environmental awareness. The first two variables are negatively and the latter ones positively related to WTP. Furthermore, the educational level has a positive influence on stated WTP. Finally, people living in rural areas were found to exhibit a higher WTP for green electricity compared to respondents in urban areas. The statistical framework

---

[94] However, there is no statistically significant difference between respondent's perception of wind power and the generic green energy source.
[95] The indicated values have been adjusted for purchasing power and translated into a common currency (Euros) as already mentioned at the beginning of this chapter.

used to figure out these causal relationships were a Tobit regression model as well as a more general double-hurdle model.[96]

Another CV application from ZOGRAFAKIS ET AL. (2010) focussed on the assessment of public WTP for renewable energy sources on the island of Crete (Greece). Currently, a major part of generated electricity arises from wind power, photovoltaic installations, small hydropower plants and biomass based on the by-products of olive oil production. Due to the prevalent geographical and climatic conditions in Crete, the potentials for further energy production from wind, solar and agricultural biomass are very high. Thus, the promotion of renewable energy sources represents an integral part of the Greek energy policy. Public WTP for "green" energy was elevated through 1,440 interviews with randomly selected households from the six major cities of Crete.[97] The elicitation format used to determine public WTP for "green" electricity supply was a double-bounded dichotomous choice (DB-DC) question. Hence, people were asked "Would you be willing to pay € x for renewable projects that create multiple environmental and social benefits to you?" Three different bids (€ 5, € 10, € 12) were used in three different questionnaire versions. Double-bounded means that people who had answered positively to the initial question were asked about a higher bid, while respondents who refused to pay the initial amount were asked a lower bid. After "free-riders" and protest votes were filtered out, a sample size of 1,235 interviews remained. Mean WTP for renewable energy based electricity calculated from the outcomes of the DB-DC answers amounts to quarterly € 17.0 (Median: € 13.5), adjusted for purchasing power. Finally, various influencing factors were identified within the scope of an econometric model. As a result, energy awareness or a high level of energy saving practice affect WTP for the further implementation of renewable energy sources positively. Another important result is the nexus between stated WTP and respondents' family income. Respondents with a

---

[96] The first step of such a double-hurdle model involves the estimation of a Probit model in order to determine which factors significantly influence people's decision to participate in a green energy scheme. In the subsequent step, a truncated regression model is used to address determinants of the level of payment for green electricity.

[97] These are Heraklion, Chania, Rethymnon, Agios Nikolaos, Ierapetra and Sitia.

high family income are on average willing to pay more for "green" energy and vice versa. This result is in line with economic theory.[98]

BOLLINO (2009) explored consumer's WTP for increasing the use of renewable energy sources in electricity production in Italy. Thereby, two elicitation formats were used, the payment card method and the referendum approach (dichotomous choice). Furthermore, the author examined which attitudinal, behavioural and demographic variables affect WTP for renewable energy. For these purposes a national online survey of 1,601 respondents was conducted in collaboration with a professional market research institute. In order to elicit WTP respondents were confronted with a price vector consisting of five bids, once in downward elicitation format (from € 30 to € 0 bimonthly) and once in upward elicitation format (from € 0 to € 20). Additionally, the WTP questions were supplemented with a "certainty correction" measuring the intensity of acceptance of the bid on a five-point scale ("definitely yes", "probably yes", "probably no", "definitely no", "not sure or don't know"). Mean WTP, corrected for the intensity of acceptance, amounts to € 8.1 for a two-month period. Furthermore, data were recoded in terms of a single referendum question.[99] Mean WTP resulting from this recoded dataset amounts to € 8.4. Applying a binary Probit (PRO) model[100], WTP was found to be positively affected by age, education, income, and whether the respondent is a house owner or not. Conversely, females and respondents with a non-accurate knowledge of renewable energy sources exhibit significantly lower WTP values.

A similar investigation for Italy was made by BIGERNA AND POLINORI (2011). The authors estimated public's WTP for renewable energy using a payment card with 17 bids ranging from € 0 to € 200. In total, the representative sample contains 1,019 interviews conducted by a professional

---

[98] ZOGRAFAKIS ET AL. (2010) determined a number of further factors influencing WTP for renewable energy sources. However, these are not explicitly explained within the scope of this review due to lack of space.

[99] This means that each answer (price) obtained from the payment card in combination with the "certainty correction" question was treated like an answer to a single referendum question. In doing so, the answers "definitely yes", "probably yes" and "don't know" were treated as "Yes", all other possible answers as "No".

[100] If these values are adjusted for purchasing power, WTP amounts to € 8.10 and € 8.35 respectively. These bimonthly estimates correspond to yearly amounts of € 48.60 and € 50.10.

survey agency throughout Italy. To correct for hypothetical bias a "certainty correction" question was included as in BOLLINO (2009). Mean WTP adjusted for purchasing power is € 10.1 (median: € 4.2) for the two-month electricity bill period. Correspondingly, annual values amount to € 60.5 and € 25.1 respectively.[101] Using a parametric interval regression model the main factors influencing WTP could be determined. First, better knowledge of renewable energy sources positively affects WTP, while negative appreciations of the future energy situation have an adverse effect on stated WTP. Additionally, the geographical location and the size of respondents' place of residence were found to be statistically significant. Hence, people in the northern and central part of Italy and respondents living in urban areas (with more than 100,000 inhabitants) exhibit a relatively higher WTP. Finally, WTP is influenced by standard socio-economic characteristics like educational level, gender, age or household size.

A CV application from the United States (ZARNIKAU, 2003) examined consumer's willingness to pay a premium for electricity generated from renewable sources like photovoltaic panels, biomass, geothermal power and wind farms ("green power").[102] The data bases for this research were the so-called Deliberative Polls conducted by the seven largest electric utilities in Texas. Within the scope of these polls a number of randomly selected customers (usually 200-250) of the electricity providers are debriefed about their attitudes towards energy-related issues such as renewable energy, energy efficiency or natural gas generation. In particular, people were confronted with a comprehensive questionnaire containing about 100 questions, both at the beginning of the poll (pre-event survey) and after the two-day event where people participated in discussions and educational seminars (post-event survey). The included WTP question used an open-ended format asking people how much more they would be willing to pay on top of their monthly electric bill for electricity generation from renewable technologies. The total sample size of the pre-event survey was 2,800 re-

---

[101] These estimates are based on the conservative interval regression model treating only "definitely yes" answers within the certainty correction as "yes".
[102] In addition, the author explored public WTP for increasing energy efficiency. However, since the topic of this review relies on renewable energy, efficiency issues are not elucidated at this point.

spondents. Mean WTP for renewable energy amounts to monthly € 6.7 in the pre-event survey and € 6.4 in the post-event sample, all values adjusted for purchasing power. This slight decrease in WTP arises from the fact that after participating in the poll more respondents were willing to pay at least a small premium for renewable energy. By contrast, the share of people willing to pay larger premiums decreased subsequently to the poll. Using a Tobit regression model, significant determinants of WTP have been identified. The first result reflects standard economic theory and shows that income is positively related to WTP. Moreover, WTP is positively affected by the educational level of respondents and their ethnic background (white or non-white). Conversely, older respondents are less willing to pay a premium for green power than younger ones. The same applies to house owners who are also willing to pay less.

WISER (2007) explored WTP for renewable energy[103] in the USA distinguishing between collective and voluntary payment vehicles on the one side and provision mechanisms on the other side. Generally, the author distinguished between four payment and provision scenarios: (1) *Mandatory* increase in electricity bill of all customers, funds collected and spent by the *government*; (2) *voluntary* increase in electricity bill, funds collected and spent by the *government*; (3) *voluntary* increase in electricity bill, funds collected and spent by *electricity suppliers*; (4) *mandatory* increase in electricity bill of all customers, funds collected and spent by *electricity suppliers*. This distinction enables to assess differences in stated WTP based on the payment method and the provision arrangement. Another issue addressed in the paper of WISER (2007) is whether expectations about the contribution of others to support renewable energy affect stated WTP. The elicitation format used to elevate WTP was a single-bounded dichotomous choice (SB-DC) question with predetermined bids amounting to $ 0.5, $ 3 and $ 8 per month. Additionally, a four-cell experimental design was used in order to address different payment vehicles (collective or voluntary) and provision methods (government or private). In total, 4,056 randomly selected individuals across the 50 states were given a mail questionnaire. The response rate was 39 % resulting in a total sample size of 1,574 completed

---

[103] In this study, wind, geothermal, solar power and biomass are subsumed under renewable energy.

CV surveys.[104] The empirical distribution of stated WTP first indicates that WTP is significantly lower when payment is voluntary compared to a collective mandatory payment. Furthermore, private provision is associated with higher WTP values than government provision. These results are confirmed in the multivariate Logit model used to identify significant determinants of the probability of willing to pay for renewable energy. Voluntary payments are associated with a lower probability, while private provision is positively related to the probability of willing to pay for renewable energy. In a more complex regression model, participation expectations were found to have a statistically significant impact on the probability of willing to pay. More precisely, respondents who expect that others also contribute to the promotion of renewable energy are more likely to be willing to pay too. Finally, several socio-demographic and attitudinal variables were found to have a significant influence on the probability of willing to pay for green electricity. These are, for instance, income, gender, educational level or environmental awareness in respondent's household.

Another CV application from the USA (MOZUMDER ET AL., 2011) refers to consumer's preferences for renewable energy in New Mexico where electricity providers are forced to generate or buy 10 % of the sold electricity from renewable sources like solar, wind or hydropower. Until 2020 the aim is to achieve a share of 20 %. Hence, the authors explored public WTP for a renewable energy programme that provides 10 % and 20 % respectively from renewable resources. For this purpose, a web-based survey was designed using an open-ended question in order to elicit monthly WTP for the provision of 10 % electricity from renewable sources. Subsequently, people were asked whether they would be willing to pay for incrementing the share of renewables from 10 % to 20 % (yes/no decision). Those who responded positively were then requested to state the monetary amount they would be willing to incur per month for the additional 10 percentage points of renewable energy. The final sample contains 367 completed surveys; the response rate of the online survey was 27 %. Average monthly WTP for

---

[104] In addition, an opinion survey containing more general questions on preferences for renewable energy (no valuation exercise) was conducted. This sample contained 202 respondents in total (response rate: 37 %). The results of this opinion survey are not discussed within the scope of this review.

generating 10 % of electricity from renewable sources is € 10.2 (converted from USD into Euros). Furthermore, about 40 % of the respondents are generally willing to pay for increasing the share of renewable energy from 10 % to 20 %. The mean WTP for this increased provision is € 3.7 per month, a much lower value compared to avarage WTP for the initial 10 % of renewable energy. Using a Tobit regression model several determinants of stated WTP have been identified including classical socio-demographic characteristics. Hence, household income and household size are positively related to WTP, while respondent's educational level has a negative impact on WTP. In addition, environmentally conscious respondents are generally willing to pay more for providing 10 % of renewable energy. The same applies to households donating a higher percentage of their income to environmental organisations. Beyond these basic Tobit models, simultaneous equation systems were estimated in order to further investigate determinants of incremental WTP (from 10 % to 20 % share) and the probability of willing to pay for this increment. As a result, income is a significant determinant of WTP for the baseline provision (10 %), but not for the increased provision (10 % to 20 %) of renewable energy and the underlying decision to pay more for this increased share of renewable energy.

IVANOVA (2005) assessed public WTP for a policy that increases the share of renewable energy sources in Queensland, Australia from 10.5 % to 12.5 %. Additionally, the study addressed differences between voluntary and mandatory payment mechanisms like in WISER (2007). To elicit WTP, the author used two open-ended questions. First, people were asked how much they are willing to pay for increasing the share of renewable energy sources in electricity generation by 2 percentage points as supposed by the Australian Mandatory Renewable Energy Target. Second, respondents were requested to state their voluntary contribution to increase the use of "green" technologies for generating electricity. The random sample from the entire population in Queensland contains 213 households. The survey was mailed out to 820 households; hence, the response rate was 26 %. Average quarterly voluntary payment people are willing to incur for an increase of green energy use is € 15.6 (adjusted for purchasing power and converted into Euros). By contrast, WTP for the policy aim of increasing the share of renewables by 2 percentage points is significantly lower with € 12.2 per quarter.[105] A distinction between technologies shows that WTP for solar and wind power is highest, while hydropower and biomass report significantly lower WTP estimates. Finally, it was found that voluntary WTP decreases with age and increases with income, while gender-related differences do not exist. Further on, people who are environmentally concerned are more likely to pay for an intensified use of renewable energy sources.

The last CV studies reviewed within the scope of this work refer to WTP for renewable energy in Asia. First, NOMURA AND AKAI (2004) explored public WTP for the promotion of renewable energy systems (photovoltaics and wind power) in Japan using a double-bounded dichotomous choice approach. Hence, people were first asked whether they would be willing to pay a premium of 500, 1000 or 2000 yen per month (yes/no decision). In a follow-up question, positive responses were confronted with a higher bid (twice of the first bid), while people responding negatively to the initial question got a lower bid (half of the first bid). In total, about 1,000 questionnaires were mailed out to a random sample of people living in one of the 11 large cities (more than 1 million inhabitants) in Japan and to a selection of residents of one middle sized city (between 150,000 and 1 million inhabitants), one small city (less than 150,000 inhabitants) and one small town or village in each of the ten Japanese areas. With a response rate of approximately 37 %, the final sample contains 379 respondents. Median estimated WTP per month for promoting renewable energy is € 12.4 (converted from yen into Euros considering purchasing power). This corresponds to a quite high annual value of € 149.3.

Another attempt to estimate WTP for renewable energy development in Korea was made by YOO AND KWAK (2009). The main objective of this study was to value Korea's policy aim of increasing the proportion of green electricity in total electricity supply from 0.2 % to 7 %. For this purpose, 800 households were interviewed in the metropolitan area of Korea (including the cities Incheon, Gyeonggi and Seoul). The face-to-face inter-

---

[105] The annual values are € 62.5 (voluntary contribution) and € 49.0 (policy aim) respectively.

views were conducted with the help of a professional polling institute. The CV question format was a double-bounded dichotomous choice situation with bids ranging from 1,000 to 10,000 Korean won (KRW) per month at intervals of 1,000 KRW.[106] In order to account for zero WTP responses, a so-called spike model was estimated in addition to a conventional parametric model. On average, people are willing to pay a premium of monthly € 1.5 for the promotion of renewable energy sources. The non-parametric estimate of mean WTP amounts to € 1.9 per month, all values adjusted for purchasing power and converted into Euros. Main determinants of the propensity to be willing to pay for renewable energy are income and the bid level. People with higher incomes are more likely to approve of the green electricity policy, while higher bids reduce the likelihood for a "yes" response to the dichotomous choice question.

---

[106] The bids for the DB-DC questions were determined within the scope of a pre-test focus group discussion.

Table 19: Overview of existing stated preference applications – renewable energy

Authors (year)	Country	SP technique (elicitation format)	Survey method	Sample (response rate)	Main objective	Model	Marginal WTP in € per year (adjusted for purchasing power)
BERGMANN ET AL. (2004)	Scotland	CE (unlabelled)[107]	Mail survey	n = 211 (43 %)	Valuing the external costs and benefits associated with renewable energy investments and the impact of these on economic welfare.	MNL	– From high landscape impact to no impact: € 44.06 – From slight increase in harm to improved wildlife: € 65.17 – Slight increase in air pollution to none: € 76.86 – Local jobs: *no significant attribute*
BERGMANN ET AL. (2008)	Scotland	CE (unlabelled)	Mail survey	n = 211 (43 %)	Quantify people's preferences for the attributes of renewable energy projects in rural areas of Scotland.	RPL	*See above*
LONGO ET AL. (2008)	England (Bath)	CE (unlabelled)	Interviews	n = 300 (–)	Investigate WTP for a hypothetical programme that promotes the production of renewable energy.	MNL	– GHG emission reduction by 1 %: € 146.27 – Decrease in electricity shortages by 1 min: € 1.78 – One additional job in the electricity sector: € 0.10
FIMERELI ET AL. (2008)	South-East England	CE (labelled)	Mail survey	n = 376 (31 %)	Analyse public preferences for the use of low-carbon energy technologies (on-shore wind power, biomass and nuclear power).	MNL	– Technology: *wind* € 122.77, *biomass* € 71.86 – Distance from home per mile: € 4.29 – Increase in biodiversity: € 28.02 – CO_2 emission reduction by 1 %: € 1.43

---

[107] With an unlabelled CE, generic titles (e.g. Strategy A, Strategy B or Alternative A, Alternative B…) are used as headings for the alternatives in the choice card. Such a generic title does not convey any information to the respondent other than that this is the first of the alternatives (HENSHER ET AL., 2005:112). In contrast, a labelled experiment attaches meaningful names to each alternative in the choice card (e.g. the names of renewable energy sources like hydropower, solar power or wind power).

Table 19 continued

Study	Country	Method	Survey	Sample	Objective	Model	Results
Navrud and Braten (2007)	Norway	CE (labelled)	Interviews	n = 189 (–)	Elicit people's preferences and WTP for different renewable energy sources and their characteristics.	MNL	– Wind power: €89.30 – Hydro power: €-167.26 – Natural gas: €-177.45 – Many small instead of few large power plants: €-42.72 – More medium sized instead of few large power plants: €-31.96
Ku and Yoo (2010)	Korea	CE (unlabelled)	Interviews	n = 774 (–)	Analyse WTP for the benefits of renewable energy projects compared to fossil fuel power plants.	MNP	– Landscape improvement: *no significant attribute* – Improvement in wildlife by 1 %: €0.075 – Reduction in air pollution by 1 %: €0.092 – One additional job: €0.119
Borchers et al. (2007)	USA	CE (unlabelled)	Interviews	n = 128 (34 %)	Elicit WTP for generic green and specific energy sources providing 25 % of electricity from the regarding source.	NL	– Generic green: €162.47 – Solar power: €205.86 – Wind power: €147.85 – Farm methane: €118.32 – Biomass: €101.21
Zoric and Hrovatin (2012)	Slovenia	CV (open-ended)	Online survey	n = 450 (–)	Analyse WTP for electricity generated from renewable energy sources.	TOBIT	– Mean WTP: €49.58
Zografakis et al. (2010)	Greece (Crete)	CV (DB-DC)	Interviews	n = 1,235 (–)	Assessment of public WTP for renewable energy development.	OLS	– Mean WTP: €67.97 – Median WTP: €53.90
Bollino (2009)	Italy	CV (payment card)	Online survey	n = 1,601 (–)	Explore consumer's WTP for increasing the use of renewable energy sources.	OLS, PRO	– Mean WTP (payment card): €48.60 – Mean WTP (referendum): €50.10
Bigerna and Polinori (2011)	Italy	CV (payment card)	Interviews	n = 1,019 (–)	Estimate consumers' WTP for the development of renewable energy sources.	OLS	– Mean WTP: €60.49 – Median WTP: €25.12

Table 19 continued

Study	Country	Method	Survey	n (response rate)	Objective	Model	Results
ZARNIKAU (2003)	USA	CV (open-ended)	Interviews	n = 2,800 (--)	Examine WTP for electric utility investments in renewable energy and energy efficiency.	TOBIT	– Mean WTP (pre-event): €80.73 – Mean WTP (post-event): €76.90
WISER (2007)	USA	CV (SB-DC)	Mail survey	n = 1,574 (39%)	Analyse payment preferences for renewable energy addressing differences between voluntary/mandatory payment vehicles and government/private provision.	LOGIT	*No mean values calculated since single-bounded dichotomous choice questions were used to elicit WTP.*
MOZUMDER ET AL. (2011)	USA	CV (open-ended)	Online survey	n = 367 (27%)	Investigate WTP for different renewable energy scenarios, once providing 10% of renewable energy supply and another time 20%.	TOBIT	Mean WTP (for 10%): €122.59 Mean WTP (for an increase from 10% to 20%): €44.91
IVANOVA (2005)	Australia	CV (open-ended)	Mail survey	n = 213 (26%)	Examine households' mandatory and voluntary WTP for promoting the use of renewable energy sources in electricity generation.	OLS	Mean WTP (voluntary payment): €62.45 Mean WTP (policy aim): €48.97
NOMURA AND AKAI (2004)	Japan	CV (DB-DC)	Mail survey	n = 379 (37%)	Explore public WTP for the promotion of renewable energy systems.	--	Median WTP: €149.32
YOO AND KWAK (2009)	Korea	CV (DB-DC)	Interviews	n = 800 (--)	Estimate WTP for Korea's policy aim to increase the share of renewable energy sources in total electricity supply from 0.2% to 7%.	Spike model	Mean WTP (spike model): €18.40 Mean WTP (non-parametric method): €22.68

Source: OWN REPRESENTATION

5.2 Wind power studies

In addition to research focussing on the multiple impacts of renewable energy sources, there exist a number of valuation studies on the topic of wind power. A selection of investigations – especially CE studies – is summarized in Table 20. In contrast to the renewable energy studies reviewed above (see section 5.1), wind power research mainly focuses on the environmental and visual impacts associated with this specific technology.

So, EK (2005) examined public preferences over the environmental effects of wind power generation in Sweden. The attributes considered in the choice experiment refer to the *location of windmills* (with the levels onshore, offshore or in the mountains), the *height* (60 or 100 meters), the *grouping of wind turbines* (separately located, small groups of less than ten windmills or large groups with ten to fifty wind turbines) and the *noise pollution* (30 or 40 decibel). The monetary attribute was introduced as a change of the electricity price per kilowatt hour (kWh) ranging from -0.10 to +0.15 Swedish krona (SEK). The whole questionnaire was mailed to 1,000 Swedish residential homeowners. The response rate was 56 % resulting in a sample size of 457 respondents. Adjusting for incomplete answers, 488 responses remained in the sample for the econometric analysis. Since people were confronted with choice tasks containing two alternatives (current situation and an alternative wind power scenario respectively), a Random Effects Probit (REP) Model was used to estimate public preferences over the environmental attributes of wind power. The results show that a reduced noise level is perceived as an environmental improvement by the respondents. Additionally, off-shore windmills are preferred over on-shore located ones, while wind turbines in mountainous areas are less preferred compared to their on-shore counterparts. Moreover, the grouping of windmills is a significant determinant of choice. Hence, people dislike large wind parks. Instead, they prefer separately located wind turbines or small groups of not more than ten windmills. By contrast, the height of wind turbines does not affect households' utility in a statistically significant manner.[108]

MEYERHOFF ET AL. (2008) made an attempt to estimate public preferences for the utilization of wind power in Germany. The methodological approach to quantify preferences over wind power was a discrete choice experiment using five attributes in total. These are *size of the wind parks* (small: 4-6, medium-sized: 10-12 or large: 16-18 turbines), *maximum height of the wind turbines* (110 m, 150 m, 200 m), *environmental impacts* (small, medium, strong impact) and *minimum distance of a wind park to residential areas* (750 m, 1100 m, 1500 m). The monetary attribute included in the CE was defined as a premium on the monthly electricity bill (€ 0, 1, 2.5, 4, 6). In total, 2,067 respondents were interviewed using a web-based survey tool; the response rate was 34 %. For statistical and econometric analyses a total sample of 1,998 respondents remained. A standard MNL model revealed that people prefer medium sized wind parks over large ones. However, small wind farms are less preferred compared to large ones. Furthermore, respondents exhibit a positive WTP for reducing the maximum height of wind turbines and increasing the distance from residential areas. Finally, the outcomes show that strong environmental impacts are valued negatively while a reduction of local nature impacts creates a positive value to people.

The investigation of MEYERHOFF ET AL. (2010) is based on the previously reviewed study of MEYERHOFF ET AL. (2008). Principally, the same CE was used to quantify landscape externalities from on-shore wind power in Westsachsen and Nordhessen (Germany). The only difference is related to the environmental impacts attribute. In contrast to MEYERHOFF ET AL. (2008), the environmental attribute now refers to the *effect on red kite population*[109] (with the levels 5 %, 10 % of 15 % reduction). Moreover, inter-

---

[108] The estimated WTP values shown in Table 20 must be treated with caution since no opt-out alternative was included in the CE. In that case implicit price can usually not be interpreted as WTP. Nevertheless, the calculated WTP values are displayed for the purpose of comparison. The indicated values have been converted from WTP per kWh to an annual WTP by using the average electricity consumption per household in 2003 which amounts to 9,240 kWh (ORF, 2006, online).
[109] The red kite is a bird that is most threatened by wind turbines in the areas of investigation.

views were conducted via telephone, in total 708, about half of them in Westsachsen (WS) and the other half in Nordhessen (NH). Prior to the main survey, the questionnaire including the CE was pre-tested within focus group meetings and a pilot study. Another extension of the current examination is that it captures preference heterogeneity by using Latent Class (LC) models in addition to a standard MNL model. LC models are based on the assumption that the population can be divided into a priori unknown segments, each having different preferences. Generally, people living in Westsachsen and Nordhessen prefer wind power development programmes that move wind turbines further away from residential areas. Additionally, local biodiversity impacts (effect on red kite population) are continuously valued negatively. By contrast, preferences for wind farm size vary significantly among the population segments identified in the LC model. Finally, the height of wind turbines does not matter for people, hence is no statistically significant determinant of choice.[110]

The study of LADENBURG AND DUBGAARD (2007) focussed on off-shore wind power in Denmark. Generally, off-shore wind farms create visual disamenities that diminish with larger distances from the coast. Hence, the aim of the study was to examine WTP for reducing the visual impacts arising from future off-shore wind parks. The problem is that placing new wind farms further away from the coast is associated with a price increase for the generated electricity. Using a choice experiment, this trade-off was evaluated. Thus, the following attributes were used to describe the visual externalities of prospectively planned wind turbines: *Distance from the shore* (with the levels 8 km, 12 km, 18 km and 50 km), *size of the wind farm* (49, 100 or 144 turbines per wind park) and finally *annual surcharge on household's electricity bill* (€ 0, 12.5, 23, 40, 80 and 175).[111] The response rate of the mail survey was 52 % leaving a sample size of overall 362 respondents. The basic fixed effect logit (FEL) model including attributes only revealed that people exhibit a positive WTP for reducing the visual impact of wind

---

[110] The WTP values indicated for Westsachsen and Nordhessen in Table 20 represent the means of the three segments identified in the LC model. Thereby, only statistically significant values are taken into account.
[111] The visual impacts from off-shore wind farms were illustrated using computer-based visualisations which have shown wind turbines at varying distances from the coast under clear whether conditions.

turbines by locating them further away from the shore. WTP thereby increases with the distance from the coast. The size of the wind parks, by contrast, has no significant influence on choice, i.e. does not matter for people. Furthermore, preferences differ substantially across subgroups. In particular, people who can see off-shore wind farms from their residence are willing to pay by far more for increasing the distance of wind turbines from the shore than respondents having no wind farms within sight. Moreover, it was shown that younger people (aged under 30 years) are not willing to pay for reducing negative landscape impacts by increasing the distance of windmills from the coast.

An attempt to quantify public preferences over the environmental impacts of wind farms in Spain was made by ALVAREZ-FARIZO AND HANLEY (2002). The study site was La Plana of Zaragoza, an important national heritage (limestone plateau) in the valley of Ebro. The authors wanted to figure out how people value the environmental effects of a potential wind farm project for La Plana. Thus, a choice experiment was developed using three environmental attributes. These attributes describe the impact of wind farms on *cliffs, fauna and flora,* as well as *landscape*, each using two levels (protection or loss). The cost attribute used in the CE referred to an annual payment in the form of a tax increase taking the levels 500, 1000 or 1500 Spanish pesetas. In total, 488 personal interviews were conducted in order to estimate preferences associated with the potential wind power project. The application of a standard MNL model indicated that all three environmental attributes represent statistically significant determinants of choice. Hence, the protection of cliffs, flora and fauna, as well as landscape is valued positively, i.e. exhibits a positive WTP. Comparing the estimated effects, the conservation of flora and fauna creates the highest value to people followed by the preservation of landscape. Protecting the unique cliffs of La Plana is ranked last as measured by the calculated WTP.[112]

DIMITROPOULOS AND KONTOLEON (2009) analysed the factors that determine local acceptance of wind power projects in two Greek Aegean is-

---

[112] The WTP values shown in Table 20 were converted from the national currency (Spanish pesetas) into Euros using *http://www.coufal.biz/euro.html*. These converted values have then been adjusted for purchasing power.

lands, Naxos and Skyros. Wind farms create externalities like impacts on landscape or wildlife (especially birdlife) and are subject to noise pollution and technical issues as for instance unreliability or high costs. For those reasons, new projects are often faced with resistance from local communities. Therefore, the authors have chosen a somewhat different approach compared to previous valuations of wind power attributes. Instead of estimating public WTP for the attributes of wind power technologies, the investigation of DIMITROPOULOS AND KONTOLEON (2009) focussed on willingness to accept (WTA) the externalities associated with wind power projects. Hence, the following attributes were used to describe wind farm projects on the inlands of Naxos and Skyros: *wind farm size* (small: 2-6, medium: 7-13, large: 14-20 or larger: 21-40 turbines), *height of the wind turbine* (50 m, 90 m), *location of the wind farm project* (in or out of a Natura 2000 protected area) and *cooperation with municipal authorities* in the course of planning a wind power project (with or without cooperation). Due to the fact that the study examines willingness to accept the physical, site-specific and institutional effects of wind power projects, the payment vehicle was defined as an annual subsidy households receive from the promoter of the project (with the levels € 50, 100, 200 or 300). In total, 212 face-to-face interviews were conducted in the areas of investigations (Naxos: 108, Skyros: 104 interviews). Response rates were quite high amounting to 67 % on average. In order to account for preference heterogeneity in the sampled population, a Random Parameter Logit (RPL) model was estimated to quantify public preferences over wind farm projects in the Aegean islands. The results indicate that people would be willing to abandon parts of their annual subsidy if the height of wind turbines is reduced from 90 m to 50 m if the project is built out of a Natura 2000 protection area and if local municipalities are involved in the planning of the wind farm project. By contrast, the wind farm size affects people's choice negatively meaning that they require a yearly subsidy in order to accept an additional windmill. Comparing the results of the econometrical analysis, it was shown that siting and institutional factors were perceived more important as against the physical attributes of wind farms like height or size.

Similar to LADENBURG AND DUBGAARD (2007), KRUEGER (2007) focussed on valuing the public preferences for off-shore wind power in Delaware, USA. Hence, a choice experiment was developed using five attributes to describe potential off-shore wind farms in Delaware. These are *location of the wind farm* (Delaware Bay, North Ocean, South Ocean), *distance from shore* (0.9, 3.6, 9 miles out of sight) and the *royalty payment* to be made by wind farm developers (USD 1, 2, or 8 million) to an associated *beach royalty fund* (Beach nourishment fund, Delaware green energy fund, Delaware general fund). The payment vehicle was a renewable energy fee added to respondent's monthly electricity bill (USD 0, 1, 5, 10, 20 or 30). The surveys were mailed to a random selection of Delaware residents. 52 % of the addressed people responded to the poll resulting in a sample size of 949 completed surveys. The sample is divided into three areas, "Bay"[113], "Ocean"[114] and "Inland"[115], whereas Bay and Ocean regions were oversampled. A RPL model was applied to account for the given preference heterogeneity in the sample. The most important outcome shows that people prefer wind farm projects located further away from the coast. Here, substantial differences between subsamples could be identified. People in the "Bay" and "Ocean" sample exhibit a higher WTP for placing wind turbines further off-shore than people living inland.[116] Hence, people living along the coastline perceive wind turbines more as a visual disamenity than inland residents. No clear results were found with respect to wind farm location and royalty fund payment.

The last two wind power studies discussed within this review refer to applications of the CV method. First, KOUNDOURI ET AL. (2009) elicited public preferences for renewable electricity generation, or more precisely, the construction of a wind farm in the island of Rhodes, Greece. Prior to the valuation question respondents were given information about the planned wind farm project like size of the wind farm, electricity that can be generated or landscape impacts associated with the wind turbines. The elicitation

---

[113] Respondents living along Delaware Bay in Kent and Sussex Counties.
[114] Respondents living along the Delaware Atlantic coastline in Sussex County.
[115] Respondents throughout Delaware not living along coastal areas.
[116] The WTP values indicated in Table 20 represent weighted averages (adjusted for sample size) of the WTP calculated for the Inland, Bay and Ocean sample.

format used to estimate WTP was a double bounded dichotomous choice (DB-DC) question using bids of € 2, 4, 6, 8 and 12 as a premium on respondent's bimonthly electricity bill. Positive responses to the initial valuation question were asked whether they would be willing to pay a higher bid. Conversely, respondents who refused to pay the stated amount in the first question were subsequently confronted with a lower bid. The survey was carried out in collaboration with a market research company using phone interviews of 200 Rhodes residents. Mean WTP calculated from the intervals defined by the two valuation questions amounts to annually € 43.9 (median: € 51.1). These values have already been adjusted for purchasing power. Using an interval regression (IR) model, the authors identified the main determinants of stated WTP. First, people who think that they are well informed about the renewable energy potentials in Rhodes are willing to pay more compared to those individuals being less informed. Furthermore, individuals thinking that the construction of the new wind farm will cause negative impacts on landscape and environment are willing to pay less for it. Finally, some socio-demographic characteristics were found to affect WTP significantly. Accordingly, the household size is positively related to WTP. The same applies to children, i.e. households with children are on average willing to pay more than childless households. However, the effect of children on WTP decreases with the number of children. Conclusively, as expected, stated WTP is positively affected by the educational level; respondents with a tertiary education are on average willing to pay more than people with a lower level of education.

GROOTHUIS ET AL. (2007) assessed, similar to DIMITROPOULOS AND KONTOLEON (2009), public willingness to accept (WTA) the construction of windmills in the mountains of North Carolina, USA. More precisely, wind turbines can cause negative externalities to people living near the windmills as for instance visual impacts on the landscape respectively mountain views, noise disturbances or shadow effects. The authors wanted to find out how much people need to be compensated in the form of an electricity bill reduction in order to accept such externalities. For that purpose a CV survey was developed using a dichotomous choice question with monthly bids ranging from USD 1 to 50. In total, 389 residents of Watauga County in

North Carolina were surveyed by mail; the overall response rate was 43 %. Prior to the CV question, people were asked whether they would participate in a green energy programme in North Carolina. Here about one third of the respondents approved to participate. As anticipated, it was shown that the percentage of "yes" votes in the referendum question increased with the compensation amount. Median WTA was found to be € 18.3 per household and year (converted from USD into Euros). The determinants of willing to accept windmills in the mountainous area have been analysed by means of a binary Probit (PRO) model. First, it was shown that people are more likely to accept the construction of windmills as compensation increases. Next, income is negatively related to WTA indicating that households with higher incomes need less compensation payment in order to accept the wind farm project. Furthermore, people who perceive wind power as a clean energy source have a higher probability to accept the construction of wind farms. In contrast, respondents with the opinion that windmills blemish the landscape are less likely to vote for. Another interesting result refers to the fact that people regularly visiting the mountains require more compensation to accept the negative externalities of windmills as compared to those who do not retire to the mountains. Finally, the outcomes of the econometrical models revealed that people who are willing to participate in a general green energy programme are also more likely to accept wind turbines in the environment.

Table 20: Overview of existing stated preference applications – wind power

Authors (year)	Country	SP technique (elicitation format)	Survey method	Sample (response rate)	Main objective	Model	Marginal WTP in € per year (adjusted for purchasing power)
Ek (2005)	Sweden	CE (unlabelled)	Mail survey	n = 547 (56 %)	Examine public preferences over the environmental attributes of wind power generation.	REP	- Reduced noise level: € 66.44 - Location in the mountains: € -21.62 - Location off-shore: € 34.41 - Small wind parks: € 15.37 - Large wind parks: € -16.26
MEYERHOFF ET AL. (2008)	Germany	CE (unlabelled)	Online survey	n = 1,998 (34 %)	Estimate public preferences for the utilization of wind power.	MNL	- Wind park size (from large to small): € -4.65 - Wind park size (from large to medium): € 6.64 - Height (from 200 m to 110 m): € -15.74 - Environmental impact (from medium to small): € 48.46 - Environmental impact (from medium to large): € -63.73 - Distance (from 750 m to 1,100 m): € 5.69
MEYERHOFF ET AL. (2010)	Germany	CE (unlabelled)	Telephone Interviews	n = 708 (--)	Quantify landscape externalities from on-shore wind power in Westsachsen and Nordhessen.	LC	- Wind park size (from large to medium): € -6.45 (WS), € -15.36 (NH) - Red kite population (from 10 % to 5 %): € 19.06 (WS), € 27.12 (NH) - Red kite population (from 10 % to 15 %): € -23.62 (WS), € -20.30 (NH) - Distance (from 750 m to 1,100 m): € 36.70 (WS), € 49.51 (NH) - Distance (from 750 m to 1,500 m): € 42.77 (WS), € 55.58 (NH)

Table 20 continued

Study	Country	Method	Survey	n (response rate)	Objective	Model	Results
LADENBURG AND DUBGAARD (2007)	Denmark	CE (unlabelled)	Mail survey	n = 362 (52 %)	Elicit public WTP for reducing the visual dis-amenities from off-shore wind farms.	REL	– Distance (from 8 km to 12 km): € 38.58 (overall), € 221.40 (see wind farms) – Distance (from 8 km to 18 km): € 80.51 (overall), € 341.32 (see wind farms) – Distance (from 8 km to 50 km): € 102.31 (overall), € 405.90 (see wind farms)
ALVAREZ-FARIZO AND HANLEY (2002)	Spain	CE (unlabelled)	Interviews	n = 488 (–)	Quantify public preferences over the environmental impacts of a wind farm project in La Plana.	MNL	– Protection of flora & fauna: € 44.28 – Protection of landscape: € 43.38 – Protection of cliffs: € 25.21
DIMITROPOULOS AND KONTOLEON (2009)	Greece	CE (unlabelled)	Interviews	n = 212 (67 %)	Assess public WTA for physical, siting and institutional effects of wind farm projects.	RPL	*Annual WTA values:* – Height (from 90 m to 50 m): €-482.02 – Location out of a Natura 2000 area: €-823.87 – Cooperation with municipalities: €-994.66 – Wind farm size (per wind turbine): € 43.38
KRUEGER (2007)	USA	CE (unlabelled)	Mail survey	n = 949 (52 %)	Valuing public preferences for off-shore wind power, especially for placing wind farms further off-shore.	RPL	– Distance (from 0.9 to 3.6 miles): € 162.09 – Distance (from 0.9 to 6 miles): € 221.82 – Distance (from 0.9 to 9 miles): € 269.23 – Distance (from 0.9 to 20 miles): € 362.50
KOUNDOURI ET AL. (2009)	Greece	CV (DB-DC)	Telephone interviews	n = 200 (–)	Explore public preferences towards the construction of a new wind farm in the South of Rhodes.	IR	– Mean WTP: € 43.90 – Median WTP: € 51.11
GROOTHUIS ET AL. (2007)	USA	CV (DC)	Mail survey	n = 389 (43 %)	Elicit public willingness to accept windmills that create negative externalities to the environment.	PRO	– Median WTA: € 18.33

Source: OWN REPRESENTATION

5.3 Hydropower studies

Stated preference or more precisely choice experiment applications assessing the public value of the multiple impacts associated with hydroelectric production are scarcely available. There exist only a limited number of studies dealing with the technology hydropower. These few investigations mainly focus on the specific environmental impacts arising from hydroelectric power. Thus, SUNDQVIST (2002a) analysed households' preferences for green electricity generated from hydropower in Sweden. In particular, the author wanted to find out how people weigh the various impacts of hydropower against one another. For that, a CE was designed using the following attributes: *downstream water level* (minimum flow, 25 % higher, 50 % higher water level), *erosion and vegetation* (-25 %, -50 % reduction) and *impacts on fish life* (harmful to certain fish species, river adapted to migratory fish species, adapted to all fish species). Beside these water-related attributes, the cost attribute was defined as an *electricity price increase per kWh* (0.05, 0.10, 0.15, 0.20, 0.25 SEK). The whole questionnaire including the CE was distributed to a random sample of 1,000 Swedish households (only house-owners). The response rate of the survey was 48 % leaving a sample of 479 respondents. Due to incomplete answers the final sample reduced to 397 observations. Applying a Random Effects Probit (REP) model, it was shown that people are willing to incur extra costs for environmental improvements, like the reduction of erosion and vegetation by 50 % or the preservation of all fish species.[117] By contrast, Swedish people are unwilling to pay for increased water levels in hydropower regulated rivers.

The same CE was applied to a random sample of small- and medium-sized firms in Sweden (SUNDQVIST, 2002b). The final sample of non-residential electricity consumers contained 243 firms after accounting for late responses and incompletely filled questionnaires. The response rate was 36 %. Again, a REP model was estimated to quantify non-residential preferences

---

[117] The estimated WTP values per kWh have been converted into yearly WTP (see Table 21) using the average electricity consumption per household in 2003 which amounts to 9,240 kWh (ORF, 2006, online).

over the environmental impacts of hydropower. This analysis principally yields the same results as in the private household sample. Small- and medium-sized enterprises are willing to incur extra costs for reduced erosion and vegetation and the adaption of hydropower regulated rivers to all inhabitant fish species. By contrast, they are not willing to pay for increasing the water level beyond minimum required flow.

An attempt to estimate WTP for environmental improvements in hydropower regulated rivers in Sweden was made by KATARIA (2009). In doing so, a CE was developed describing the environmental improvements that can be achieved by adopting remedial measures.[118] Hence, the attributes of the CE are *increased fish stock* (0 %, 15 %, 25 % increase), *improved conditions for bird life* (yes, no), *species richness of benthic invertebrates* (high, moderate, considerably reduced) and *river-margin vegetation and erosion* (high amount, somewhat reduced amount, considerable reduced amount of plant species and biomass growth). The payment vehicle was defined as an additional annual electricity cost for the household (with the levels 0, 200, 375, 600, 850, 1175 or 1400 SEK). The whole survey was mailed to 1,475 randomly selected Swedish households. The questionnaire was returned by 638 respondents resulting in a response rate of 43 %. However, due to incomplete answers, not all the questionnaires could be used for analysis. In total, 568 completely filled questionnaires were available for econometric analysis. In order to account for preference heterogeneity in the sampled population, a RPL model was estimated. The outcome revealed that the Swedish population is in general willing to pay for environmental improvements in hydropower regulated rivers. More precisely, people exhibit a positive WTP for increasing the fish stock, improving the living conditions for birds and increasing species richness. The same applies to an improvement of river-margin vegetation and erosion.

A contingent valuation (CV) study addressing the trade-off between hydropower use and the aesthetical and recreational value of a natural site was made by EHRLICH AND REIMANN (2010). More specifically, at the Jägala

---

[118] These measures are assumed to contribute to an attainment of the good ecological river status required by the European Water Framework Directive (WFD).

Waterfall in Estonia a small-scale hydropower plant is planned to be built. However, this project would lead to significant impacts regarding the nature values of the waterfall since most of the water past the waterfall will be redirected in order to operate the turbines of the hydropower station. The aim of the study was to evaluate how much people would be willing to pay per year for keeping the Jägala Waterfall in its natural state. For that purpose, an open-ended CV question format was applied using photos to visualise the impact of the hydropower plant on landscape and nature. In total, 950 Estonian residents were surveyed in spring 2009. 60 % of the respondents were hypothetically willing to incur extra costs for preserving the natural state of the waterfall. A two-step procedure was used to identify determinants of stated WTP. First, a Logit model was estimated revealing that female respondents and people with higher income are more likely to state a positive WTP while the probability of willing to pay decreases with age. In a second step, a standard OLS regression was used to determine the factors that influence the amount of WTP. Here, only income was found to be a statistically significant determinant. So, the payment decision is affected by various socio-demographic indicators while the payment amount is simply influenced by income. On average, Estonian residents are willing to pay € 20.5 annually for keeping the Jägala Waterfall in its natural state.[119]

---

[119] This value has already been adjusted for purchasing power. Furthermore, mean WTP was calculated by dividing the overall consumer surplus derived from the fitted demand curve (€ 10 million) by the number of people exhibiting a positive WTP (567,000).

Table 21: Overview of existing stated preference applications – hydropower

Authors (year)	Country	SP technique (elicitation format)	Survey method	Sample (response rate)	Main objective	Model	Marginal WTP in € per year (adjusted for purchasing power)
SUNDQVIST (2002a)	Sweden	CE (unlabelled)	Mail survey	n = 397 (48 %)	Analyse households' preferences over the environmental impacts of hydroelectric power.	REP	− Erosion and vegetation (-50 %): € 14.58 − River adapted to all fish species: € 16.46
SUNDQVIST (2002b)	Sweden	CE (unlabelled)	Mail survey	n = 243 (36 %)	Examine preferences of small- and medium-sized firms over the environmental impacts of hydroelectric power.	REP	− Erosion and vegetation (-50 %): € 13.98 − Erosion and vegetation (-25 %): € -12.20 − River adapted to all fish species: € 17.45
KATARIA (2009)	Sweden	CE (unlabelled)	Mail survey	n = 568 (43 %)	Measure WTP for reducing the impacts on the environment in hydropower regulated rivers.	RPL	− Increased fish stock (by 1 %): € 1.68 − Improved conditions for bird life: € 10.44 − Species richness (moderate compared to considerably reduced): € 24.58 − Species richness (high compared to considerably reduced): € 29.80 − Vegetation and erosion (somewhat reduced compared to considerably reduced): € 44.37 − Vegetation and erosion (high compared to considerably reduced): € 39.84
EHRLICH AND REIMANN (2010)	Estonia	CV (open-ended)	Mail survey	n = 950 (−)	Estimate public WTP for preserving the natural state of the Jägala Waterfall in a natural state compared to a usage for hydroelectric generation.	LOGIT, OLS	− Mean WTP: € 20.47

Source: OWN REPRESENTATION

5.4 Comparison of previous findings

From the comprehensive review of existing stated preference applications it can be concluded that there exists a lot of international research on assessing the public value of renewable energy investments in general, using both choice experiments as well as contingent valuation methods. Most of the work in this field of research has been done in the United Kingdom and the USA. Few studies – especially CV applications – have also been carried out in other European countries (Italy, Greece and Slovenia) and Asia (Korea and Japan). In contrast, stated preference studies addressing specific renewable technologies are scarcely available. Especially for hydropower, there exists only a very limited number of studies using the CE approach to value the multiple impacts of the technology. The few investigations carried out in the past are primarily from Sweden where hydropower traditionally plays an important role in the electricity sector. Moreover, stated preference studies often focus on wind power. Relevant investigations have been implemented in Germany, Spain, Greece, Scandinavian countries like Sweden or Denmark and also in the USA. In these countries wind power potentials are typically high. To conclude, renewable energy sources generally play an important role all over the world, especially in view of climate change issues. Hence, various attempts have been made to estimate the public value of renewables. Specific technologies like hydropower or wind power may play a more crucial role in one country compared to another. This is why studies focussing on specific renewable technologies have only occasionally been conducted, mainly in those countries that exhibit significant potentials or where the particular technology currently plays a bearing role in the energy sector.[120] For Austria, however, no existing valuation studies were found although the intensified use of renewable energy sources, especially hydropower, is an integral part of the national climate and energy strategy.

---

[120] Further attempts to estimate the public value of specific renewable energy sources have been made for biomass and tidal power technologies (see for instance LEE AND YOO, 2009; SOLOMON AND JOHNSON, 2009; HITE ET AL., 2008 or JENSEN ET AL., 2004).

What can we learn from previous studies? First, the attributes used to describe renewable energy investment strategies or specific technologies appeared to be of main interest. Here, a clear trend was identified. Beside the type of renewable technology which was integrated at least in some of the past applications, mainly two types of attributes have been used in previous choice experiments. These are *nature-related attributes* referring to the impacts of renewable energy on landscape and wildlife as well as *climate- and economic-related attributes* like the reduction of air pollution or employment effects. Previous investigations assessing public preferences towards specific technologies (wind and hydro power) focussed on the negative environmental externalities associated with electricity generation from the particular technology rather than on global impacts like climate change prevention or economic effects. The environmental impacts of hydroelectric power assessed within past investigations mainly refer to erosion, vegetation and fish population. The negative externalities evaluated within the scope of previous wind power studies were noise level, height, environmental impacts (especially the effects on bird life) and landscape impacts. Landscape externalities have usually been specified by the size of the wind park and the location (on-shore, off-shore, mountainous). Another attribute often used in wind power research was distance, both from the residential area or the coast in case of off-shore wind farms. The cost attribute used in each of the reviewed choice experiments[121] mainly referred to households' electricity bill.[122] This is particularly advantageous since people are usually familiar with such a payment vehicle. Few of the studies, by contrast, used a premium on the electricity price per kWh. However, this raises the problem of correctly perceiving the real financial burden associated with such a price increase.[123]

Concerning the elicitation format of previous CE studies, a clear tendency towards unlabelled experiments was found. Unlabelled experiments use

---

[121] As mentioned before, the inclusion of a price or cost factor among the attributes of a CE is necessary for the calculation of implicit prices respectively WTP.
[122] This means that the cost attribute was defined as a surcharge to monthly, bimonthly, quarterly or yearly electricity bill.
[123] In particular, people would have to know at least roughly about their monthly or yearly electricity consumption in kWh in order to capture the effect of such a price increase on the amount of their electricity bill.

generic titles (e.g. Alternative A, Alternative B,...) for the headings in the choice sets that do not convey any information to the respondent other than the order of the alternatives (HENSHER ET AL., 2005). Labelled experiments have especially been applied in those investigations that distinguished between renewable technologies. According to that, the different technologies served as the labels in the CE representing the headings of the alternatives in the choice sets. When looking at previous CV studies, a clear preference for certain elicitation formats was recognized as well. Most of the studies used referendum formats (single-bounded or double-bounded dichotomous choice questions), as recommended by the NOAA Panel of Contingent Valuation (ARROW ET AL., 1993). Beside that, the open-ended question format was found to be very popular in CV studies exploring the monetary value of renewable energy sources. This result coincides with LIEBE AND MEYERHOFF (2005) who stated that the open-ended elicitation format experiences a "renaissance" to some extent. Payment cards and bidding game approaches, by contrast, appeared to be less popular in previous research on renewable energy valuation.

The surveys carried out within the scope of previous studies were conducted via mail, telephone, face-to-face or online. Although no clear preference for one of these data collection methods can be distinguished, mail surveys and face-to-face interviews appeared to be the most widely used methods. In contrast, online surveys are still less prevalent in stated preference applications. The survey sample sizes of previous CE studies range between 128 and 1,998 respondents; the mean sample size is 528 (see Table 22). Response rates achieved within previous CE applications range between 31 % and 67 % (mean response rate: 45 %). In past CV studies the average sample size amounts to 921, ranging from 200 to maximum 2,800 respondents. Compared to CE studies, the average sample size of previous CV applications is significantly different as shown by a "mean comparison" test[124] (t=1.750, p-value=0.092).[125] The response rates of the CV studies reviewed

---

[124] More on mean comparison tests can be found in HARTUNG ET AL. (1991:505ff).
[125] In addition to that, the alternative hypothesis H_a: $mean_{CV} - mean_{CE} > 0$ was tested yielding a statistically significant result (p-value=0.046).

within this work lie between 26 % and 43 %; the mean value is 34 %.[126] In comparison, past CE studies achieved on average significantly higher response rates than previous CV applications (t=1.937, p-value=0.073).[127]

Table 22: Sample sizes and response rates of previous CE and CV studies

Statistical parameter	Sample size	Response rate
Mean – CE studies	528	45 %
Mean – CV studies	921	34 %
Mean comparison test	t = 1.750 p = 0.092*	t = -1.937 p = 0.073*
Minimum – CE studies	128	31 %
Minimum – CV studies	200	26 %
Maximum – CE studies	1,998	67 %
Maximum – CV studies	2,800	43 %
*significant at 10 % level		

Source: OWN CALCULATIONS AND REPRESENTATION

Further methodological issues worth to be discussed, relate to the complexity of the choice exercises and the econometrical models used to quantify people's preferences. First, in order not to overextend people, the choice exercises were tried to keep as simple as possible. Most of the studies used four choice cards; the total number of choice sets presented to respondents did not exceed six choice exercises.[128] Additionally, the number of alternatives included in the choice cards was limited to three alternatives plus a status-quo or opt-out option. Regarding the econometric models used in previous CE studies, a clear preference for MNL models was observed; the MNL model is the simplest econometric approach used to quantify people's preferences over the attributes of a CE. Some of the studies considered unobserved preference heterogeneity ("taste differences") by estimating RPL or LC models or, in the binary case, REP or REL models. The main determinants of stated WTP in CV applications have usually been identified using Tobit models for censored data. Some of the studies used two-step procedures estimating first the probability of willing to pay (with

---

[126] However, a significant portion of the previous CV studies used personal interviews as the data collection method. For these investigations, no response rates have been reported.
[127] Testing the hypothesis H_a: mean$_{CV}$ – mean$_{CE}$ < 0 yields a significant result (p=0.037).
[128] Only one study (DIMITROPOULOS AND KONTOLEON, 2009) used eight choice sets to be presented to respondents.

Probit or Logit models) and subsequently the determinants of the stated WTP amount (using standard OLS for truncated data).

As far as the results of previous research are concerned, it was shown that people exhibit in general a positive WTP for renewable energy. However, investments in renewable technologies are often affiliated to negative externalities like impacts on landscape or wildlife. Although people adopt in general a positive attitude towards renewable energy use, these externalities are consistently valued negatively, i.e. people wish to be compensated for them. In addition, the impacts on nature and wildlife were commonly found to be important compared to other attributes. The "Not in my backyard" (NIMBY) phenomenon has also been addressed by several studies revealing that people prefer larger distances between power plants and their residential areas. This especially turned out to be true when referring to wind turbines. Furthermore, most of the studies that distinguished between different technologies concluded that solar power is the most preferred renewable energy source followed by wind power. Biomass, by contrast, was less preferred, while hydropower was only used by NAVRUD AND BRATEN (2007) as one of the technologies at choice. Turning to the results of previous CV studies, it was found that standard socio-demographic characteristics like gender, age or income represent the main determinants of stated WTP. In addition, preceding investigations revealed that increasing influence on stated WTP arises from the general attitude towards the environment and renewable energy. Hence, environmentally aware people are usually willing to pay (more) for renewable energy than their non-environmentally conscious counterparts.

6 Study design and implementation

6.1 The general approach

Generally, the implementation of a stated preference (SP) study assessing the public value of future hydropower and renewable energy expansion in Austria involves several steps. This is shown in Figure 36. The structure of tasks applies to both, contingent valuation (CV), as well as choice experiment (CE) studies. The first step of this multi-stage procedure comprises some initial research defining the research question to be answered, as well as the object or impact being valued. This is usually done by means of literature research, expert interviews or focus groups. The main task is to characterize the decision problem in a way that people understand it (ADAMOWICZ ET AL., 1998a:12; PEARCE ET AL., 2002:28). Once the decision problem has been determined, the researcher has to decide on the valuation technique used to assess the economic value of the regarding object or impact. As a general rule, if the SP study focuses on examining the economic value of the different characteristics or attributes of an environmental good, then choice modelling is preferable. In contrast, if the focus of interest is on the good as a whole (e.g. the economic value of a river landscape), the contingent valuation approach is more likely to be relevant. Sometimes it may also be useful to use contingent valuation and choice experiment methods simultaneously, especially in order to increase the robustness and to check the validity of the results (PEARCE ET AL., 2002:32f).

Another important issue of the second stage of the procedure is the choice of the survey method. In general, SP studies can be conducted by mail, telephone, online or face-to-face. Each survey method features advantages and disadvantages. Accordingly, mail or postal surveys are relatively cheap compared to other data collection methods, can easily handle sensitive questions, do not suffer from potential interviewer bias and put less time pressure on respondents. However, mail surveys are subject to low response rates and self-selection bias. Moreover, they are time-consuming, restrict the use of visual aids, and the researcher has little control over who fills the questionnaire.

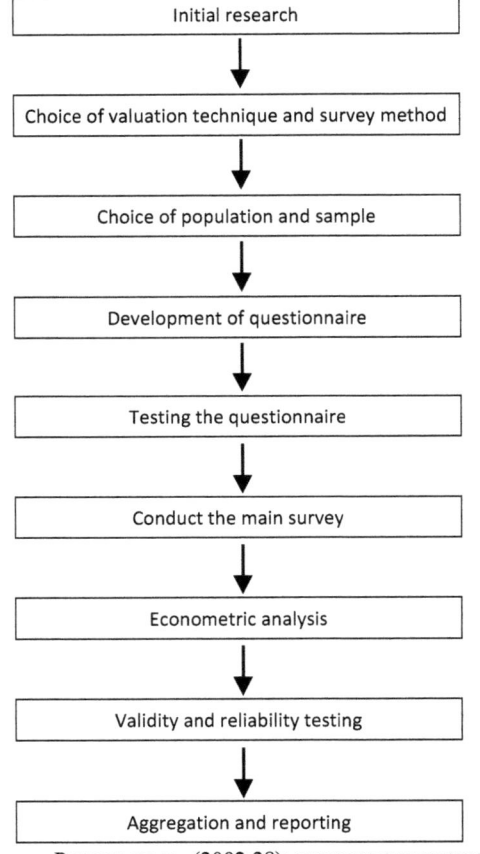

Figure 36: Multi-stage procedure of a stated preference study

Source: PEARCE ET AL. (2002:28); OWN REPRESENTATION

Face-to-face interviews, by contrast, are highly flexible and allow complex questionnaire structures, as well as the use of visual and demonstration aids. Furthermore, personal interviews provide the possibility for sample control and usually obtain high response rates of 70 % and more. Nevertheless, face-to-face surveys are quite expensive, give rise to interviewer and self-selection bias and are normally not representative. Further weaknesses include the limited sample size, as well as geographic limitations. Telephone interviews are generally cheaper than face-to-face surveys, relatively quick to administer and easy to monitor. In addition, telephone surveys include the possibility of random sampling and provide good geographic coverage. Response rates are usually also very high amounting to 60-75 %. Major disadvantages of telephone surveys include interviewer bias, the inability to use visual aids, the necessity to be brief, the refusal of respondents to answer sensitive questions and the risk that respondents get tired (EVANS AND MATHUR, 2005:203ff; PEARCE ET AL., 2002:42).

Online surveys have gained increasing importance since the first web-based applications in the 1990s. Web-based surveys are rather flexible and can easily be tailored to costumer demographics by putting multiple versions of a questionnaire online. The speed and timeliness are other advantages of online surveys meaning that the period of time to get a survey into the field and to collect the data is minimized. Moreover, no bias can be caused by the interviewer and respondents can take as much time as they need to fill in the questionnaire. The ease of data entry and analysis is another benefit of online surveys since data inputting stage is not necessary and the administrative burden of sending and receiving questionnaires is reduced. Usually, online surveys are relatively cheap in their administration because the surveys are self-administered and no interviewers are required.[129] Moreover, web-based surveys enable to obtain large samples, can minimize non-responses by imposing an obligation to answer a question before advancing to the next one, and finally, permit the use of complex questionnaire structures without confusing respondents.[130] A major disadvantage of online surveys is usually the low response rate. Furthermore, the sample may be skewed towards internet population. Typically, internet users are not truly representative of the general population precluding random samples. However, in modern times the internet population is becoming more representa-

---

[129] Due to the technological and programming requirements the preparation of online surveys could theoretically be costly. However, the availability of advanced survey software and specialized market research institutes put downward pressure on the costs for preparing a web-based survey (EVANS AND MATHUR, 2005:199).

[130] More precisely, online surveys can be tailored in a way that respondents answer only those questions that pertain specifically to them. Thus, complicated instructions (e.g. if your answer is yes please continue with question x, if your answer is no go to question y) become redundant.

tive. Hence, the gap between offline and online populations is quickly closing. Although more and more people are using the internet, web-based surveys may still cause difficulties due to the lack of familiarity with online questionnaires. Finally, self-administered online surveys require extremely clear instructions without overloading respondents (EVANS AND MATHUR, 2005:197ff).

Once the valuation technique and survey method has been decided on, the subsequent step involves the definition of the target population and the kind of sample drawn from this population. Generally, the target population can comprise the whole population of a country or a region or simply a defined group of people (e.g. green electricity consumers). The selection of the target population depends on the good to be valued and the kind of value that is of interest.[131] Next, the sample frame population from which the sample will finally be drawn has to be identified. The sample frame is usually an explicit list such as the list of registered voters or the commercial costumers of an electricity utility. Finally, the researcher has to select a sample from the frame population. The sample can be purely random, systematic, stratified or clustered multi-stage[132] (PEARCE ET AL., 2002:43f).

The main and most important part of a SP study is the development of the questionnaire and the included CE or CV task. The design of a CE or CV involves, in turn, several steps. In case of a CV study, the researcher has to decide on the kind of value to be assessed (WTP or WTA), the elicitation format and the payment vehicle used to assess the economic value of the good in question (PEARCE ET AL., 2002:28). ARROW ET AL. (1993:32ff) recommend the investigation of WTP instead of WTA values, and the use of referendum formats for the CV question.[133] The design of a CE is somewhat more elaborate compared to a CV study. First, the attributes and lev-

---

[131] For instance, if non non-use values are to be assessed, wide geographic coverage is required.
[132] With a stratified sample the frame population is divided into several sub-populations. Then a separate and independent random sample is drawn from each sub-group. Cluster sampling means that the population is divided into a set of groups or "clusters". Subsequently, a set of clusters is randomly selected and each individual within these randomly selected clusters is surveyed. In a multi-stage sample only a sample of individuals within the selected clusters is surveyed.
[133] See also LIEBE AND MEYERHOFF (2005:9). A detailed description of the referendum or dichotomous question format is given in section 4.2 of this work.

els used to describe the good in question have to be selected. This is usually done through literature reviews (see chapter 5), focus group discussions or direct questioning. Then, the analyst has to assign levels to each of the chosen attributes (ADAMOWICZ ET AL., 1998a:13). Generally, the identification of the set of attributes and the levels represents the key phase in choice experiment design. Attributes and levels must be chosen so that several requirements are fulfilled. First, the attributes must be relevant to the problem which is being analysed. Second, the attributes must be credible, realistic and consistent with government policies. Third, the selected attributes are required to be relevant and easily understandable for the sample population. Finally, the attributes should be applicable to policy analysis (BERGMANN ET AL., 2004:5; LONGO ET AL., 2008:143). A rule thumb is not to choose more than four or five attributes in order to ensure that the choice situation can be handled by respondents (PEARCE ET AL., 2002:55). Once the attributes and the corresponding levels have been determined, choice sets need to be constructed. This is formally known as the experimental design of the CE generating different combinations of the attribute levels, i.e. alternatives. The generated alternatives are then grouped into choice sets, also known as choice cards, to be presented to respondents. A choice set usually contains a baseline scenario (status quo) and several alternative options in which the specified attributes differ in quantity. The number of choice cards should typically not exceed eight; a working rule is to use four to six choice sets in a CE (ADAMOWICZ ET AL., 1998a:13; PEARCE ET AL., 2002:55). Usually, a CV or CE task is introduced by an introductory section describing the general context for the decision to be made. Moreover, the introductory section contains a detailed description of the good to be valued and, in case of a CE survey, an accurate description of the attributes and levels used to describe the good in question. Finally, the introduction to a CE or CV should include a description of the institutional setting in which the good will be provided, and a characterisation of the payment vehicle, i.e. how will the good be paid for. It is extremely important to ensure that respondents understand the attributes and levels used in a CE study, as well as the initial situation in a CV application. Nevertheless, people should not become over informed, because then they are no longer representative of the total population. After the CE or CV task, debriefing questions may give some indication of why respondents answered certain questions the way they did (ALRIKSSON AND ÖBERG, 2008:248; CARSON, 1999:11; LIEBE AND MEYERHOFF, 2005:9).

After the CE or CV tasks have been designed, the whole exercise must be embedded into a comprehensive questionnaire. Beside the valuation scenario including an introductory section and follow-up questions, the questionnaire should additionally contain attitudinal questions and socio-economic characteristics (PEARCE ET AL., 2002:47ff).

Before conducting the main survey, the draft questionnaire has to be tested within focus group sessions and pilot surveys in order to ensure that the design is appropriate for the research question and gives the desired results. If the CE or CV design works well on small groups (focus groups) or in a pilot survey, it is reasonable to assume that it will also give reliable and valid results using the full sample. The design and testing of questionnaires should be closely connected, and iteration between the two may be needed. What is learnt from testing can be fed back to improve the design of the questionnaire. It is very important to start with the full survey only if the questionnaire has performed satisfactorily in focus groups and pilot surveys (CARSON, 1999:11f; PEARCE ET AL., 2002:56ff). If this is ensured one can proceed to the next step and conduct the main survey. The final steps of a SP study involve the econometric analysis, some kind of statistical tests, as well as the aggregation and reporting of the study results. These tasks are covered explicitly in chapters 9 to 11 of this work.

6.2 The choice experiments

6.2.1 Choice experiment for hydropower expansion

As investments into hydropower are expected to grow in the future, it is important for policy makers to know how people perceive different hydropower expansion strategies and their potential related impacts. The positive and negative effects associated with future hydropower investments (see section 3.7) can be considered as characteristics, criteria or attributes of hydropower expansion. It is useful and policy relevant to gain knowledge on the relative economic values of these attributes in order to determine which hydropower expansion strategy would be preferred by the population.[134] Based on previous research and two discussion rounds with external experts from the field of environmental economics research, the electricity industry, and different federal and provincial government departments, the five attributes presented in Table 23 have been selected to describe future hydropower expansion. Previous investigations on hydropower exclusively focussed on the specific impacts of hydropower use on the river ecosystem. Within the scope of the present investigation, by contrast, a broader perspective has been chosen considering also climate-related, as well as economic effects in addition to the impacts on landscape and ecosystem.

Employment

According to that, the first attribute used in the CE for hydropower expansion refers to employment creation. As shown in section 3.7, future hydropower investments are accompanied by positive impacts for the local economy, and thus lead to job creation. This is why employment was considered to be a relevant attribute and it was included in the CE with four levels: 10, 50, 100 or 500 jobs. In particular, the indicated number of jobs refers to jobs created in the residential area of the respondent. Employment effects have also been used in previous studies such as BERGMANN ET AL. (2004) and (2008) or KU AND YOO (2010) as part of the impacts caused by renewable energy investments.

---

[134] As shown in chapter 2, there is a general need to gain information on people's preferences in order to determine the optimal provision of a public good. Moreover, the monetary valuation of external costs and benefits provides an attempt to internalize them when energy policy decisions are made.

Table 23: Attributes and levels used in the CE for hydropower expansion

Attribute	Description	Level 1	Level 2	Level 3	Level 4
Jobs	Number of created jobs in the residential area of the respondent.	10 jobs	50 jobs	100 jobs	500 jobs
CO_2 reduction	Attainable reduction of CO_2 emissions in the electricity sector.	-10%	-20%	-40%	-60%
Nature and landscape	Impact of new hydropower plants on the natural environment (ecosystem) and the landscape.	Small impact	Strong impact		
Distance	Distance of a new hydropower plant from the respondent's home.	2 km	4 km	8 km	20 km
Cost	Increase in monthly electricity bill (6 levels).	€3, €6, €9, €12, €15, €18			

Source: OWN REPRESENTATION

Reduced air emissions

According to the EU climate and energy package, Austria is committed to reduce its GHG emissions by 16 % in 2020 compared to 2005 levels (EC, 2008b:15). The necessity to reduce emissions and to prevent climate change is one of the main reasons why there is such a strong need for the expansion of renewable energies like hydropower. The electricity sector is one of the key elements for moving towards a low carbon economy (EC, 2011b:2). The major part (70-80 %) of total GHG emissions in developed countries stems from fossil fuel combustion such as coal, oil or gas in the energy sector (CAPROS ET AL., 2008:2). Hence, the increase of renewable energy sources in the electricity sector may contribute significantly to the reduction of climate-damaging GHG emissions. Depending on the amount of electricity generated from hydropower, different levels of CO_2 emission reductions in the electricity sector can result, amounting to -10 %, -20 %, -40 % or -60 %. The effect of renewable energy investments on emission levels have also been used in different forms in previous research as for instance BERGMANN ET AL. (2004) and (2008), FIMERELI ET AL. (2008) or LONGO ET AL. (2008).

Impact on nature and landscape

Although the expansion of hydropower utilisation contributes to job creation and the prevention of climate change, the construction of new hydropower stations is subject to negative environmental side effects.[135] These include visual impacts on the landscape, negative consequences for the ecosystem of the water body[136], and correspondingly the impact of these changes on fish and other water-dependent wildlife. The environmental consequences associated with hydropower can take on different dimensions, namely a small and a strong impact. This definition meets the special requirement to keep the environmental impact attribute as simple as possible so that it is valid for all types of hydroelectric development (SUNDQVIST, 2002a:6). With a strong environmental impact only the mini-

---

[135] The trade-off between positive effects on the labour market and air pollution and the negative environmental side effects is described in detail in section 3.7 of this work.
[136] For instance, the inability of fish to migrate up the river due to the existence of dams, alteration of flow conditions or reduced water level downstream of the hydropower plant (see also section 3.7).

mum requirements predetermined by the WFD are fulfilled. These include the installation of fish ladders and a minimum water flow. A small impact, by contrast, implies that the ecological integrity of new hydropower stations is improved by adopting higher environmental standards. Measures to minimize the impact of new hydropower stations on the landscape and natural environment may include the restoration of riverbanks in a near-natural state or the connection of subsidiary water bodies. Wildlife and biodiversity impacts have also been considered in previous studies valuing the multiple impacts arising from hydropower or renewable energy (see for instance BERGMANN ET AL., 2004; FIMERELI ET AL., 2008; KATARIA, 2009 or SUNDQVIST, 2002a and 2002b).[137]

Distance to home

The fourth attribute used in the CE describes the distance of the nearest power plant to the respondent's home. With the distance attribute the degree to which individuals are affected by an expansion of hydropower is taken into account. Previous studies like FIMERELI ET AL. (2008) or MEYERHOFF ET AL. (2010) have shown that people prefer energy sources to be placed at a larger distance from their home. Hence, the acceptance of hydropower expansion is assumed to decrease as the distance between the closest new hydro plant and respondent's home declines and vice versa. This relationship is better known as the "Not in my backyard" (NIMBY) phenomenon for which evidence should be found by the inclusion of the distance attribute. The levels of the distance attribute amount to 2 km, 4 km, 8 km and 20 km.

Increase in monthly electricity bill

Finally, one of the attributes used in a CE should typically be a price or cost factor in order to obtain monetary values associated with the other attributes (BOXALL ET AL., 1996:244; PEARCE ET AL., 2002:55). Accordingly, the payment vehicle was specified as an increase in household's monthly

---

[137] However, previous analysis used the attribute describing environmental impacts in a somewhat different form. On the one hand, landscape and wildlife impacts were often used as two separate attributes. On the other hand, environmental impacts associated with hydroelectric production were described in greater detail referring precisely to downstream water level, erosion and vegetation or impacts on fish life (see also section 5.3).

electricity bill with levels ranging from € 3 to € 18. Here, it is extremely important that people are familiar with the payment vehicle. This is usually the case when referring to utility bills (CARSON, 1999:13).

The payment levels have been determined within the scope of a pre-test[138] by asking people directly what they would be willing to pay for green electricity from increased hydropower (CV). These data have then been taken as the basis for defining the bandwidth of the monetary attribute. As can be seen from Table 24, mean WTP for hydropower in the pre-test sample was € 8.8 with a standard deviation of € 8.3. Median WTP amounted to € 8.5; the minimum was € 0.0 and the maximum € 50.0. These analyses are based on a sample size of 54 observations.

Table 24: **Directly stated WTP for hydropower in the pre-test**

Statistical key figure	WTP for hydropower
Mean	€ 8.8
Standard deviation	€ 8.3
Median	€ 8.5
Minimum	€ 0.0
Maximum	€ 50.0
Observations	54

Source: OWN CALCULATIONS AND REPRESENTATION

Looking at Figure 37, it can be seen that most of the people (49 or 89.1 %) in the pre-test sample exhibit a WTP of not more than € 15. Only a small proportion (9.3 %) indicated a WTP of more than € 15 pointing towards a small number of outliers. Hence, the levels of the payment vehicle ranging from € 3 to € 18 seem to cover people's appreciation of hydropower expansion fairly well.

---

[138] More on pre-testing can be found in section 6.6 of this chapter.

Figure 37: Distribution of directly stated WTP for hydropower in the pre-test (in %)

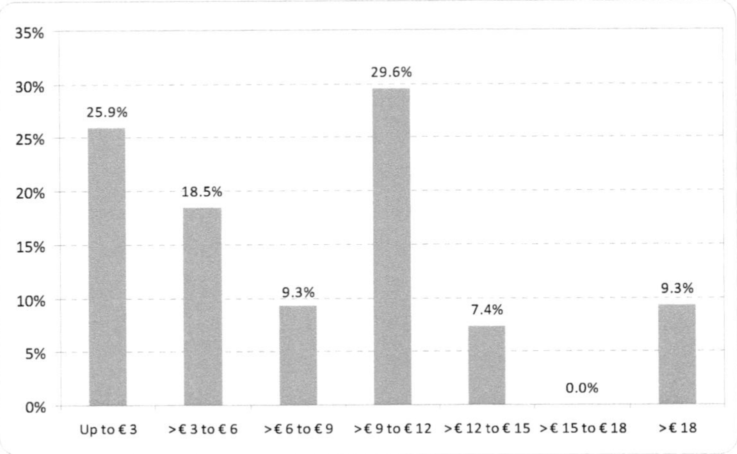

[Bar chart showing:
- Up to €3: 25.9%
- >€3 to €6: 18.5%
- >€6 to €9: 9.3%
- >€9 to €12: 29.6%
- >€12 to €15: 7.4%
- >€15 to €18: 0.0%
- >€18: 9.3%]

Source: OWN CALCULATIONS AND REPRESENTATION

6.2.2 Choice experiment for renewable energy expansion

Although the expansion of hydropower use represents an important strategy to meet future climate and energy goals like the reduction of GHG emissions, it is only part of a broad strategy aiming at expanding the use of renewable energy sources. Beside hydropower, there are also significant future potentials for the use of biomass, wind and solar power (see for instance BGBL, 2011b or ÖSTERREICHISCHER BIOMASSE-VERBAND, 2008). The focus on hydropower may cause "super informing" and draw peoples' attention to this special technology leading to upward biased results. This problem is formally known as framing bias. Inadequate framing occurs when "the choices made in a valuation experiment do not reflect the choice framework existing in a real world situation" (ROLFE ET AL., 2000:293). Accordingly, in real life respondents may not choose to consume hydropower but rather another type of electricity as for instance green electricity from wind power. Hence, if people are given the option they would probably buy green electricity from other power sources (SUNDQVIST,

2002a:19).[139] For that reason, hydropower expansion was embedded into a broader strategy promoting an expansion of renewable energy sources. The renewable energy sources investigated in this context are – beside hydropower – biomass, solar (photovoltaics) and wind power representing a more realistic decision situation (see Table 25).

As shown in Table 25, the attributes and levels used to describe the different renewable energy expansion strategies are the same as in the case of hydropower, since employment creation, reduced air emissions, nature and landscape impacts, as well as the NIMBY phenomenon may also play a role when referring to an expansion of biomass, solar power or wind power (see section 3.7).[140] In contrast to the hydropower CE, respondents now have the possibility to decide between different renewable technologies. This approach enables to test for the impact of framing hydropower demand in the context of alternative renewable energy sources. More precisely, it is possible to find out whether people's preferences for the attributes of hydropower expansion are affected by the context in which respondents made their choices.[141]

---

[139] The negligence of the potential significance of other renewable energy sources has also been criticised in MEYERHOFF ET AL. (2008:18).

[140] As in the case of hydropower, the bandwidth of the payment vehicle was determined by means of a pre-test containing 51 observations. The mean WTP in the pre-test sample on renewable energy was € 8.5 with a standard deviation of € 5.9; median WTP amounted to € 7.0. Moreover, it was shown that the bulk of respondents (92.2 %) indicated a WTP smaller than € 18. Correspondingly, the payment levels ranging from € 3 to € 18 seemed to be appropriate.

[141] Additionally, the CE on renewable energy expansion enables to gain information on the implied ranking of different renewable energy sources.

Table 25: Attributes and levels in the CE for renewable energy expansion

Attribute	Description	Level 1	Level 2	Level 3	Level 4
Technology	Label of the choice experiment indicating which technology will be expanded.	Biomass	Hydropower	Solar power	Wind power
Jobs	Number of created jobs in the residential area of the respondent.	10 jobs	50 jobs	100 jobs	500 jobs
CO_2 reduction	Attainable reduction of CO_2 emissions in the electricity sector.	-10%	-20%	-40%	-60%
Nature and landscape	Impact of new renewable energy plants on the natural environment (ecosystem) and the landscape.	Small impact	Strong impact		
Distance	Distance of a new renewable energy plant from the respondent's home.	2 km	4 km	8 km	20 km
Cost	Increase in monthly electricity bill (6 levels).	€3, €6, €9, €12, €15, €18			

Source: OWN REPRESENTATION

6.2.3 Choice experiment for a specific hydropower project

Beside the valuation of the Austrian strategy to expand hydropower and renewable energy use, we go a level deeper. Hence, a specific hydropower project is considered within the scope of this work. This regional hydropower case study refers to a project in the province of Styria in Graz-Puntigam along the river Mur. Further details of this project are elucidated in chapter 10.

As in the previous choice experiments, it is assumed that a new hydropower plant can be described by its characteristics or attributes. However, the attributes used to describe the planned hydropower station fairly differ from those in the previous choice experiments referring to hydropower and renewable energy. Accordingly, the new hydropower plant is characterised by the following four attributes (see Table 26).

Electricity generation amount
The main advantage of the installation of a new hydropower plant is the emission-free generation of electricity for local consumers and the associated greenhouse gas reduction. According to the project operator, the number of households able to be provided with electricity is estimated at approximately 20,000. The attainable reduction of CO_2 emissions amounts to 60,000 tonnes per year (DOBROWOSKI AND SCHLEICH, 2009:10; ENERGIE STEIERMARK, 2010a, online). In view of a conservative estimate the levels of electricity generation were fixed to 5,000, 10,000 and 15,000 households.

Impact on nature and landscape
As shown in section 3.7, the development and construction of a new hydropower plant definitely cause impacts on the landscape and the ecosystem. Generally, the damming of a river and the associated loss of vegetation causes adverse effects on the landscape. Moreover, the visual barrier effect that is associated with dams has a negative influence on the appearance of the natural landscape (PISTECKY, 2010:28). Other environmental concerns related to the new hydropower project involve biodiversity impacts and a change in water quality. The construction and operation of the

hydropower station is suspected to lead to a deterioration of water quality from good to moderate status (ORF, 2012, online).[142] Furthermore, the damming of the river and vegetation clearance will lead to a loss of habitats along the river banks. Additionally, fish will be negatively affected by the hampered ability to pass the dam (PISTECKY, 2010:18ff). This is only a brief summary of the environmental impacts associated with the planned hydropower project. A detailed description of the consequences on human and ecological systems is usually given in the environmental impact assessment. Such an assessment is obligatory for projects that are expected to cause substantial environmental effects.[143] In view of the requirement to keep attributes as simple as possible, the nature and landscape attribute was included in the CE with the same levels as before, namely a small and a strong impact. With a strong impact, the natural habitats of flora and fauna, as well as the landscape are severely affected. A small environmental impact, by contrast, means that a strong emphasis is put on the preservation and protection of flora, fauna and landscape. By means of a near-natural design of the power plant and the implementation of extensive ecological accompanying measures, the hydropower plant is likely to merge harmoniously with its surroundings.

Recreational activities

The third attribute included in the CE describes possible future recreational activities along the riverside. Generally, the power plant is expected to upgrade the urban area of Graz by creating leisure space and recreational areas. This includes the linking of existing foot and cycling paths, as well as the provision of leisure activities like boating or canoeing. Additionally, the commercial benefit of the hydropower project can be enhanced by the es-

---

[142] Other investigations, by contrast, indicate that the construction of the new hydropower plant will be associated with a considerable improvement of the water quality. This may be due to the simultaneous realisation of important infrastructure projects like the installation of a central sewage collector (ENERGIE STEIERMARK, 2010b, online).

[143] The Environmental Impact Assessment Act became Austrian law in 1993 and has been repeatedly refined over the course of the following years. The general aim of the mandatory environmental impact assessment is to assess, describe and evaluate the direct and indirect consequences of a project on human beings, plants, animals, soil, water, air and climate, landscape, as well as natural and cultural heritage. Moreover, environmental impact assessments aim to examine measures that avoid or reduce the harmful effects on the environment (BGBL, 1993, § 1). For the current hydropower project in Graz-Puntigam, a positive decision on the environmental impact assessment has been issued (ORF, 2012, online).

tablishment of riverside localities like cafés or restaurants (DOBROWOLSKI AND SCHLEICH, 2009:14; ENERGIE STEIERMARK, 2010b, online; PISTECKY, 2010:12).[144] The attribute has two levels. First, the new hydropower plant extends the possibilities for recreation. Second, the hydropower project creates adverse effects on public recreation. In this case no additional measures aiming at improving the possibilities for public recreation are adopted.

Increase in monthly electricity bill

Finally, the monetary attribute was specified as an increase in respondent's monthly electricity bill. As before, the six payment levels range between € 3 and € 18. That this is an appropriate bandwidth can again be shown by the results of a pre-test based on 103 observations. In the pre-test sample, the mean of directly stated WTP was € 9.5 with a rather broad distribution (standard deviation: € 14.0); median WTP amounted to € 5.0. Furthermore, the bulk of the respondents, namely 88.3 %, indicated a WTP of less than € 18.

---

[144] Generally, there are ambiguous empirical results regarding the impact of hydro-electric power plants on recreational activities. For further details please refer to section 3.7.

Table 26: Attributes and levels in the CE for the specific hydropower project

Attribute	Description	Level 1	Level 2	Level 3
Households	Number of households that can be provided with green electricity from the new hydropower plant.	5,000 households	10,000 households	15,000 households
Nature and landscape	Impact of the new hydropower plant on the natural environment and the landscape.	Small impact	Strong impact	
Recreational activities	Impact of the new hydropower plant on the possibilities for recreation.	Extended	Restricted	
Cost	Increase in monthly electricity bill (6 levels).	€3, €6, €9, €12, €15, €18		

Source: Own representation

6.3 Experimental design

Once the attributes and associated levels have been determined, the next important step in conducting a choice experiment is to create the experimental design. "Experimental design is concerned with how to create the choice sets in an efficient way, i.e. how to combine attribute levels into profiles of alternatives[145] and profiles into choice sets" (ALPIZAR ET AL., 2001:15). Usually, some kind of orthogonal design is used to generate the profiles or alternatives (ADAMOWICZ ET AL., 1998a:13). Orthogonality means that the combinations of the attributes of the alternatives are uncorrelated in all choice sets. The experimental design procedure generally involves two steps. First, the analyst has to attain the optimal combinations of attribute levels to be included in the CE and second, translate these combinations into choice sets. The starting point of this design procedure is usually a full factorial design (ALPIZAR ET AL., 2001:15f). Generally speaking, in factorial designs each attribute level is combined with every level of all the other attributes. A full or complete factorial design contains all possible combinations of the attribute levels. However, full factorial designs are often impracticable because the number of possible alternatives resulting from the combination of the attribute level with each other may be too large to be handled within the experiment. For that reason, one might use only a subset of all possible combinations (ALPIZAR ET AL., 2001:16; LOUVIERE ET AL., 2000:84ff). This attempt is formally known as fractional factorial design. "Fractional factorial designs involve the selection of a particular subset or sample (i.e. fraction) of complete factorials, so that particular effects of interest can be estimated as efficiently as possible (LOUVIERE ET AL., 2000:90).[146] The problem is that "traditional fractional factorial designs were designed for creating sets of single profiles, so they need to be adapted if they are to be used to generate sets of choice sets" (CHRZAN AND ORME, 2000:4). Consequently, if respondents are to be confronted with a

---

[145] A profile is a single attribute level combination, also called "treatment combination" in the statistical literature (ADAMOWICZ ET AL., 1998a:13). Hence, a profile represents a single choice alternative presented to the respondents. For that reason, profiles are referred to as alternatives in the following remarks.

[146] A more detailed statistical elaboration of factorial designs can be found in MONTGOMERY (1984:261ff).

sequence of choices, alternative designs are required. One possibility is to create randomized designs. In a randomized design respondents are randomly selected to receive different versions of the choice sets.[147] The choice sets can be created using different approaches. In the present context the method of "complete enumeration" has been used. With this approach, the generated alternatives or profiles fulfil the efficiency requirement of orthogonality within respondents as effectively as possible. In addition, two-way frequencies of level combinations between attributes are equally balanced, a principle called level balance. Finally, the complete enumeration approach is in line with the principle of minimal overlap meaning that attribute levels do not repeat themselves in a choice set (CHRZAN AND ORME, 2000:6; ALPIZAR ET AL., 2001:16).

The experimental designs of this work were generated with the software package Sawtooth. The randomized CE design on hydropower expansion involves 900 profiles or alternatives of attribute level combinations including an opt-out alternative.[148] These profiles were grouped into 300 choice sets consisting of two alternatives plus an opt-out. The inclusion of an opt-out alternative in all choice sets avoids that people are forced to (hypothetically) buy electricity from hydropower which would depict an unrealistic decision situation (DIMITROUPOULOS AND KONTOLEON, 2009:1846) The design, i.e. the 300 choice sets, was finally blocked into 50 versions each containing six choice sets (see Table 27).

Table 27: Components of the randomized CE designs

Hydropower	Renewable energy	Specific hydropower project
900 alternatives (incl. opt-out)	900 alternatives (incl. opt-out)	540 alternatives (incl. opt-out)
300 choice sets each with	300 choice sets each with	180 choice sets each with
2 alternatives plus opt-out	2 alternatives plus opt-out	2 alternatives plus opt-out
↓	↓	↓
50 blocks each with	50 blocks each with	30 blocks each with
6 choice sets	6 choice sets	6 choice sets

Source: OWN REPRESENTATION

---

[147] Randomized designs enable to systematically control for the variability arising from extraneous sources. Further reading on randomized designs can be found in MONTGOMERY (1984:123ff).
[148] The opt-out alternative is always defined in the same way as "none of the two strategies". Thus, only 600 "true" profiles with varying attribute levels have been generated.

An example of a choice set is given in Figure 38. As can be seen from the example card, the opt-out alternative is referred to as "none of the two strategies". Such an experiment is called an unlabelled experiment, since it uses generic titles (Strategy A and Strategy B) for the alternatives or profiles presented in the choice set. These generic titles do not convey any information to the respondent other than strategy A is the first of the alternatives and Strategy B the second (HENSHER ET AL., 2005:112).

Figure 38: Choice set example for hydropower

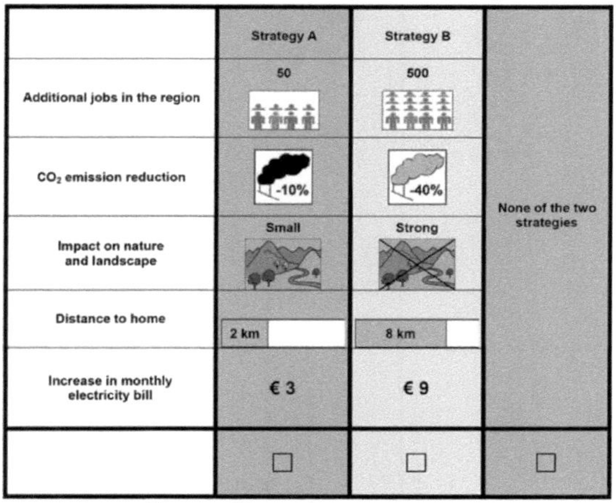

	Strategy A	Strategy B	
Additional jobs in the region	50	500	
CO_2 emission reduction	-10%	-40%	None of the two strategies
Impact on nature and landscape	Small	Strong	
Distance to home	2 km	8 km	
Increase in monthly electricity bill	€ 3	€ 9	
	☐	☐	☐

Source: OWN REPRESENTATION

As shown in Table 27, the components of the randomized CE design on renewable energy expansion are in principle the same as in the case of hydropower. Again, 900 alternatives or profiles were generated in total, including the opt-out which is always defined equivalently as "none of the two strategies". These profiles have then been grouped into 300 choice sets each with two alternatives plus the opt-out. The final design consisted of 50 blocks each containing six of the generated choice sets. An example is shown in Figure 39. Such an experiment is called a labelled experiment because it attaches meaningful names, i.e. labels, to each of the alternatives presented in the choice card. In the present context, the different renewable energy sources represent the labels of the CE. The main advantage of using alternative specific labels lies in the possibility to reduce the cognitive burden of the experiment by inducing familiarity with the context. However, the use of labels also bears the risk that respondents focus only on the labels and do not consider trade-offs between the attributes. In an unlabelled experiment, by contrast, respondents are likely to place greater emphasis on the attributes because there are no labels to focus on (ALPIZAR ET AL., 2001:21).

Figure 39: Choice set example for renewable energy

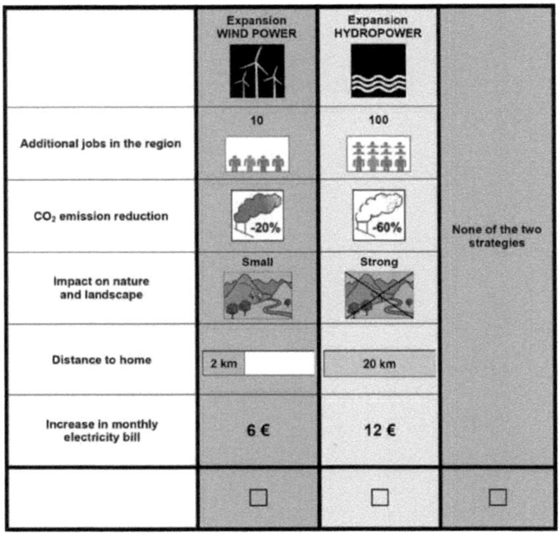

	Expansion WIND POWER	Expansion HYDROPOWER	
Additional jobs in the region	10	100	
CO_2 emission reduction	-20%	-60%	None of the two strategies
Impact on nature and landscape	Small	Strong	
Distance to home	2 km	20 km	
Increase in monthly electricity bill	6 €	12 €	
	☐	☐	☐

Source: OWN REPRESENTATION

Finally, the CE design for the regional hydropower case study is somewhat smaller due to the reduced number of attributes.[149] Accordingly, 540 profiles of different attribute level combinations were generated including the opt-out alternative, referred to as "none of the two alternatives". As before, each choice set consisted of two alternatives plus the opt-out resulting in a total of 180 choice sets. The generated choice sets were then grouped into 30 blocks each containing six choice tasks. An example for such a choice set is given in Figure 40. Again, this represents an unlabelled experiment.

---

[149] In particular, the CE for the specific hydropower project uses only four attributes rathern than five as in the other experiments (see also section 6.2.3).

Figure 40: Choice set example for the specific hydropower project

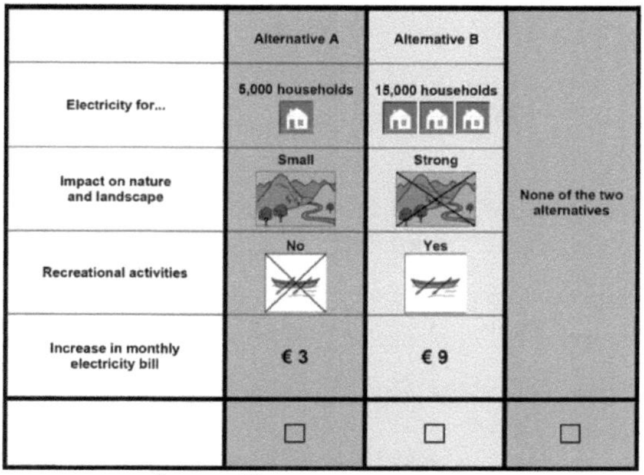

	Alternative A	Alternative B	
Electricity for...	5,000 households	15,000 households	
Impact on nature and landscape	Small	Strong	None of the two alternatives
Recreational activities	No	Yes	
Increase in monthly electricity bill	€ 3	€ 9	
	☐	☐	☐

Source: OWN REPRESENTATION

Once a CE design has been generated, it is extremely important to test whether the design is appropriate or efficient. Inappropriate designs may produce biased parameter estimates or unidentifiable models (KERR AND SHARP, 2009:2). Generally, efficiencies are measures of design goodness. A common measure of efficiency is the concept of D-efficiency. D-efficiency is typically stated in relative terms, hence, always refers to the efficiency of one design relative to another (CHRZAN AND ORME, 2000:10). The measure of D-efficiency is given by equation 6.1:

$$D-efficiency = \frac{1}{|\Omega|^{\frac{1}{K}}} \quad (6.1)$$

where Ω is the asymptotic covariance matrix for the design and K the number of parameters or estimated coefficients. Generally, a design is D-efficient when the D-errors are minimized (ALPIZAR ET AL., 2001:16; KERR AND SHARP, 2009:2; KUHFELD, 1997:7).

An advanced test for design efficiency involves the generation of dummy data or dummy responses. These dummy responses are then used to estimate the effects of the attributes using (multinomial) logit (SAWTOOTH SOFTWARE, 2008:9). In doing so, there are generally two possibilities. First, the analyst might consider only main effects, i.e. the effects of the attributes used in the CE allowing the estimation of a strictly additive "main effects only" utility specification. Such designs are, not surprisingly, called "main effects designs (ADAMOWICZ ET AL., 1998a:14; CHRZAN AND ORME, 2000:7). However, main effects are not the only effects that may be of interest in practical SP studies. In addition to the main effects, "interaction effects" suggested by theory or previous empirical evidence might be of interest (LOUVIERE ET AL., 2000:87). Generally, main effects account for 70 % to 90 % of the explained variance in a choice model while two-way interactions typically account for only 5 % to 15 % of explained variance (LOUVIERE ET AL., 2000:94). Therefore, interaction terms are often neglected within the scope of experimental design procedures. According to that, traditional CE analysis typically produces only main effects (CHRZAN AND ORME, 2000:7).

Consequently, only main effects (i.e. no interaction terms) were considered in the present context. The tests for design efficiency were performed using the software package Sawtooth. For the CE design on hydropower the estimation of the main effects plan was carried out using 455 dummy responses[150], and assuming that about 15 % of the respondents chose the opt-out alternative in some of the choice sets. The results of this test procedure are given in Table 28. Particular attention must be drawn to the standard errors of the estimated effects. Very large standard errors are indicative for inappropriate or deficient designs. A rule of thumb, required to get coefficient estimates in the actual model significant, is to have standard errors below a level of 0.05 in the simulated model (SAWTOOTH SOFTWARE, 2008:10).[151] As can be seen from the table below, the calculated standard errors correspond in principle to this rule of thumb, except for the standard errors of the cost attribute which were slightly above the threshold value of 0.05. However, this did not cause any problem in the actual model as can be seen later in chapter 9. According to these results, the CE design for hydropower seemed to be highly efficient.

---

[150] Usually, the sample size in the simulated model should be similar to the number of respondents that is actually planned to obtain (SAWTOOTH SOFTWARE, 2008:9).
[151] Regarding interaction terms, this threshold value for standard errors is 0.10. However, because only main effects are considered this does not apply to the present context.

A measure of relative D-efficiency is reported by the "strength of design". The design efficiency is computed relative to an "ideal" hypothetical orthogonal design. Efficiencies that are far away from a value of 100 may be perfectly satisfactory, a requirement that is satisfied in the CE design for hydropower (see Table 28). However, the "ideal" hypothetical orthogonal design used as a reference is far from possible making the efficiency measure less convincing. For that reason, it would be useful to compare the reported strength of the design relative to the efficiency of an alternative design (KUHFELD, 1997:7; SAWTOOTH SOFTWARE, 2008:9).

Table 28: Efficiency of the CE design for hydropower expansion

Attribute level	Effect	Standard Error	t Ratio
EMPLOYMENT			
10 jobs	-0.021	0.043	-0.490
50 jobs	0.000	0.043	0.000
100 jobs	-0.063	0.043	-1.472
500 jobs	0.084	0.043	1.971
CO_2 REDUCTION			
-10 %	0.037	0.043	0.869
-20 %	-0.053	0.043	-1.241
-40 %	-0.037	0.043	-0.860
-60 %	0.053	0.042	1.240
ENVIRONMENTAL IMPACT			
Small	0.024	0.021	1.153
Strong	-0.024	0.021	-1.153
DISTANCE			
2 km	0.049	0.043	1.159
4 km	-0.009	0.043	-0.219
8 km	-0.002	0.043	-0.055
20 km	-0.038	0.043	-0.875
COST			
€ 3	0.060	0.058	1.038
€ 6	-0.030	0.057	-0.522
€ 9	-0.004	0.058	-0.076
€ 12	-0.025	0.058	-0.442
€ 15	0.030	0.058	0.522
€ 18	-0.030	0.058	-0.523
NONE (Opt-out)	-1.124	0.055	-20.346
Strength of design = 586.936			

Source: OWN CALCULATIONS AND REPRESENTATION

Table 29: Efficiency of the CE design for renewable energy expansion

Attribute level	Effect	Standard Error	t Ratio
TECHNOLOGY (LABEL OF THE CE)			
Hydropower	0.048	0.029	1.680
Biomass	-0.023	0.029	-0.801
Solar power	0.003	0.029	0.116
Wind power	-0.028	0.029	-0.986
EMPLOYMENT			
10 jobs	-0.031	0.029	-1.059
50 jobs	-0.033	0.029	-1.131
100 jobs	0.032	0.029	1.114
500 jobs	0.031	0.029	1.083
CO_2 REDUCTION			
-10 %	-0.002	0.029	-0.061
-20 %	-0.022	0.029	-0.792
-40 %	0.011	0.029	0.366
-60 %	0.014	0.029	0.489
ENVIRONMENTAL IMPACT			
Small	-0.009	0.014	-0.654
Strong	0.009	0.014	0.654
DISTANCE			
2 km	0.004	0.029	0.140
4 km	-0.021	0.029	-0.736
8 km	-0.048	0.029	-1.676
20 km	0.066	0.029	2.288
COST			
€ 3	0.057	0.039	1.470
€ 6	-0.039	0.039	-1.000
€ 9	-0.001	0.039	-0.032
€ 12	0.010	0.039	0.267
€ 15	-0.012	0.039	-0.313
€ 18	-0.015	0.039	-0.384
NONE (Opt-out)	-1.112	0.037	-29.974
Strength of design = 1,316.508			

Source: OWN CALCULATIONS AND REPRESENTATION

Table 29 shows that the CE design for renewable energy expansion is highly efficient as well. The simulated model is based on 1,000 random responses and assumes a proportion of people choosing the opt-out of approximately 15 %. All standard errors are significantly below the required value of 0.05 even for the cost attribute which seemed to create some kind of problem in the previous hydropower CE. Also the strength of the design is extremely high compared to an ideal hypothetical orthogonal design.

Finally, the advanced efficiency test was carried out for the CE design on the specific hydropower project, providing a less satisfactory result. The test is based on 110 dummy responses; again 15 % of the respondents were assumed to choose the opt-out alternative in some of the choice sets. As shown in Table 30, the standard errors for the environmental and recreational effects conform to the requirement of lying below 0.05. By contrast, the standard errors of the electricity generation levels are slightly above the threshold value of 0.05 but still acceptable. The result causing the relative poor performance of the design refers to the cost attribute with standard errors lying considerably above 0.05. However, in the actual model this deficiency did not cause any severe problems as can be seen later in chapter 10 of this work.

Table 30: Efficiency of the CE design for the regional hydropower case study

Attribute level	Effect	Standard Error	t Ratio
ELECTRICITY GENERATION AMOUNT			
5,000 households	0.056	0.068	0.810
10,000 households	-0.014	0.069	-0.205
15,000 households	-0.041	0.069	-0.600
ENVIRONMENTAL IMPACT			
Small	-0.016	0.043	-0.377
Strong	0.016	0.043	0.377
RECREATIONAL ACTIVITIES			
Yes	0.036	0.043	0.828
No	-0.036	0.043	-0.828
COST			
€ 3	-0.056	0.120	-0.470
€ 6	0.112	0.118	0.949
€ 9	0.275	0.116	2.369
€ 12	-0.274	0.122	-2.240
€ 15	-0.133	0.120	-1.112
€ 18	0.076	0.118	0.648
NONE (Opt-out)	-0.877	0.104	-8.474
Strength of design = 162.111			

Source: OWN CALCULATIONS AND REPRESENTATION

6.4 Contingent valuation exercise

Despite the high potential and popularity of the CE technique, as well as its flexibility to obtain preferences for any good that can be subdivided into discrete attributes, it may be beneficial to validate the results of a CE on the basis of another stated or revealed preference technique (VAN DER POL ET AL., 2008:2043). Since there are no revealed preferences to study, contingent valuation (CV) may be a valuable complement to the CE method (ADAMOWICZ ET AL., 1998b:65). In order to increase the robustness of a CE study and check the estimated values for consistency it would be useful to use both, CV and CE (PEARCE ET AL., 2002:32).

For those reasons, a CV exercise was included in addition to the CE within the scope of the present investigation. This CV task takes an open-ended form. Although the open-ended elicitation format is associated with several disadvantages like a large number of non-responses, zero bids and outliers, or the attraction of strategic behaviour (see section 4.2), the open-ended format has recently experienced a so-called "renaissance".[152] Open-ended methods provide individual specific estimates of WTP enabling a more robust comparison with the results of a CE. Additionally, open-ended formats alleviate the calculation of mean WTP compared to referendum formats and generate more conservative estimates of WTP since closed-ended methods usually produce higher estimates of value (VAN DER POL ET AL., 2008:2044). Hence, people were asked directly for the monthly amount they would be at maximum willing to pay to support the expansion of hydropower. That is:

> **Welchen Aufschlag zu Ihrer monatlichen Stromrechnung würden Sie maximal für den weiteren Ausbau der <u>Wasserkraft</u> bezahlen, damit Ihr Haushalt Ökostrom bekommt?**

---

[152] This can be seen from previous SP studies. For instance, ZORIC AND HROVATIN (2012), EHRLICH AND REIMANN (2010), IVANOVA (2005) or ZARNIKAU (2003) used open-ended formats to elicit WTP for hydropower or renewable energy sources respectively (see chapter 5).

The CE on renewable energy expansion was complemented by the same CV task, with the sole difference that it asks for maximum WTP to support the expansion of renewable energy sources. That is:

> **Welchen Aufschlag zu Ihrer monatlichen Stromrechnung würden Sie maximal für den weiteren Ausbau erneuerbarer Energiequellen bezahlen, damit Ihr Haushalt Ökostrom bekommt?**

Usually, a CV exercise should be introduced by a detailed description of the general context, the good to be valued and the payment vehicle (CARSON, 1999:11). Since the CV task represents a supplement to the CE carried out within the scope this work, and hence, the same good is to be valued, no separate description of the general valuation context was included. All necessary information is provided by the introductory section of the CE. For further details see section 6.5.

6.5 The questionnaire

The previously presented CE and CV tasks were embedded in a comprehensive questionnaire on hydropower and renewable energy comprising three parts. The general structure of the questionnaire is shown in Figure 41.

Figure 41: General structure of the questionnaire

```
┌─────────────────────────────────────────┐
│         Attitudinal questions           │
└─────────────────────────────────────────┘
                    │
┌─────────────────────────────────────────┐
│   Choice tasks & debriefing questions   │
└─────────────────────────────────────────┘
                    │
┌─────────────────────────────────────────┐
│ Socio-economic characteristics & CV task│
└─────────────────────────────────────────┘
```

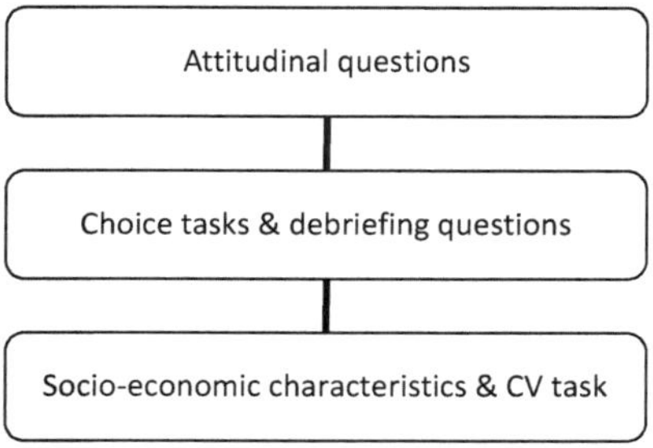

Source: OWN REPRESENTATION

Attitudinal questions

The first section of the questionnaire contains questions about the respondent's electricity use behaviour, as well as their general perception and attitude towards renewable energy and hydropower use. The extent of this part of the questionnaire depends on the type of the choice experiment. As we have previously seen, three differing experiments have been designed to evaluate people's preferences for hydropower expansion, renewable energy expansion, and a specific regional hydropower project. According to which CE is used, more or less questions are contained in the first part of the questionnaire (see Table 31).[153]

Table 31: Extent of attitudinal section of the questionnaire

Renewable energy	*Hydropower*	*Regional case study*
– Questions on electricity use – Questions on the expansion of renewable energy sources	– Questions on electricity use – Questions on the expansion of renewable energy sources	– Questions on electricity use – Questions on the expansion of renewable energy sources
	– Additional questions referring to hydropower and its intensified use	– Additional questions referring to hydropower and its intensified use
		– Additional questions referring to the specific hydropower project

Source: OWN REPRESENTATION

Respondents are, for instance, asked whether or not they do consciously obtain electricity from renewable energy sources, what they think from which energy sources electricity in Austria should be derived in the future or how important they think it is to increase the share of renewable energy. These general questions are included in each version of the questionnaire. The hydropower questionnaire contains, on top of that, some additional questions referring to hydropower in general and its intensified use. For instance, people are asked whether they have heard of the plan to expand hydropower utilisation in Austria, what their attitude towards the construc-

---

[153] Full versions of the questionnaires can be found in the appendix of this work.

tion of new hydropower stations is or how far away the next hydropower station is from their home. In the regional case study questionnaire, general questions on hydropower are partly substituted by questions referring to the specific hydropower project. Accordingly, respondents are for instance asked whether they have heard from the plan to build the new hydropower plant, what their attitude towards the new hydropower project is or if they feel positively or negatively affected by the project.

Choice experiment

The second part of the questionnaire is made up of six choice experiment questions. The choice tasks are introduced by an explanatory text, familiarizing respondents with the relevant attributes or impacts associated with an expansion of hydropower or renewable energy use. Here, a general problem of complexity arises because impacts on CO_2 emission levels, landscape or biodiversity are complex issues that involve uncertainty and often generate significant levels of scientific debate. Hence, it may be difficult for laypeople to obtain a clear understanding of the particular impacts. As a rule of thumb, attributes should be kept as simple as possible in order not to "over-inform" people.[154] "The challenge is to delineate which are the core layers of information that need to be transferred to respondents" (ROLFE ET AL., 2000:293).

According to that, the explanatory text gives a short overview of the current situation regarding renewable energy and hydropower use in Austria. In this context it is mentioned that currently more than half of domestically generated electricity comes from renewable energy sources, especially hydropower, and that there is still substantial potential for new facilities. Subsequently, the impacts associated with these future hydropower or renewable energy investments are presented to respondents in plain words. Here, the real task is to translate complex scientific issues in a form that is understandable for the general public. Attribute levels can be generally communicated textually or via pictures. The visual (non-textual) representation of attribute levels may contribute to a more homogeneous perception of the levels (ADAMOWICZ ET AL., 1998a:13; CARSON, 1999:11). However, photo-

---

[154] "Super-informed" people may no longer be representative of the general population.

graphs can give very different impressions of an impact, depending for instance on the angle from which a photo is taken (MEYERHOFF ET AL., 2010:87). In order not to influence people's perception of one attribute or level compared to another caused by the attractiveness of a picture, simple pictograms in black and white colour shades have been used to communicate the levels of the attributes used in the current experiments. These pictograms were included in the choice cards as well, so as to improve the comprehensibility of the decision situations (see section 6.2 and 6.3).

In order to understand the motives behind the stated choices, the CE is followed up by a number of debriefing questions. On the one hand, these follow-up questions are related to the perceived complexity of the experiment. On the other hand, the debriefing section aims to find out which attributes respondents perceived as more important compared to others and to reveal the possible presence of protest responses.

Socio-economic characteristics & CV

In the final part of the questionnaire, information on respondents' demographic and socio-economic background like household size, number of children, profession, educational level, household income or environmental organisation participation is elicited. Furthermore, this section contains a number of questions referring to people's current electricity bill. For instance, people are asked to state the amount of their monthly electricity and whether this amount is paid by them or another household member.

A very important task included in the last part of the questionnaire is the open-ended CV question asking directly for the monthly amount that people are at maximum willing to pay to support the expansion of renewable energy or hydropower. The CV task was included subsequently to the questions on household income and current electricity bill. This order of questions may induce people to carefully think about how much disposable income they have available and how much they are already paying for electricity before stating their WTP. Without elucidation of the budget constraint, people may state too large WTP bids (ARROW ET AL., 1993:14).

To conclude, the questionnaire structure consisting of attitudinal questions, the valuation task and socio-economic questions is in line with general recommendations for SP studies (see for instance PEARCE ET AL., 2002:47ff) and past CE applications (see chapter 5). The total number of questions included in the final questionnaire on renewable energy expansion was 31. Due to the greater extent of the attitudinal section, the hydropower questionnaire contained 44 questions in total. Overall, 43 questions were included in the questionnaire referring to the specific hydropower project in Styria (see Table 32).

Table 32: Extent of the designed questionnaires

Questionnaire on...	*Number of questions*
...renewable energy expansion	31 questions
...hydropower expansion	44 questions
...the specific hydropower project in Styria	43 questions

Source: OWN REPRESENTATION

6.6 Pre-Testing

The choice experiment and questionnaire designs presented in the previous sections are the result of a series of pre-tests carried out over a period of several months. That questionnaires need to be tested in order to generate valid and reliable results has already been emphasised in section 6.1 of this work. Correspondingly, the developed questionnaires have been tested several times. In a first pre-test, using face-to-face interviews and a mail survey, 41 respondents got the draft questionnaire on hydropower and 50 the questionnaire on renewable energy,. Special attention was given to the choice experiment and how understandable the experiment and its design were for laypeople. After this first pre-test, the questionnaire and CE design was revised in light of people's responses, to eliminate any problems and maximise the amount of information that can be gathered.

The revised questionnaires were retested within a second pre-test round. The second pre-test sample contained 109 respondents in total. 55 received the questionnaire on hydropower, 54 the questionnaire on renewable energy expansion. First, a web-based survey was implemented on a small sample of respondents. Furthermore, three focus group discussions were conducted revealing useful information regarding comprehensibility and plausibility of the questionnaire, especially the choice experiment. After this test round, the questionnaire was slightly modified once again and pilot-tested on a sample of 290 respondents within a web-based survey.[155] This final pre-test showed that the questionnaire performs satisfactorily and was ready to be used on the full sample in the final survey.

Table 33: Conducted pre-tests of the questionnaires

Pre-test & questioning technique	Sample size & results
First pre-test round: face-to-face interviews and mail survey	Renewable energy: 50 respondents Hydropower: 41 respondents ↓ Revision of the questionnaires
Second pre-test round (re-test 1): web-based survey and focus group discussions	Renewable energy: 54 respondents Hydropower: 55 respondents ↓ Revision of the questionnaires
Third pre-test round (re-test 2): web-based survey	Renewable energy: 290 respondents ↓ Slight modification of the questionnaire
Pre-test of the regional case study: web-based survey	Regional case study: 103 respondents ↓ No further modifications required

Source: OWN REPRESENTATION

After testing and modifying the questionnaires on hydropower and renewable energy, the questionnaire referring to the regional hydropower case study was directly tested within a pilot online-survey of 103 respondents.[156] This was especially important, because a large number of protest responses were suspected in the regional sample. However, this has not yet been proven. Quite contrary to previous expectations, the questionnaire performed satisfactorily in the pre-test and, correspondingly, there was no need for further modifications.

---

[155] This pilot survey was conducted by a professional market research institute.
[156] This pilot survey was again carried out by an external survey agency.

6.7 Data collection

As we have seen in section 6.1, there exist various methods for conducting a survey as for instance face-to-face interviews, mail or online surveys, each with advantages and disadvantages. For the current purpose, an online survey has been used, although web-based surveys are usually associated with low response rates and a potential for skewed samples. The best way to minimize the possibility of skewed samples is to use demographically balanced online panels, because quotas and screening can help to target the proper respondents in a demographically balanced manner (EVANS AND MATHUR, 2005:209). The professional market research institute MARKETAGENT uses such an online panel for its market research. More precisely, the institute is active in Austria, Germany, Slovenia and Switzerland and provides an online access panel containing about 340,000 participants.[157] The implementation of the online surveys of this work has therefore been carried out in collaboration with MARKETAGENT. The programming of the web-based surveys was done ourselves using the software package Sawtooth, a special tool for online interviewing and conjoint analysis. The survey agency MARKETAGENT only delivered the address data and was responsible for the distribution of the survey across respondents. The online research institute guaranteed the requested number of filled questionnaires, as well as the representativeness of the sample.

Since more than one CE had to be conducted (see section 6.2), three random samples were drawn from different target populations. In case of the renewable energy survey, the target population was the whole population of Austria. For the hydropower survey, by contrast, only people living in Carinthia, Salzburg, Styria, Tyrol, Vorarlberg and Vienna were among the target group because these are the states where most of the future hydropower stations are planned to be built.[158] Finally, the target population of the regional hydropower case study were people living around the project,

---

[157] For more information see *http://www.marketagent.com/customer/ma_customer_main_free.aspx*.
[158] This selection of federal states is based on the list of planned hydropower stations issued by the UMWELTDACHVERBAND (2010, online). Vienna was included simply due to the fact that it represents the capital of Austria and therefore cannot be ignored when conducting a nationwide survey.

thus people from Graz (capital of Styria) and the surrounding communities.[159]

Generally, it is possible to determine the required size of the sample drawn from the appropriate target population dependent on the error probability and the deviation from true value (precision). The method of calculation is given in equation 6.2, where z is the quantile of the standard normal distribution for error probability α (see HARTUNG ET AL., 1991:891), p the proportion in true sample which is usually unknown prior to the sampling procedure, and L refers to the precision or deviation from true value.[160]

$$n \geq \frac{4 * z_{1-\alpha/2}^2 * p(1-p)}{L^2} \qquad (6.2)$$

For the unknown p usually the "worst case" of $p=0.5$ is assumed which reduces equation 6.2 to:

$$n \geq \frac{z_{1-\alpha/2}^2}{L^2} \qquad (6.3)$$

The higher the precision and the lower the error probability should be, the larger the required sample must be (HUDEC AND NEUMANN, n.d.:23). As can be seen from Table 34, a precision level of +/- 1 % from the true value and an error probability of 1 % would require a sample size of 16,587 observations. The necessary sample size declines with increasing error probabilities and decreasing precision. Accordingly, an error probability of $\alpha=0.05$, which is a standard assumpiton, and a precision level of +/- 5 % from the true value would require a sample of 384 respondents. In order to get results with a confidence level of 99 % and a precision of +/- 5 %, 663 respondents must be surveyed. The choice of the appropriate sample size generally entails a trade-off between cost and precision of estimate because larger samples are usually more costly (PEARCE ET AL., 2002:45).

---

[159] The general sample frame from which the samples were ultimately drawn, corresponds to Austrian people registered in the online access panel of MARKETAGENT.
[160] More precisely, L = (deviation from true value + deviation from true value)².

Table 34: Required sample size depending on error probability and level of precision

Error probability → ↓ Precision	1 %	5 %	10 %
+/- 1 %	16,587	9,604	6,764
+/- 5 %	663	384	271
+/- 10 %	166	96	68

Source: OWN CALCULATIONS AND REPRESENTATION

The online surveys of the present work were conducted between the beginning of June and mid of July 2011 with the sample sizes shown in Table 35. The different samples were started subsequently, beginning with the regional case study in Styria. Afterwards the survey referring to hydropower expansion, and lastly, the survey on renewable energy were implemented. Regarding the hydropower sample, 4,892 people were invited to participate in the survey. The response rate was 18.5 % meaning that 905 respondents completed the survey. This is the main sample of the investigation and its size is quite above the number of respondents required to provide results with +/- 5 % precision and an error probability of 1 %. In addition to the hydropower sample, two smaller samples have been drawn. In total, 1,781 people were invited to participate in the survey on renewable energy expansion. 301 respondents completed the survey corresponding to a response rate of 16.9 %. Finally, the survey on the specific hydropower project was distributed to 959 people living in Graz and its surrounding communities. Here, the highest rate of return has been achieved (22.0 %) resulting in a sample size of 211 respondents. The sample sizes of these surveys are significantly lower compared to the main survey on hydropower. Smaller samples may be preferred from a cost perspective, but usually go at the expense of precision of estimate. Nevertheless, the obtained samples with 200 to 300 observations lie well within the range of required sample sizes associated with commonly assumed error probability and precision levels.

Table 35: The different samples of the online survey

Sample	Region/target population	Sample size	Response rate
(1) Hydropower	Austria (only Carinthia, Salzburg, Styria, Tyrol, Vorarlberg and Vienna)	n = 905	18.5 %
(2) Renewable energy	Austria	n = 301	16.9 %
(3) Regional hydropower	Graz and surrounding communities	n = 211	22.0 %

Source: OWN CALCULATIONS AND REPRESENTATION

6.8 Identifying non-valid responses

Prior to the analysis of the collected data it is necessary to filter out non-valid responses. Generally, non-valid responses refer to people that refuse to answer the valuation question or do not reveal their true WTP but respond with a zero value instead (PEARCE ET AL., 2002:65). Accordingly, choosing the opt-out alternative in each of the six choice tasks may be an expression of protest. In the sample referring to hydropower expansion, 73 respondents chose the opt-out alternative in each of the six choice cards. In terms of the total sample size this corresponds to a proportion of 8.1 %. In the renewable energy sample the amount of people choosing the opt-out six times was 13, a share of 4.3 % based on the total sample. Finally, in the sample relating to the regional hydropower project altogether 18 respondents chose the opt-out alternative in each of the six choice tasks. This corresponds to a share of 8.5 % of the total sample (see Figure 42). However, not all of these respondents may represent protest bidders. In order to identify protest responses, people choosing six times the opt-out alternative, were confronted with a series of statements shown in Table 36. The first statement was assumed to indicate protest. The others were taken as valid responses, thus representing "true" zeros. In addition, respondents were given the opportunity to state individual reasons for choosing the opt-out in an open-ended task. These individual answers were analysed carefully and some of them have been categorised as protest votes. More precisely, the open-ended statements categorised as protests especially refer to respondents' unwillingness to pay due to the currently high electricity price level.

Table 36: Categorisation of protest responses

Response	Valid (✓) / Protest (✗)
I am strictly against the expansion of renewable energy sources / the expansion of hydropower / the construction of the hydropower plant.	✗
I am not interested in this matter	✓
The current situation is already satisfactory.	✓
I cannot afford additional monthly payments.	✓
The additional monthly payments are too high.	✓
I think other issues are more important.	✓
Other (open-ended)	✓ / ✗

Source: OWN REPRESENTATION

According to that, 13 protest responses were identified in the sample referring to the expansion of hydropower. This is a proportion of 1.4 % of the total number of respondents in the sample. In the sample on renewable energy expansion the number of protest votes is extremely low (2), accounting for only 0.7 % of the total sample. In the regional sample, the number of protest responses is – as previously expected – somewhat higher and amounts to 12, a proportion of 5.7 % based on the total sample. Another distinctive factor concerning the regional sample is that the major part of the respondents (12 out of 18) choosing the opt-out in every choice set was able to be categorised as "true" protest. In the other samples, by contrast, only a small fraction of the respondents who chose six times the opt-out were "true" protests. This is shown in Figure 42.

Figure 42: Number of protest responses in the different samples

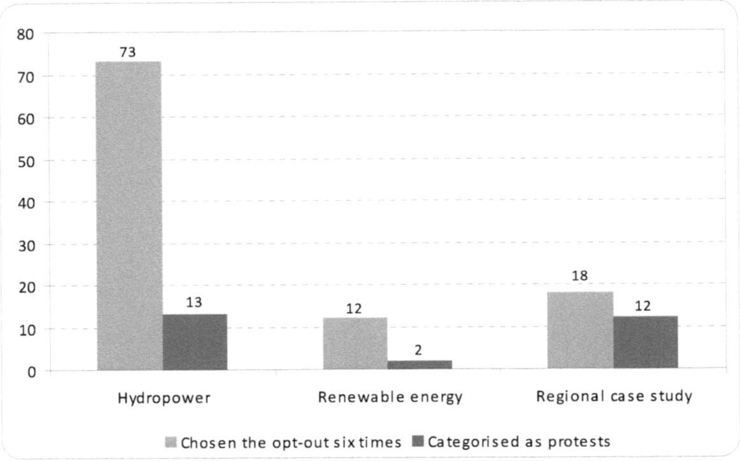

	Hydropower	Renewable energy	Regional case study
Chosen the opt-out six times	73	12	18
Categorised as protests	13	2	12

Source: OWN CALCULATIONS AND REPRESENTATION

Once protest responses have been identified, the sample must be adjusted. One possibility would be to omit "true" protests from the sample and the subsequent analysis. This is a valid approach as long as the characteristics of the reduced sample do not differ significantly from the unadjusted sample (PEARCE ET AL., 2002:66). Since the proportion of protest responses is in each of the three samples relatively small compared to the overall sample size, this requirement can be expected to hold. Consequently, the identified protest responses were excluded from the following analysis resulting in the following sample sizes.

Table 37: Sample sizes after adjustment for protest responses

Sample	Unadjusted sample	Reduced sample
Hydropower	n = 905 →	n = 892
Renewable energy	n = 301 →	n = 299
Regional hydropower	n = 211 →	n = 199

Source: OWN REPRESENTATION

6.9 Representativeness of the samples

The first step in data analysis is to ensure that the samples are representative of the population in general (PEARCE ET AL., 2002:66). This is why the survey sample characteristics have been compared with the characteristics of the corresponding target population. First, Table 38 shows that representativeness is given with respect to gender. The gender of the respondents is in both samples, hydropower and renewable energy, very close to the Austrian average with 48.7 % men and 51.3 % women. In the regional sample, however, there is a slight surplus of male respondents compared to the total Styrian population.[161]

Table 38: Gender or respondents compared to total population

Gender	absolute	in %	Total population
HYDROPOWER SAMPLE			
Male	435	48.8 %	48.7 %
Female	457	51.2 %	51.3 %
RENEWABLE ENERGY SAMPLE			
Male	146	48.8 %	48.7 %
Female	153	51.2 %	51.3 %
REGIONAL HYDROPOWER SAMPLE (STYRIA)			
Male	103	51.8 %	48.9 %
Female	96	48.2 %	51.1 %

Source: STATISTIK AUSTRIA (2011a:48); OWN CALCULATIONS AND REPRESENTATION

The age structure corresponds in principle to that of the total population in Austria (see Table 39). However, in the hydropower sample people aged between 18 and 59 years are slightly overrepresented, while elderly people (60 years and older) are proportionally low compared to the total population. The under-representation of older people may be due to the data collection method, since the older population is usually less familiar with online surveys or the internet in general (see also section 6.1). Mean age amounts to 42.3 years with a standard deviation of 15.1 years; median age is 42 years.

---

[161] Due to a lack of reliable data for the area of Graz and surroundings, the regional hydropower sample is compared to the whole province of Styria.

The age structure of the renewable energy sample lies, by contrast, well within the distribution of the population in Austria. Contrary to expectations, older people aged between 70 and 75 years are even over proportional compared to the total population. Mean age is slightly higher than in the hydropower sample with 44.4 years (standard deviation: 15.5 years, median: 43 years).

In the regional sample, the age category older than 59 years is proportionally low compared to the total Styrian population which may again be attributable to the online survey method. The same applies to the age group 18-19 years which is also slightly underrepresented in the sample. In contrast, respondents aged between 20-29 years are stronger represented with a proportion of 26.1 % in the sample compared to 17.5 % in the total population. The mean age in the regional sample is slightly lower compared to the other samples and amounts to 40.9 years (standard deviation: 14.2 years; median: 41 years; see Table 39).

Table 39: Age of the respondents compared to total population

Age	HYDROPOWER SAMPLE		RENEWABLE ENERGY SAMPLE		Austrian population	REGIONAL HYDROPOWER SAMPLE		Styrian population
	absolute	in %	absolute	in %		absolute	in %	
18-19 years	50	5.6 %	12	4.0 %	3.3 %	4	2.0 %	3.3 %
20-29 years	165	18.5 %	52	17.4 %	17.4 %	52	26.1 %	17.5 %
30-39 years	179	20.1 %	55	18.4 %	18.2 %	38	19.1 %	17.8 %
40-49 years	194	21.7 %	64	21.4 %	22.7 %	46	23.1 %	22.4 %
50-59 years	162	18.2 %	53	17.7 %	17.8 %	36	18.1 %	18.2 %
60-69 years	107	12.0 %	43	14.4 %	14.7 %	19	9.5 %	14.6 %
70-75 years	35	3.9 %	20	6.7 %	5.9 %	4	2.0 %	6.2 %
In total	**892**	**100.0 %**	**299**	**100.0 %**	**100.0 %**	**199**	**100.0 %**	**100.0 %**
Mean age	42.3 years		44.4 years			40.9 years		
Median age	42 years		43 years			41 years		

Source: STATISTIK AUSTRIA (2011a:48), (2011b:316) and (2011c:72); OWN CALCULATIONS AND REPRESENTATION

The geographical distribution of the nationwide samples is given in Table 40. As far as the renewable energy sample is concerned, the distribution among Austrian states corresponds almost perfectly to the distribution in the total population. The survey related to the expansion of hydropower has only been conducted in certain states, namely those that are actually affected by the Austrian hydropower expansion plans. As already mentioned before, these are Carinthia, Salzburg, Styria, Tyrol, Vorarlberg and Vienna.[162] In each of these states, approximately 150 respondents have been surveyed. As a consequence, the sample on hydropower may not be representative with respect to the population distribution but with respect to hydropower expansion in Austria.

Table 40: Geographical distribution of the samples on hydropower and renewable energy

State	HYDROPOWER SAMPLE		RENEWABLE ENERGY SAMPLE		Total population
	absolute	in %	absolute	in %	
Burgenland	0	0.0 %	10	3.3 %	3.4 %
Carinthia	150	16.8 %	17	5.7 %	6.7 %
Lower Austria	0	0.0 %	58	19.4 %	19.2 %
Upper Austria	0	0.0 %	53	17.7 %	16.8 %
Salzburg	147	16.5 %	14	4.7 %	6.3 %
Styria	150	16.8 %	46	15.4 %	14.4 %
Tyrol	150	16.8 %	25	8.4 %	8.4 %
Vorarlberg	149	16.7 %	14	4.7 %	4.4 %
Vienna	146	16.4 %	62	20.7 %	20.3 %
In total	**892**	**100.0 %**	**299**	**100.0 %**	**100.0 %**

Source: STATISTIK AUSTRIA (2011a:48); OWN CALCULATIONS AND REPRESENTATION

The target population of the regional hydropower case study involved only people living around the project, particularly in Graz and the directly surrounding communities. In total, Graz and its surrounding area[163] have about 338,000 inhabitants. 21.6 % of them are living in one of the surrounding communities and 78.4 % have their residence in the city of Graz. This dis-

---

[162] Vienna is not directly affected by the hydropower expansion plans, however, cannot be neglected since it represents the capital of Austria.
[163] Graz has in total 19 directly surrounding communities, namely Feldkirchen, Fernitz, Gössendorf, Grambach, Hausmannstätten, Kalsdorf, Raaba and Seiersberg in the south. In the north of Graz are the communities Deutschfeistritz, Eisbach, Gratkorn, Gratwein, Judendorf-Straßengel, Peggau, Stattegg and Weinitzen. Finally, in the east and west, Graz is surrounded by the communities Hart, Kainbach and Thal.

tribution is roughly reflected in the sample with 75.2 % of the respondents living within the city limits of Graz and 24.8 % living in one of the surrounding communities (see Table 41). The respondents from the area around Graz are thereby equally allocated among all surrounding communities. Accordingly, from each community on average 2-3 inhabitants have been surveyed.

Table 41: Geographical distribution of the regional hydropower sample

Region	absolute	in %	Total population
Graz	151	75.2 %	78.4 %
Surrounding communities	48	24.8 %	21.6 %
In total	**199**	**100.0 %**	**100.0 %**

Source: LAND STEIERMARK (2012a); OWN CALCULATIONS AND REPRESENTATION

With respect to the educational situation, the samples referring to hydropower and renewable energy are slightly higher educated than the average Austrian population. The secondary level (higher school certificate) is rather overrepresented (21.7 %/20.5 % compared to 13.6 % in the Austrian population), while people with compulsory schooling as the highest completed level of education are underrepresented with 12.3 %/15.8 % compared to 19.5 % in the total population. With respect to tertiary education (college of education or university degree), as well as apprenticeship and professional school, the samples are fairly representative (see Table 42).

A similar result shows up in the regional hydropower sample. As can be seen from Table 42, the sample is somewhat higher educated than the total population. As an aside, sample characteristics are here compared with the population of the district "Graz-Stadt". This appeared to be an appropriate approach because the sample consists mainly of people from this area (see also Table 41). Respondents with a higher school certificate are considerably overrepresented while lower educated people (compulsory school, apprenticeship and professional school) are significantly underrepresented compared to the total population of Graz. The proportion of respondents obtaining a university degree lies only slightly above the population share (26.3 % compared to 24.0 % in the total population). These deviations from the population distribution must be seen in light of the difficulty for the market research institute to draw a representative sample from such a small geographical area.

The professional situation of the respondents is given in Table 43. There are marginal deviations from the distribution in the total population. In the hydropower and regional sample, employed people are significantly overrepresented with 66.0 % or 62.8 % compared to 53.8 % in the total population.[164] Furthermore, in each of the three samples unemployed people are proportionally low compared to the total population. The same applies to the share of retired people which is disproportionately small as well. As in the case of older people being underrepresented in the samples, this may be due to the online survey vehicle. Retired people usually represent the older fraction of the population and are often not familiar with online panels. Finally, in the regional sample people in education are represented more than in the Austrian population (15.6 % compared to 7.3 %). This may be motivated by the fact that Graz represents a major university town in Austria. Comparable data on provincial level are not available.

---

[164] Due to the absence of appropriate data at provincial level, the professional situation in the regional sample is compared with data for the entire Austrian population.

Table 42: Educational level of respondents compared to total population

Educational level	HYDROPOWER SAMPLE		RENEWABLE ENERGY SAMPLE		Austrian population	REGIONAL HYDROPOWER SAMPLE		Graz
	absolute	in %	absolute	in %		absolute	in %	
Compulsory school	110	12.3 %	47	15.8 %	19.5 %	10	5.1 %	15.8 %
Apprenticeship, professional school	464	52.0 %	155	52.0 %	52.3 %	58	29.3 %	37.0 %
Higher school certificate	194	21.7 %	61	20.5 %	13.6 %	74	37.4 %	20.0 %
College of education	20	2.2 %	5	1.7 %	3.5 %	3	1.5 %	3.2 %
University (of applied sciences)	102	11.4 %	28	9.4 %	11.1 %	52	26.3 %	24.0 %
Other	2	0.2 %	2	0.7 %	0.0 %	1	0.5 %	0.0 %
In total	**892**	**100.0 %**	**299**	**100.0 %**	**100.0 %**	**198**	**100.0 %**	**100.0 %**

Source: STATISTIK AUSTRIA (2012a:432ff); OWN CALCULATIONS AND REPRESENTATION

Table 43: Professional situation of respondents compared to total population

Professional situation	HYDROPOWER SAMPLE		RENEWABLE ENERGY SAMPLE		Austrian population	REGIONAL HYDROPOWER SAMPLE		Austrian population
	absolute	in %	absolute	in %		absolute	in %	
Employed	589	66.0 %	167	55.9 %	19.5 %	125	62.8 %	53.8 %
Unemployed	19	2.1 %	9	3.0 %	52.3 %	7	3.5 %	5.0 %
In education	74	8.3 %	27	9.0 %	13.6 %	31	15.6 %	7.3 %
Retired	155	17.4 %	68	22.7 %	3.5 %	28	14.1 %	25.5 %
Housewife	51	5.7 %	27	9.0 %	11.1 %	8	4.0 %	7.8 %
Other	4	0.4 %	1	0.3 %	0.0 %	0	0.0 %	0.6 %
In total	**892**	**100.0 %**	**299**	**100.0 %**	**100.0 %**	**198**	**100.0 %**	**100.0 %**

Source: STATISTIK AUSTRIA (2012b:261); OWN CALCULATIONS AND REPRESENTATION

The distribution of disposable monthly household income is shown in Table 44. In the hydropower sample 50.0 % of the respondents have a household income less than € 2,000 per month. In the regional hydropower sample this proportion is slightly higher. Accordingly, 53.0 % of the respondents have less than € 2,000 available per month. The median income category in these two samples therefore corresponds to € 1,501-2,000 which is considerably below median household income in Austria of approximately € 2,490. Consequently, the income distribution in two samples referring to hydropower is slightly skewed towards those with lower incomes. In the renewable energy sample, the income distribution differs marginally. Accordingly, the share of people with a monthly available household income less than € 2,000 is only 48.2 %. Thus, median income falls into the next higher income category, i.e. € 2,001-2,500 and therefore corresponds more appropriately to median household income in the total population.

Table 44: Distribution of disposable monthly household income in the samples

Income category	HYDROPOWER SAMPLE		RENEWABLE ENERGY SAMPLE		REGIONAL HYDROPOWER SAMPLE	
	absolute	in %	absolute	in %	absolute	in %
Up to € 1,000	144	16.1 %	42	14.0 %	35	18.9 %
€ 1,001-1,500	151	16.9 %	45	15.1 %	27	14.6 %
€ 1,501-2,000	152	17.0 %	57	19.1 %	36	19.5 %
€ 2,001-2,500	130	14.6 %	52	17.4 %	26	14.1 %
€ 2,501-3,000	128	14.3 %	40	13.4 %	29	15.7 %
€ 3,001-3,500	72	8.1 %	28	9.4 %	12	6.5 %
More than € 3,500	115	12.9 %	35	11.7 %	20	10.8 %
In total	**892**	**100.0 %**	**299**	**100.0 %**	**199**	**100.0 %**
Median category	€ 1,501-2,000		€ 2,001-2,500		€ 1,501-2,000	
Median (population)			€ 2,487			

Source: STATISTIK AUSTRIA (2011a:248); OWN CALCULATIONS AND REPRESENTATION

7 Public attitude towards hydropower in Austria

Beside the valuation tasks (CE and CV), the developed questionnaire a section relating to people's electricity consumption patterns, as well as their general attitude towards renewable energy and hydropower use. These questions are statistically evaluated in the following sections providing a clear insight into respondents' perception of renewable energy and hydropower use.[165]

7.1 Descriptive analysis of public attitude

First, there is a general agreement upon the importance of renewable energy use. The major part of the respondents appreciates electricity from renewable sources. More specifically, 86.9 % of the respondents indicate utmost importance that their obtained electricity stems from renewable sources. Of those, 36.5 % state that they consciously choose an electricity supplier that provides them exclusively with electricity from renewable energy. Finally, 39.9 % of the respondents that consciously buy green electricity are willing to accept a higher price in order to obtain electricity from renewable sources.

The perceived importance of future renewable energy expansion is shown in Figure 43. The majority of the respondents (75.6 %) regard the intensified use of renewable energy sources in the future as very important; another 21.9 % think at least that it is rather important to increase the use of renewable energy sources. Only 2.6 % of the respondents consider the expansion of renewable energy as rather or totally unimportant.

---

[165] The analyses are based on the main hydropower sample containing 892 observations (see section 6.7).

Figure 43: Perceived importance of future renewable energy expansion

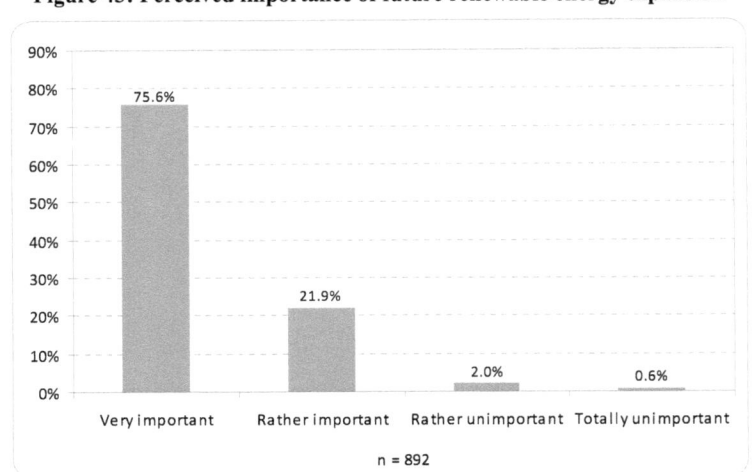

Source: OWN CALCULATIONS AND REPRESENTATION

The meaning of renewable energy sources further arises when asking people from which energy sources electricity in Austria should come from in the future. In particular, people were allowed to choose several randomly ordered options including fossil fuels. The results of this task are shown in Figure 44. What becomes immediately apparent is the prioritisation of renewable energy sources as opposed to fossil fuels. Most people (716 or 80.3 %) think that hydropower should be an important future energy source. Solar and wind power were selected by 709 (79.5 %) respectively 666 (74.7 %) respondents. Biomass is significantly less preferred being selected by only 320 respondents or 35.9 %. Furthermore, 108 respondents (12.1 %) consider natural gas as a preferred energy source for future electricity generation. Fossil fuels (coal and oil) and nuclear power are to be found at the end of the ranking with only a low number of respondents choosing these technologies. The category "other" which has been selected by 7 respondents mainly contains geothermal energy.

Figure 44: Preferred energy sources for future electricity generation in Austria

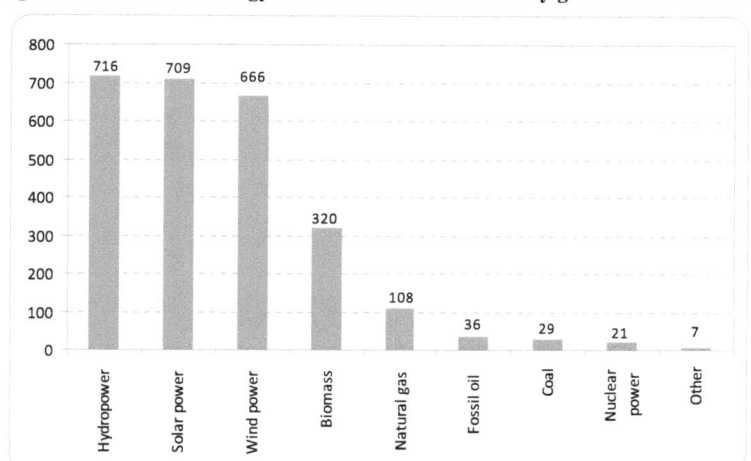

Source: OWN CALCULATIONS AND REPRESENTATION

Another question refers to renewable energy sources only. In particular, people were asked to rank the two most preferred renewable energy sources to be promoted in the future. In particular, respondents were allowed to choose rank one for the most preferred renewable energy source and rank two for the second most preferred option. The outcome is shown in Figure 45. Solar power (photovoltaics) is the first preferred energy source for 39.2 % of the respondents, closely followed by hydropower which is the first choice of 39.0 % of the respondents. Additionally, solar power was placed second by 31.8 % of the respondents, while the share of respondents assigning the second place to hydropower was 26.3 %. For 12.8 % of the respondents wind power is the first preferred renewable energy source. At the same time, nearly one third (32.6 %) thinks that wind power should be the second most preferred energy source to be promoted in the future. Finally, biomass is the least preferred energy source for future renewable energy expansion with 9.0 % of the respondents giving the technology rank one and 9.2 % assigning rank two. This result may be due to the fact that biomass is generally less associated with renewable energy as compared to solar, hydro or wind power.

Figure 45: Preferred renewable energy sources for future expansion in Austria

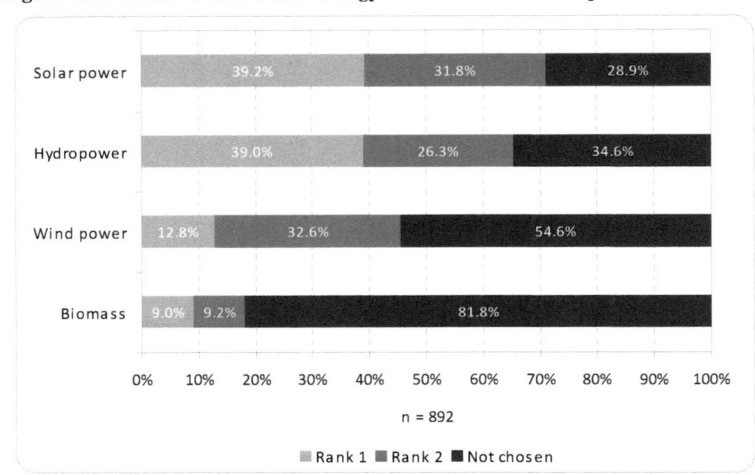

Source: OWN CALCULATIONS AND REPRESENTATION

The following assessments particularly refer to hydropower and propose to gain a clear insight into people's attitude towards the most important renewable energy source in Austria. First, the public's general attitude towards hydropower use in Austria is basically positive with 45.5 % of the respondents having a very positive and 50.2 % a rather positive attitude. In contrast, 4.3 % of the survey participants are negatively confronted with hydropower use in Austria (see Figure 46).

Despite this positive result, a significant lack of information was identified. As can be obtained from Table 45, more than one third of the respondents (38.1 %) feel rather bad and another 4.9 % very bad informed about hydropower use in Austria. In addition, only 58.6 % of the respondents heard about the concrete plans to build new hydropower stations along Austrian rivers. Hence, a significant part of the respondents (41.4 %) has little information about the Austrian hydropower expansion plans.

Table 45: Perceived state of information regarding hydropower use in Austria

State of information	absolute	in %
Very good	116	13.0 %
Rather good	392	43.9 %
Rather bad	340	38.1 %
Very bad	44	4.9 %
In total	**892**	**100.0 %**

Source: OWN CALCULATIONS AND REPRESENTATION

Asking people about their general attitude towards the construction of new hydropower plants, it was found that 43.8 % are very well disposed towards hydropower expansion. The share of people with a rather positive attitude is 48.3 %. Only a minority of 7.9 % is in principle against the construction of new hydropower plants along Austrian rivers (see Figure 46).

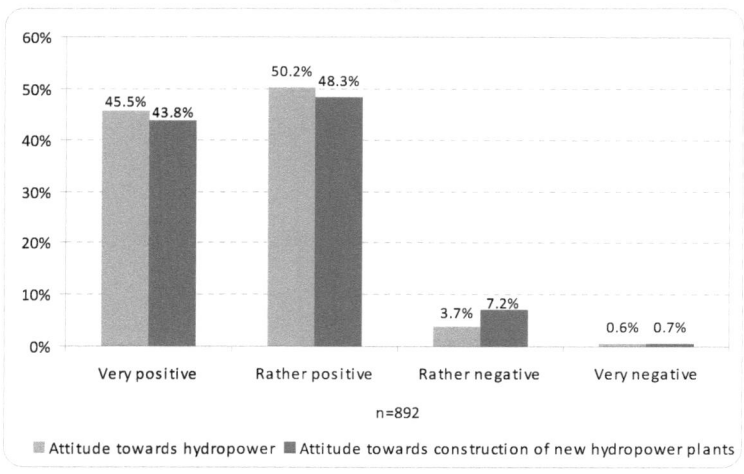

Figure 46: Attitude towards hydropower use and the construction of new plants in Austria

Source: OWN CALCULATIONS AND REPRESENTATION

In addition to the attitude and information status towards hydropower use and its expansion, the questionnaire aimed to elicit people's current experience with the technology. First, it can be shown that the majority of the respondents (84.4 %) are living in a distance of maximum 10 kilometres to a river which might be affected by future hydropower expansion. For those respondents recreational activities along the riverside may play an important role. For that reason, people were confronted with a question asking for the frequency certain recreational activities are explored along the riverside. In particular, respondents were asked to indicate whether they exert ten different recreational activities "frequently" (1), "sometimes" (2) or "never" (3). For each leisure activity an index value (mean) has been calculated according to equation 7.1. The lower the index, the more frequently a recreational activity is practised. The outcomes of these calculations are shown in Table 46.

$$Index_j = \frac{\sum_{i=1}^{n} frequency_i}{n} \quad for\ all\ activities\ j = 1,...,10 \quad (7.1)$$

The most frequently exerted recreational activities along Austrian rivers are walking along the riverside, recreating/enjoying the landscape and sportive activities. Boating and fishing are on average the least frequently practised leisure activities. Generally, the possibilities for recreation may be affected by the construction of new hydropower plants. Whether this impact will be positive or negative depends on the individual design of the hydropower station.

Table 46: Recreational activities along Austrian rivers

Recreational activity	Index value
Walking along the riverside	1.63
Recreating/enjoying the landscape	1.66
Sportive activities	1.78
Family trip	1.98
Animal watching	2.20
Swimming	2.24
Visiting a restaurant/café	2.24
Picnic on the riverside	2.46
Boating	2.62
Fishing	2.74
Observations	n=753

Source: OWN CALCULATIONS AND REPRESENTATION

Another question aimed to get a perception of how many hydropower stations are in the respondents' surrounding area. 18.7 % of the respondents were not able to make an assumption about the number of hydropower stations near their home. The majority of the respondents (63.7 %) has at least some hydropower stations in the neighbourhood. 7.5 % stated that there are many hydropower plants in the close vicinity. In contrast, a share of 10.1 % indicated that there are no hydroelectric facilities in the environment (see Figure 47).

Figure 47: Perceived number of hydropower plants in respondents' environment

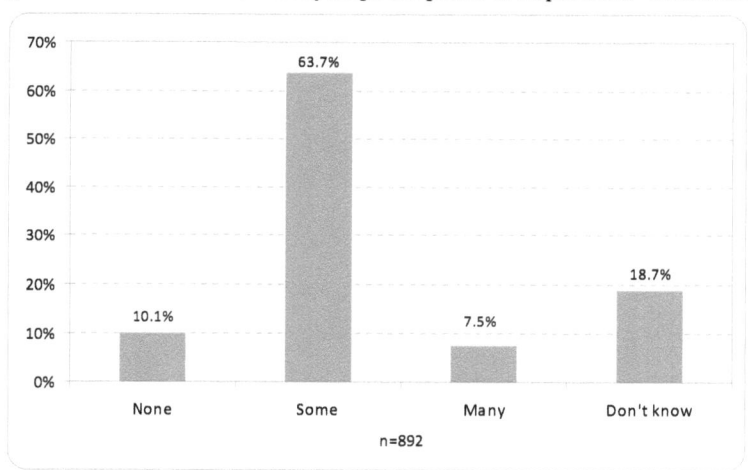

Source: OWN CALCULATIONS AND REPRESENTATION

Subsequently, people were asked to state the approximate distance between the nearest hydropower station and their home in the form of an open-ended question. As shown in Figure 48, about one quarter of the respondents (25.5 %) is living very close to a hydropower station at a maximum distance of 5 km.[166] The major part of the respondents, namely 46.8 %, stated that the next hydropower facility is placed at a distance between 5 km and 25 km. The remaining 27.6 % indicated a distance of more than 25 km. The mean distance obtained from respondents' statements is 27.9 km;

---

[166] As an aside, the distance question was voluntary. Accordingly, as people were not obliged to state the approximate distance, 27 respondents refused to answer the question (missing values). This is why the analysis of distance is based on only 865 as compared to 892 observations in the full sample.

the standard deviation of the distribution is very high (53.5 km) indicating the presence of outliers. Hence, median distance is significantly lower and amounts to 15 km.

Figure 48: Stated distance between the nearest hydropower station and respondents' home

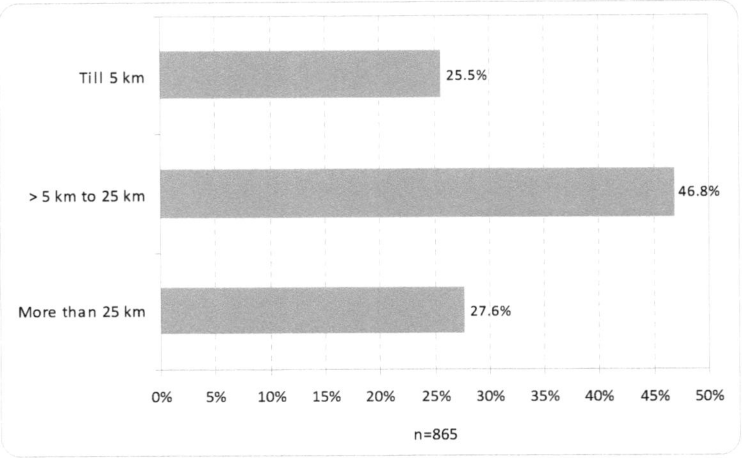

Distance	Percentage
Till 5 km	25.5%
> 5 km to 25 km	46.8%
More than 25 km	27.6%

n=865

Source: OWN CALCULATIONS AND REPRESENTATION

A follow-up question is related to the degree people are affected by existing hydropower schemes. As a result, the main part of the respondents (70.3 %) feels generally unaffected by the nearest existing hydropower plant. By contrast, about one quarter of the respondents (25.9 %) indicated to feel positively affected by the next hydropower station. The share of people feeling negatively affected is considerably low and amounts to merely 3.8 %. However, the individual concern associated with the closest hydropower station varies with distance. In the group of respondents living at a distance of less or equal than 5 km to a hydropower station, the proportion of people feeling unaffected (63.3 %) is significantly lower as compared to the subgroup living more than 5 km away (71.7 %). By contrast, people living closer to a hydropower station are more often positively affected (34.8 %) than people living further away (23.6 %). Finally, the subgroup living at a distance of maximum 5 km to a hydropower station appeared to be less frequently affected in a negative way (see Figure 49). The correlation between distance and individual concernment is statistically significant at the 1 % level as shown by the contingency analysis (Pearson-χ^2=12.912, p-value=0.002; Cramer's V=0.122).[167]

Figure 49: Individual concernment of people by the closest hydropower station

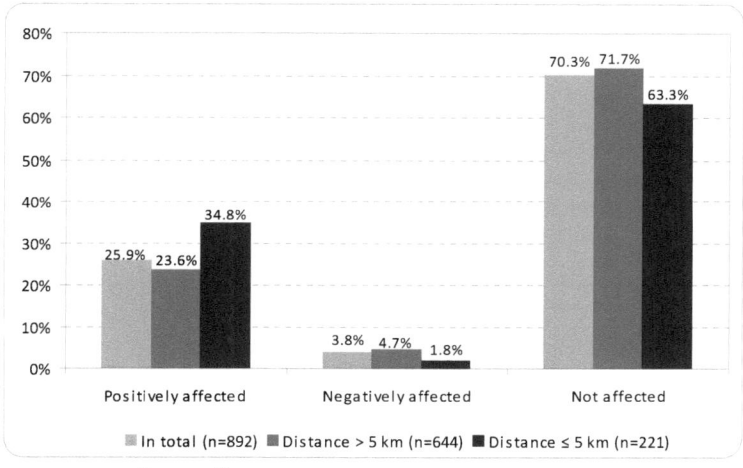

Source: OWN CALCULATIONS AND REPRESENTATION

Subsequently, people were asked for the reasons why they feel positively or negatively affected by the closest hydropower station. As can be seen from Table 47, positive concernments are mainly related to the possibility of getting electricity from a renewable energy source and the basically positive attitude towards hydropower. The primary reasons associated with a perceived negative derogation are adverse affects on landscape and biodiversity (flora and fauna) imposed by the use of hydropower.

---

[167] The higher Cramer's V, the stronger the correlation between the two variables. More on calculation and interpretation of Cramer's V can be found in HARTUNG ET AL. (1992:452).

Table 47: Reasons for the individual concernment imposed by a hydropower plant

Why positively affected		
Reasons	Sample (n=231)	in %
I can get electricity from a renewable energy source.	118	51.1 %
The hydropower plant has a positive effect on the landscape.	36	15.6 %
The hydropower plant facilitates new recreational activities.	31	9.1 %
I am principally pro hydropower.	55	23.8 %
Other reasons	1	0.4 %
Why negatively affected		
Reasons	Sample (n=34)	in %
The hydropower plant limits the possibilities for recreation.	7	20.6 %
The hydropower plant has a negative effect on the landscape.	15	44.1 %
The hydropower plant has a negative effect on flora and fauna.	10	29.4 %
I am principally against hydropower.	1	2.9 %
Other reasons	1	2.9 %

Source: OWN CALCULATIONS AND REPRESENTATION

In addition, people were asked whether a new hydropower station is built or planned near to their home, or more precisely, at a distance of maximum 10 km to their residence. 18.3 % of the respondents replied to this question with "yes", while 37.6 % think that there are no plans to build a new hydropower plant near their residence. The remaining 44.2 % don't know whether a new hydropower facility is planned to be built near their home (see Figure 50).

This result illustrates once again the prevalent information deficit throughout the Austrian population. Accordingly, respondent's knowledge about the plans to build a new hydropower station in close vicinity to their residence depends on the level of information. Generally, well informed people better know about the plans to build or not to build a new hydropower facility close to their residence. This is shown in Figure 50. As can be seen, 23.8 % of the respondents feeling well informed about hydropower use in Austria explicitly know that a new hydropower scheme will be built at a distance of maximum 10 km to their home. In the subgroup of badly informed respondents this proportion is only 10.9 %. A similar result is given for response category "no". While 41.1 % of the well informed respondents distinctly know that there aren't any plans to build a new hydropower facility in close proximity, only 32.8 % of the badly informed respondents were able to indicate this with certainty. In the response category "don't know" the opposite case applies. Correspondingly, the proportion of "don't knows" is significantly higher in the subgroup of badly informed respondents as compared to their well informed counterparts (56.3 % versus 35.0 %). The correlation between respondent's knowledge of the plans to build a hydropower station and their level of information is statistically significant at the 1 % level (Pearson-χ^2=46.172, p-value=0.000; Cramer's V=0.228).

Figure 50: Respondent's knowledge of hydropower construction plans near their home

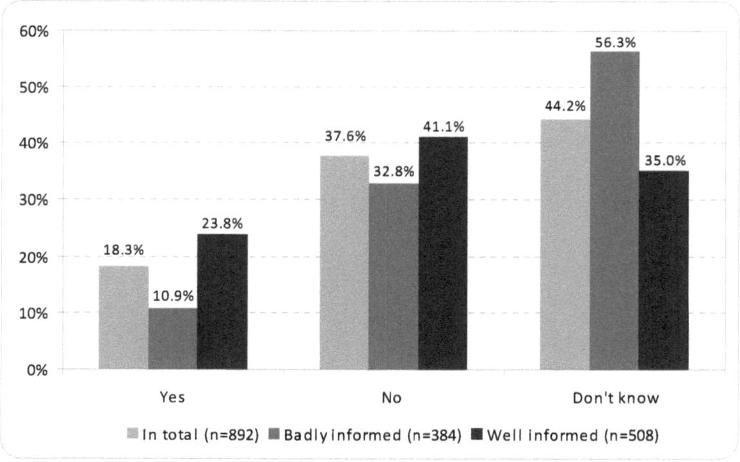

Response	In total (n=892)	Badly informed (n=384)	Well informed (n=508)
Yes	18.3%	10.9%	23.8%
No	37.6%	32.8%	41.1%
Don't know	44.2%	56.3%	35.0%

Source: OWN CALCULATIONS AND REPRESENTATION

Another important result of the descriptive statistical analysis refers to the perceived impact of a new hydropower station on the possibilities for recreational use of the river. In total, 34.5 % of the respondents think that the construction of a hydropower plant would improve the possibilities for recreational activities. By contrast, 18.4 % are of the opinion that a hydropower station would restrict the opportunities for recreation. Finally, nearly one half of the respondents (47.1 %) were unable to judge the impact of a hydropower facility on recreation (see Figure 51). Furthermore, the appreciation of the impact of a new hydropower station on recreational activities depends on the degree to which people are currently affected by hydropower. Respondents who feel positively affected by an existing hydropower plant are much more likely to think that a new hydropower facility will improve the possibilities for recreation as compared to the subgroup of negatively affected people (53.7 % versus 35.3 %). This result is contrary to the response category "deterioration". While half of the respondents feeling negatively affected by existing hydropower schemes indicate that the construction of a new hydropower station would lead to a deterioration of recreational possibilities, this applies to a proportion of only 13.9 % in the subsample of positively affected respondents. Finally, the share of people being unable to judge the impact of new hydropower stations on recreation is highest in the group of respondents that is not affected by current hydropower. The statistical correlation is highly significant at the 1 % level (Pearson-χ^2=79.596, p-value=0.000; Cramer's V=0.211).

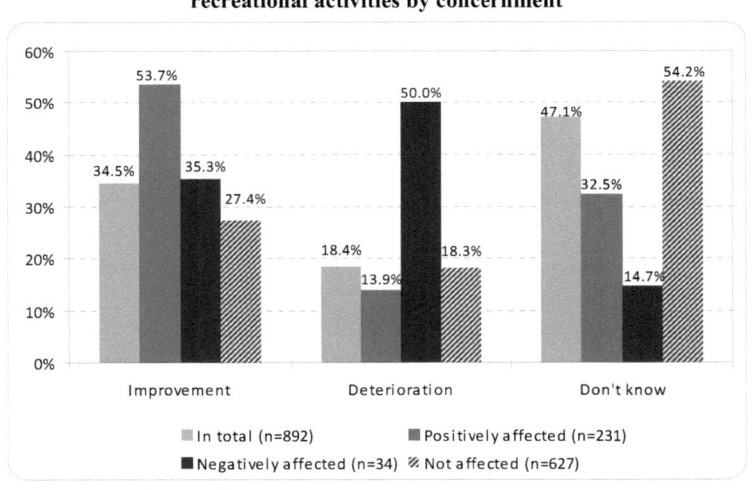

Figure 51: Perceived impact of hydropower on recreational activities by concernment

Source: OWN CALCULATIONS AND REPRESENTATION

In addition, there is a statistically significant correlation between the perceived impact of hydropower on recreational activities and people's attitude towards the construction of new hydropower plants (Pearson-χ^2=49.116, p-value=0.000; Cramer's V=0.235). Hence, people with a positive attitude towards the expansion of hydropower more often hold the opinion that a new hydropower station improves the possibilities for recreation (36.7 %) compared to respondents with a negative attitude (8.6 %). Conversely, the proportion of people thinking that a hydropower plant restricts the possibilities for recreation is significantly smaller in the group of people exhibiting a positive attitude (15.9 %) than in the group of respondents having a negative attitude (47.1 %; see Table 48).

Table 48: Perceived impact of hydropower on recreational activities by attitude

Impact on recreation	Positive attitude (n=822)	Negative attitude (n=70)
Improvement	36.7 %	8.6 %
Deterioration	15.9 %	47.1 %
Don't know	47.3 %	44.3 %

Source: OWN CALCULATIONS AND REPRESENTATION

Finally, people were confronted with a series of statements to which they had to attach their personal agreement. The highest commitment was found with respect to the importance of increased hydropower use on satisfying the growing demand for electricity in Austria. Here, 61.0 % of the respondents totally agree and 34.0 % rather agree. By contrast, only 4.5 % rather disagree and 0.6 % totally disagree to this statement.

The second statement is related to the importance of an intensified hydropower use for reducing climate-damaging CO_2 emissions. The outcome shows that 53.7 % of the respondents totally agree and 37.6 % rather agree that the increased use of hydropower is an important measure to reduce CO_2 emissions. The share of those who rather or totally disagree is 8.7 % in total.

A similar result is given with respect to the impact of hydropower expansion on the dependence on imports. Accordingly, 54.7 % of the respondents totally agree that the intensified use of hydropower contributes to reduce the necessity of electricity imports. Furthermore, 34.8 % rather agree and 10.5 % in principle disagree to this statement.

The last two statements people were confronted with relate to the negative environmental impacts associated with hydropower use. Although people's attitude towards hydropower use is generally positive as previously shown, 30.5 % of the respondents think that hydropower plants cause adverse affects on the landscape. Moreover, nearly one half (46.3 %) indicate that hydropower schemes are associated with negative wildlife impacts (see Figure 52).

Figure 52: Respondents' agreement to several statements on intensified hydropower use

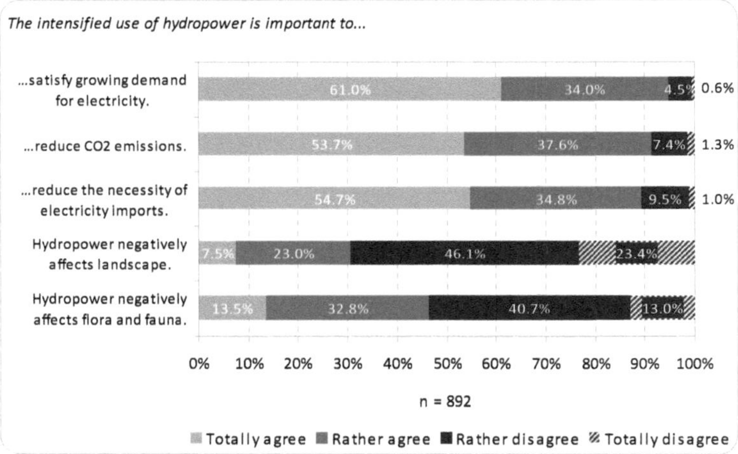

The intensified use of hydropower is important to...

Statement	Totally agree	Rather agree	Rather disagree	Totally disagree
...satisfy growing demand for electricity.	61.0%	34.0%	4.5%	0.6%
...reduce CO2 emissions.	53.7%	37.6%	7.4%	1.3%
...reduce the necessity of electricity imports.	54.7%	34.8%	9.5%	1.0%
Hydropower negatively affects landscape.	7.5%	23.0%	46.1%	23.4%
Hydropower negatively affects flora and fauna.	13.5%	32.8%	40.7%	13.0%

n = 892

Source: OWN CALCULATIONS AND REPRESENTATION

However, people's agreement to the several statements presented above varies with respondents' attitude towards the expansion of hydropower. Accordingly, the share of people coinciding with the statements that the intensified use of hydropower contributes to satisfy growing electricity demand, to reduce CO_2 emissions, and to reduce the necessity of electricity imports is significantly higher in the group of respondents with a positive attitude towards hydropower expansion.[168] The opposite applies to the agreement on landscape and wildlife impacts (see Figure 53). While only

---

[168] Satisfy growing electricity demand: Pearson-χ^2=172.080, p-value=0.000; Cramer's V=0.439. Reduce CO_2 emissions: Pearson-χ^2=75.100, p-value=0.000; Cramer's V=0.290. Reduce dependency on imports: Pearson-χ^2=60.651, p-value=0.000; Cramer's V=0.261.

26.8 % of the respondents having a positive attitude towards the expansion of hydropower think that the technology creates adverse effects on the landscape, this share amounts to 74.3 % in the group of respondents with a negative attitude. Conversely, for respondents with a positive attitude the disagreement rate is significantly higher as opposed to those with a negative attitude. A similar outcome is given for the expected impacts of increased hydropower use on flora and fauna. Thus, in the group of respondents with a negative attitude the proportion of people believing that the intensified use of hydropower causes negative wildlife impacts is twice as high as compared with the sample fraction having a positive attitude. On the contrary, disagreement is higher for people with a positive attitude towards the intensified use of hydropower. The correlation between people's attitude towards hydropower expansion and their agreement to landscape and wildlife impacts is highly significant at the 1 % level (landscape impacts: Pearson-χ^2=80.168, p-value=0.000; Cramer's V=0.300; wildlife impacts: Pearson-χ^2 =106.765, p-value=0.000, Cramer's V=0.346).

Figure 53: Agreement to landscape and wildlife impacts by attitude

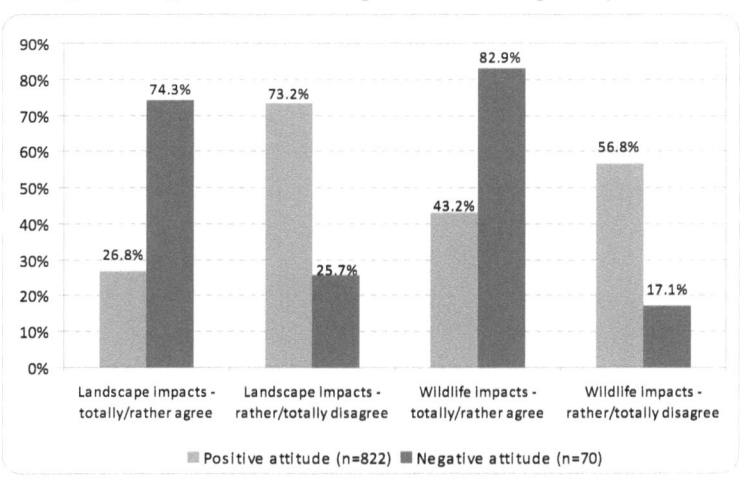

Source: OWN CALCULATIONS AND REPRESENTATION

7.2 Determinants of public attitude towards hydropower expansion

As we have seen in the previous section, people have in general a positive attitude towards hydropower use in Austria, as well as the construction of new plants. Additionally, it would be interesting to find out which factors are the main determinants of this positive attitude. For this purpose, an econometric model using a binary dependent variable[169] has been estimated. The dependent variable describes people's attitude towards an intensified use of hydropower and takes the value $y=1$ in case of a very or rather positive attitude, and the value $y=0$ when the attitude is rather negative or very negative (see equation 7.2).

$$\left.\begin{array}{l}\text{Very positive}\\ \text{Rather positive}\end{array}\right\} \rightarrow y = 1$$

$$\left.\begin{array}{l}\text{Rather negative}\\ \text{Very negative}\end{array}\right\} \rightarrow y = 0$$

(7.2)

A first option would be to estimate a "Linear Probability Model" (LPM) which generally represents an application of a classical linear regression model on a model with binary dependent variable. According to equation 7.3, the LPM estimates the probability of success or "response probability" for $y=1$ as a linear function of the independent variables x_j.

$$P(y = 1 \mid \bar{x}) = \beta_0 + \beta_1 x_1 + \ldots + \beta_k x_k \tag{7.3}$$

Such models are estimated by "Ordinary Least Squares" (OLS) and possess a useful interpretation. The estimated coefficients of a LPM measure the change in the probability of success when any of the explanatory variables, say x_j, changes, given that all other factors are constant. That is:

$$\Delta P(y = 1 \mid \bar{x}) = \beta_j \Delta x_j \tag{7.4}$$

---

[169] Binary variables are usually referred to as Dummy variables.

Nevertheless, the LPM has major limitations. Since the model estimates "response probabilities", the outcome should strictly lie between 0 and 1. However, in a LPM the resulting values may be below 0 or greater than 1 representing an invalid result. Additionally, the marginal effects resulting from a variation of x_j are independent of the level of x_j which may be an improper outcome. Finally, LPM models are associated with a violation of the "Gauss Markov assumption" of homoscedasticity leading to biased standard errors, t- and F-statistics (WOOLDRIDGE, 2000:232ff).[170] For those reasons, more advanced estimation procedures are required.

In a "Binary Response Model" (BRM) the response probability is not linear in β_j as in the LPM. Instead, $\beta_0+\beta_1x_1+...+\beta_kx_k$ is transformed by a function $G(\cdot)$. This transformation ensures that for all real numbers (and combinations of $x_1,...,x_k$) values or probabilities between 0 and 1 will result. That is:

$$P(y=1\mid \bar{x}) = G(\beta_0 + \beta_1 x_1 + ... + \beta_k x_k) \qquad (7.5)$$

$$\begin{aligned}&0 < G(z) < 1\\ &\text{with } z = \beta_0 + \beta_1 x_1 + ... + \beta_k x_k\end{aligned} \qquad (7.6)$$

If the logistic distribution function shown in equation 7.7 is used for $G(\cdot)$ the BRM is called *Logit* model. The *Probit* model uses, by contrast, the distribution function of the standard normal distribution (WOOLDRIDGE, 2000:530ff).

$$G(z) = \frac{\exp(z)}{1+\exp(z)} \qquad (7.7)$$

Generally, both transformations, *Logit* and *Probit*, provide similar results. For the current application of identifying the main determinants of people's

---

[170] As a consequence, the statistical significance of the estimated coefficients cannot be assessed reliably.

attitude towards hydropower expansion, the *Logit* transformation has been used.[171]

The parameters of *Logit* and *Probit* models are estimated by Maximum-Likelihood. Generally, the Maximum Likelihood Estimator (MLE) for β_j maximizes the Log Likelihood Function. However, the estimated coefficients cannot be calculated directly. Instead, they are determined by an iterative procedure which provides consistent, asymptotically efficient and normally distributed estimators (KOHLER AND KREUTER, 2006:285; WOOLDRIDGE, 2000:533ff).

The variables used in the estimated models are described in Table 49. For the variables "Positive impacts index" and "Negative impacts index" further explanation is required. The "Positive impacts index" represents a measure of people's perception of the positive impacts associated with an expansion of hydropower. More precisely, the variable refers to people's agreement to several statements indicating that the intensified use of hydropower contributes to satisfy increasing electricity demand (statement 1=S_1), to reduce CO_2 emissions (statement 2=S_2), and to reduce the dependency on electricity imports (statements 3=S_3). As we have seen in the previous section, people were allowed to choose on a four-point scale ranging from (1) "totally agree" to (4) "totally disagree". The index was calculated as shown in equation 7.8. It represents the mean of people's agreement to statements 1 to 3.

$$Positive\ impacts\ index_i = \frac{S_1 + S_2 + S_3}{3} \quad \forall i = 1,...,n \quad (7.8)$$
$$with\ \ 1 < index < 4$$

---

[171] Usually, *Probit* coefficients must be multiplied by a factor of 1.8 in order to be comparable with the coefficients estimated by *Logit*. That a *Probit* transformation provides similar results compared to *Logit* is shown in Table A2 in the Appendix.

Table 49: Description of the variables used in the Logit model

Variable	Type	Description	Relative frequency
Attitude expansion	Dummy variable	1 = Very/rather positive attitude 0 = Rather/very negative attitude	1 = 92.2 % 0 = 7.8 %
Age	Continuous variable	Age of the respondent in years	Mean = 42.3
Gender	Dummy variable	1 = Female 0 = Male	1 = 51.2 % 0 = 48.8 %
Education	Dummy variable	1 = Tertiary level 0 = Below tertiary level	1 = 13.7 % 0 = 86.3 %
Income	Dummy variable	1 = Income > € 3500 0 = Income ≤ € 3500	1 = 12.9 % 0 = 87.1 %
Hydropower	Dummy variable	1 = Hydropower first preferred energy source for future expansion. 0 = Other renewable energy source.	1 = 39.0 % 0 = 61.0 %
Positive impacts index	Continuous variable	Index of people's agreement to the positive impacts of hydropower (between 1 and 4)	Mean = 1.5
Negative impacts index	Continuous variable	Index of people's agreement to the negative impacts of hydropower (between 1 and 4)	Mean = 2.7
Distance	Dummy variable	1 = Distance between next hydropower station and respondent's residence is less or equal 2 km. 0 = Distance is more than 2 km.	1 = 10.9 % 0 = 89.1 %
New plant	Dummy variable	1 = New hydropower station is planned to be built near residence. 0 = Otherwise	1 = 18.3 % 0 = 81.7 %
Recreation impact	Dummy variable	1 = New hydropower plant expected to improve recreational activities. 0 = Otherwise	1 = 34.5 % 0 = 65.5 %

Source: OWN CALCULATIONS AND REPRESENTATION

Accordingly, high index values indicate a low agreement to the fact that intensified hydropower use has positive effects on the satisfaction of increasing demand, the reduction of CO_2 emissions and the dependency on imports. On the contrary, low index values denote high agreement. The same approach has been applied to create a measure of respondents' perception of the adverse effects associated with the expansion of hydropower. In particular, the "Negative impacts index" describes respondents' agreement to the fact that hydropower causes adverse affects on landscape (statement 4=S_4) and wildlife (statement 5=S_5). As before, high index values indicate high disagreement. Low index values imply high agreement to hydropower having adverse effects on landscape and wildlife (see equation 7.9).

$$\text{Negative impacts index}_i = \frac{S_4 + S_5}{2} \quad \forall i = 1,...,n \tag{7.9}$$
$$\text{with} \quad 1 < \text{index} < 4$$

7.2.1 Results of the estimated models

The estimated *Logit* models are shown in Table 50.[172] Principally, *Logit* or in the most general sense *Binary Response* models are, compared to *Linear Probability Models*, difficult to interpret. This difficulty is due to the non-linearity in parameters β_j imposed by the transformation of the estimation equation. Consequently, the estimated coefficients of a *Logit* model cannot be interpreted directly. Instead, the statistical significance and the sign of the estimated parameters, i.e. the direction of the relationship can be assessed as a first step. The parsimonious model specification (model 1) shown in Table 50 includes socio-demographic characteristics (SDC) only. As can be seen, the educational and income level have a statistically significant negative influence on people's attitude towards the expansion of hydropower. Age and gender are statistically not significant. Based on the specification using SDC only, a variety of variables including other socio-demographics, as well as attitudinal measures were included in the model specification. This set of models is not presented here. Generally, not all of these variables showed up to be statistically significant in the estimated models. The result of this iterative estimation procedure is the statistically best fit (SBF) model shown in the second column of Table 50 (model 2). Beside the socio-demographic characteristics, model 2 includes a set of variables describing people's current attitude towards hydropower, as well as their experience with the technology.[173]

---

[172] Intercooled Stata 9.2 econometric software was used to estimate the models.
[173] As 27 respondents refused to state the approximate distance between the closest hydropower station and their residence, the number of observations in model 2 amounts to merely 865 compared to 892 in model specification 1.

Table 50: Logit models explaining people's attitude towards hydropower expansion

Variable	MODEL 1 coefficients	MODEL 2 coefficients
Dependent variable: attitude expansion		
Constant	3.055*** (0.000)	2.952*** (0.001)
Age	-0.006 (0.518)	-0.017* (0.095)
Gender	-0.128 (0.610)	0.392 (0.199)
Tertiary education	-0.739** (0.013)	-0.684** (0.033)
Income	-0.843*** (0.005)	-0.596* (0.090)
Hydropower		1.167*** (0.008)
Positive impacts index		-1.728*** (0.000)
Negative impacts index		1.272*** (0.000)
Distance		-0.935** (0.044)
New plant		0.910* (0.058)
Recreation impact		0.800* (0.081)
Observations	892	865
Log likelihood	-237.627	-151.539
Wald χ^2 (p-value)	16.6 (0.002)	86.6 (0.000)
Adjusted McFadden Pseudo-R^2	0.011	0.325
AIC	485.254	325.078
BIC	509.221	377.468
Robust p-values in parentheses		
Significance: ***1 % level **5 % level *10 % level		

Source: OWN CALCULATIONS AND REPRESENTATION

In the SBF model, age is statistically significant at the 10 % level and affects respondents' attitude towards hydropower expansion negatively. This means that older people are usually more negative on the construction of new hydropower stations along Austrian rivers than their younger counterparts. Gender is still statistically insignificant indicating that there are no significant gender gaps regarding the attitude towards hydropower. Furthermore, the model shows that higher educated people holding a university degree have a more negative attitude towards hydropower expansion than lower educated ones. The same applies to household income. Accordingly, people with a high household income (above € 3,500) are more likely to exhibit a negative attitude than people with lower incomes.

As we have seen in the previous section, respondents' were asked to rank their two most preferred renewable energy sources for future electricity generation. People who chose hydropower as their first preferred energy source have a more positive attitude towards hydropower expansion than people ranking another renewable energy source (biomass, solar or wind power) as the first preferred choice.

As already explained above, the variable "Positive impacts index" describes respondents' perception of the positive effects associated with hydropower. Since high index values represent a low agreement to the assertion that increased hydropower use has positive effects on the satisfaction of electricity demand, the reduction of CO_2 emissions and the dependency on imports, the sign of the coefficient is negative. Hence, people who do not think that the intensified use of hydropower has positive effects are more likely to have a negative attitude towards its expansion. The opposite applies to the variable "Negative impacts index". The higher the index value, the higher people's disagreement to the proposition that the intensified hydropower use is associated with adverse effects on landscape and wildlife, and, consequently, the more favourable people are with respect to the expansion of hydropower.[174]

Another factor that plays a role for explaining public attitude towards the expansion of hydropower may be people's previous experience with the technology. For that reason, the variable Distance was included in the estimation model. This variable describes whether a respondent actually lives close to a hydropower station (distance ≤ 2 km) or not. The negative sign of the estimated coefficient indicates that people living currently close to a

---

[174] A statistically significant correlation between people's agreement to the positive and negative impacts of an intensified hydropower use and their attitude towards an expansion has already been found in the previous section 7.1 of this work.

hydropower plant have a less positive attitude towards the construction of further hydropower schemes than people living further away. Conversely, if a new hydropower plant is indeed under construction or planned to be built near respondent's residence (at a distance of maximum 10 km), he or she is more likely to have a positive attitude towards hydropower expansion. This result may be attributable to the fact that those respondents are actually confronted with the construction of new hydropower plants and are therefore better informed about hydropower expansion in general, or more specifically, about the effects a new hydropower station will imply.

Finally, respondents holding the opinion that the construction of a new hydropower station would improve the possibilities for recreational activities have a more positive attitude towards the construction of new hydropower plants as compared to respondents who think that hydro developments would limit recreational activities along a river.[175] This relationship is captured by the positive sign of the coefficient of the variable "Recreation impact".[176]

7.2.2 Comparison of the models

The overall fit of the estimated models can be evaluated on the basis of various statistical indicators and test procedures. A first indicator is the so-called Pseudo-R^2 by McFadden which is calculated analogous to the R^2 in a linear regression model. That is:

$$R_{MF}^2 = 1 - \frac{\ln \hat{L}(M_\beta)}{\ln \hat{L}(M_\alpha)} \qquad (7.10)$$

where $\hat{L}(M_\alpha)$ is the likelihood function of the constant only model and $\hat{L}(M_\beta)$ is the likelihood function of the extended model (KOHLER AND KREUTER, 2006:286). However, the McFadden Pseudo-R^2 increases by def-

---

[175] As an aside, the reference group of the variable "recreation impact" includes people that were unable to judge the expected impact of new hydropower plants on recreational activities in the survey.
[176] This correlation has already been elucidated in section 7.1 and was found to be statistically significant.

inition with the inclusion of further explanatory variables. Hence, when comparing two models with a differing number of explanatory variables, it would be better to use the measure of adjusted R². As can be seen from equation 7.11, this measure is adjusted for the number of explanatory variables K (LONG, 1997:104).

$$Adjusted - R_{MF}^2 = 1 - \frac{\ln \hat{L}(M_\beta) - K}{\ln \hat{L}(M_\alpha)} \qquad (7.11)$$

The interpretation of a Pseudo-R² measure, which is lying strictly between 0 and 1, is somewhat difficult as compared to the standard R² in a linear regression model. The only thing we can say is that the higher the Pseudo-R² the better the model fit (KOHLER AND KREUTER, 2006:286). Usually, a Pseudo-R² in the 0.2 to 0.4 range is a pretty good model fit and comparable to an OLS adjusted R² of 0.7 to 0.9 (BERGMANN ET AL., 2004:9). As can be seen from Table 50, the adjusted Pseudo-R² of the first model specification is very low in. By contrast, the measure of adjusted Pseudo-R² has risen considerably from 0.011 to 0.325 by the inclusion of additional (attitudinal) variables. This is within the range required for a good model fit.

Another standard test procedure for the comparison of two or more models is the calculation of information criteria according to Akaike (AIC) and Schwarz (BIC). These criteria are calculated as follows:

$$\begin{aligned} AIC &= -2 \cdot \ln \hat{L}(M_\beta) + 2 \cdot K \\ BIC &= -2 \cdot \ln \hat{L}(M_\beta) + K \cdot \ln N \end{aligned} \qquad (7.12)$$

where K represents the number of parameters in the model and N the number of observations. Generally, lower values of AIC and BIC are preferred (CAMERON AND TRIVEDI, 2010:359). Comparing the two models presented above, it can be seen that the extended model exhibits significantly lower values of AIC and BIC as compared to the model specification including SDC only. Hence, model 2 is preferable.

In addition, the quality of an estimated model can be evaluated my means of a Likelihood Ratio test (LR-Test). In particular, the LR-test examines whether the model fit improves by the inclusion of additional explanatory variables. The test statistic is calculated as follows:

$$\chi^2_{L(Diff)} = -2 \cdot (\ln L_{restricted} - \ln L_{unrestricted}) \quad (7.13)$$

$L_{restricted}$ and $L_{unrestricted}$ are the likelihood functions of the reduced and the extended model respectively (KOHLER AND KREUTER, 2006:299). As can be seen from Table 51, the two estimated models are highly significant. As compared to a base model including only a constant, the null hypothesis (H_0: $\beta_j=0$ for all $j=1,...K$) can be rejected at the 1 % significance level in both models. Starting from model 1, the inclusion of further attitudinal variables increases the model fit substantially. Accordingly, the null hypothesis that the additionally included variables are jointly not significantly different from zero can be clearly rejected.

Table 51: Results of the Likelihood Ratio Tests

Comparison of...	LR statistic	p-value
...constant only and model 1	15.40	0.004
...constant only and model 2	187.57	0.000
...model 1 and model 2	172.18	0.000

Source: OWN CALCULATIONS AND REPRESENTATION

Usually, the fitted value, i.e. the predicted outcome of an estimated *Logit* model is classified as 1 if the model predicts a probability of greater or equal 0.5. Otherwise a value of 0 is assigned. The predicted outcomes can then be compared with the actual outcomes of the dependent variable. According to that, the proportion of correctly specified answers can be calculated. For model specification 1 this proportion amounts to 92.2 %.[177] In the extended model (model 2), 94.9 % of the answers can be predicted correctly. Considered separately, 99.6 % of the observations with the characteristic $y=1$ (positive attitude towards hydropower) can be correctly speci-

---

[177] Although 92.2 % of the answers are correctly classified in total, model 1 is unable to predict any of the observations classified as $y=0$ (negative attitude towards hydropower) correctly. This indicates a bad model fit.

fied by the model. This is formally known as "sensitivity". The "specificity" refers to the proportion of correctly specified observations in the group of $y=0$ (negative attitude). Here, the share of correctly classified answers is 40.6 %. Translating the values of sensitivity and specificity into a coordinate system gives the so-called "ROC-curve" (Receiver Operating Characteristic).

Figure 54: ROC-curve of model specification 1

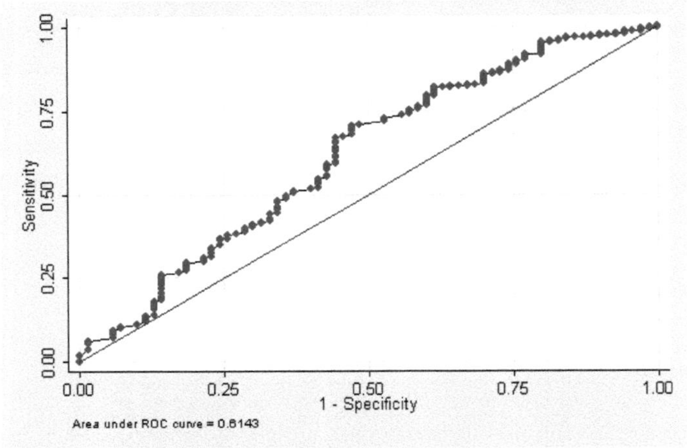

Source: OWN CALCULATIONS

The closer the ROC-curve to the diagonal in the coordinate system, the poorer the model fit. An optimal ROC-curve would correspond to the upper right angle triangle of the coordinate system. This optimal situation would be associated with an area under the ROC-curve of 1. Correspondingly, the closer the calculated area under the ROC-curve is to 1, the better the model fit.

As previously mentioned, specification 1 represents a relatively poor model with a low Pseudo-R^2 and high values of AIC and BIC. This poor model fit is also reflected by the ROC-curve. As can be seen from Figure 54, the ROC-curve is rather close to the diagonal which is associated with a small area under the curve (area=0.614). By contrast, the ROC-curve of the extended model looks significantly better (see Figure 55). The shape of the curve is closer to the optimal right angle triangle of the coordinate system, and, correspondingly, the area under the curve is with a value of 0.885 closer to the optimal case of 1.[178]

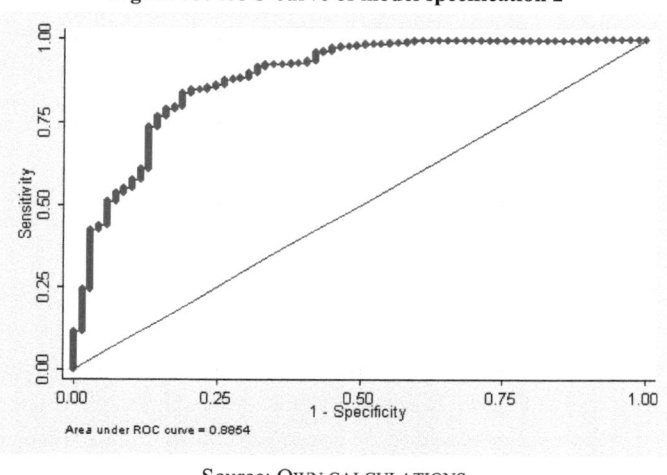

Figure 55: ROC-curve of model specification 2

Source: OWN CALCULATIONS

Based on the statistical measures and test procedures presented above, the extended model specification (model 2) is preferable to the reduced specification including socio-demographic characteristics only (model 1).

7.2.3 Odds-Ratios and probabilities

In order to give the estimated coefficients of the preferable *Logit* model more meaningfulness, Odds-Ratios have been calculated. The Odds-Ratio refers to the chance of having a positive attitude towards hydropower ($y=1$), when x_j changes by one unit. Odds-Ratios are calculated according to equation 7.14 (KOHLER AND KREUTER, 2006:283; LONG, 1997:74ff).

$$Odds - Ratio = \frac{e^{\beta_0 + \beta_j(x_j+1)}}{e^{\beta_0 + \beta_j x_j}} = \frac{e^{\beta_0 + \beta_j x_j} e^{\beta_j}}{e^{\beta_0 + \beta_j x_j}} = e^{\beta_j} \qquad (7.14)$$

---

[178] For further details on the interpretation of the ROC-curve see FAWCETT (2003).

Generally, an Odds-Ratio greater than 1 refers to a higher chance for an occurrence of $y=1$, while Odds-Ratios smaller than 1 point to a less chance.[179] The Odds-Ratios for the statistically best fit model are given in Table 52. The significance of the estimated coefficients (p-values), as well as measures of model fit such as Pseudo-R^2, log likelihood, AIC or BIC remained unchanged as compared to Table 50. The interpretation of the Odds-Ratios is based on a ceteris paribus assumption, that is, all other effects are held constant except for the variable for which the Odds-Ratio is being calculated.

Regarding age, each additional year reduces the chance of having a positive attitude towards hydropower expansion by a factor of 1.02.[180] People with a tertiary education are – ceteris paribus – twice as less likely to have a positive attitude than people with a lower educational level. A similar outcome is given for high income households. Accordingly, people with a household income higher than € 3,500 have a 1.8 times lower chance to exhibit a positive attitude towards increased hydropower use. Furthermore, the chance of having a positive attitude towards the construction of new hydropower plants is 3.2 times higher for people who ranked hydropower as their first preferred energy source. A relatively high effect originates from people's perception of the positive impacts of hydropower. Correspondingly, a one point increase of the "Positive impacts index" reduces the chance for a positive attitude by 5.6. Conversely, a one point increase in the "Negative impacts index" increases the chance of having a positive attitude by a factor of 3.6. As we have previously seen, attitude towards hydropower expansion is negatively affected by the distance between respondents' home and the closest currently existing hydropower plant. This effect is reflected by the Odds-Ratio smaller 1. Accordingly, people living at a distance of maximum 2 km to a hydropower station are 2.5 times less likely to have a positive attitude towards expansion than people living further away. The last two effects are positive and refer to the impact of the actual plans to build a hydropower station and the perceived impact of increased hydropower use on recreational activities. First, people who are indeed confronted with hy-

---

[179] As an aside, negative coefficients are associated with Odds-Ratios smaller than 1, positive coefficients with an Odds-Ratio greater than 1.
[180] This value results from the inverse of the Odds-Ratio shown in Table 52.

dropower expansion[181], exhibit a 2.5 times higher chance to be well disposed towards hydropower expansion. Finally, respondents thinking that the construction of a new hydropower plant will improve the possibilities for recreation along the regarding rivers are 2.2 times more likely to have a positive attitude.

Table 52: Statistically best fit model – Odds-Ratios

Variable	ODDS-RATIOS
Dependent variable: Attitude expansion	
Age	0.983* (0.095)
Gender	1.480 (0.199)
Tertiary education	0.504** (0.033)
Income	0.551* (0.090)
Hydropower	3.212*** (0.008)
Positive impacts index	0.178*** (0.000)
Negative impacts index	3.567*** (0.000)
Distance	0.393** (0.044)
New plant	2.484* (0.058)
Recreation impact	2.225* (0.081)
Observations	865
Log likelihood	-151.539
Wald χ^2 (p-value)	86.6 (0.000)
Adjusted McFadden Pseudo-R^2	0.325
AIC	325.078
BIC	377.468
Robust p-values in parentheses	
Significance: ***1 % level **5 % level *10 % level	

Source: OWN CALCULATIONS AND REPRESENTATION

---

[181] These are people with a new hydropower station planned to be built near their home.

Finally, it is possible to calculate the probability of having a positive attitude towards hydropower expansion ($y=1$) for different individuals. This can be done as follows:

$$P(y=1) = \frac{e^z}{1+e^z}$$

(7.15)

$$\text{with } z = \beta_0 + \beta_1 x_1 + ... + \beta_k x_k$$

In equation 7.15, z represents the so-called Logit which has, considered separately, little informational value. The values of the Logits are then converted into probabilities using Euler's number e (KOHLER AND KREUTER, 2006:274). This has been done for two different individuals. The outcome is shown in Table 53. First, the probability of having a positive attitude towards hydropower expansion is 99.9 % for an individual of mean age, with an educational level below tertiary, a monthly household income below € 3,500, hydropower as the first preferred energy source, a positive index value of 1 (complete agreement), a negative index value of 4 (complete disagreement), a distance of more than 2 km between residence and the next hydropower station, having a new hydropower scheme planned to be built near home, and holding the opinion that new hydropower plants improve recreational activities.

Conversely, another individual is considered, exhibiting the same socio-demographic characteristics but differing with respect to hydropower related attitudinal variables. Accordingly, the individual prefers another renewable energy source for future electricity generation, exhibits a positive impacts index of 4 (complete disagreement), a negative impacts value of 1 (complete agreement), lives currently close ($\leq$ 2 km) to a hydropower station, has no new hydropower scheme planned to be built near home, and thinks that intensified hydropower use restricts recreational activities. For this individual, the probability of having a positive attitude is only 0.13 %, an extremely low value compared to the probability of the other case. From the consideration of these two extreme examples, we can conclude that the probability of a positive attitude towards the construction of new hydropower plants along Austrian rivers is mainly determined by people's current attitude towards hydropower and their experience with the technology rather than socio-demographic characteristics.

Table 53: Probability of a positive attitude for two different individuals

Individual characteristics	$P(y=1\|x)$
– Mean age = 42.3 years – Education = 0 – Income3500 = 0 – Hydropower = 1 – Positive impacts index = 1 – Negative impacts index = 4 – Distance = 0 – New plant = 1 – Recreation impact = 1	99.9 %
– Mean age = 42.3 years – Education = 0 – Income3500 = 0 – Hydropower = 0 – Positive impacts index = 4 – Negative impacts index = 1 – Distance = 1 – New plant = 0 – Recreation impact = 0	0.13 %

Source: OWN CALCULATIONS AND REPRESENTATION

8 Willingness to pay for hydropower expansion

Within the scope of a Contingent Valuation (CV) task people were asked directly how much they would be willing to pay for an expansion of hydropower in order to get green electricity for their household. The outcomes of this valuation exercise are presented in the following sections, starting with the descriptive statistical analysis of people's WTP followed by an investigation of the factors that influence stated WTP.

8.1 Descriptive statistical analysis of WTP

To recap, respondents' WTP for the intensified use of hydroelectric power in Austria has been elevated through an open-ended CV question. Accordingly, people had to state the amount of money they would be willing to pay in addition to their monthly electricity bill directly. Typically, open-ended questions are suspected to elicit a large number of protest zeros and a small number of extremely high responses, so-called outliers (CARSON, 1999:14). In the present case, 60 respondents stated zero WTP which might be an expression of protest. In relation to the total number of observations, the percentage of zero bids is 6.7 %. Moreover, the number of people indicating extremely high WTP values[182] (more than € 100 per month) amounts to 15, which corresponds to a proportion of 1.7 % as related to the total sample. [183]

As shown in Table 54, mean WTP for increasing the use of hydropower is € 15.3. Looking at the calculated standard deviation (€ 29.4), it can be seen that stated WTP fluctuates substantially around its mean. Usually, "the mean is extremely sensitive to the right tail of the distribution; that is to the responses of higher bidders" (HANEMANN, 1994:25). Hence, outliers may considerably influence mean WTP (CARSON, 1999:14). By contrast, the median is a measure that is commonly very robust (HANEMANN, 1994:25).

---

[182] The indication of extremely high WTP values may also be a form of protest (see also section 4.2).
[183] The presence of non-responses has been prevented since people were obliged to answer the WTP question in the online survey.

Following this, it is recommended to calculate median in addition to the mean. In the present case, median WTP is with € 10 per month significantly lower as compared to mean WTP. This divergence is attributable to the existence of outliers.[184] The lower limit of stated WTP is € 0.0; maximum stated WTP is € 300.0.

Table 54: Stated WTP for the expansion of hydropower and current electricity bill (per month)

Statistical key figure	WTP	Current electricity bill
Mean	€ 15.3	€ 69.7
Standard deviation	€ 29.4	€ 53.1
Median	€ 10.0	€ 55.0
Minimum	€ 0.0	€ 10.0
Maximum	€ 300.0	€ 700.0
Observations	892	884

Source: OWN CALCULATIONS AND REPRESENTATION

Additionally, it might be of interest to compare the stated WTP with the amount people currently pay for electricity. The responses to a question on current electricity bill allowed the calculation of average electricity costs in Austrian households.[185] As indicated in Table 54, the average monthly electricity costs per household are € 69.7 with a standard deviation of € 53.1. So, in relation to monthly electricity bill people are willing to pay 22.0 % on top of their monthly costs in order to get electricity from intensified hydropower use. If median values are used, respondents are on average willing to pay 18.2 % on top of their current electricity bill. The premiums, people are on average willing to pay for the expansion of hydropower approximately correspond to the existing mark-up of 15.4 % accounted for the promotion of green electricity in Austria ("Ökostromförderbeitrag"; see section 3.5).

---

[184] Correspondingly, the divergence between mean and median WTP is significantly lower when excluding respondents with a WTP of more or equal € 50. In this case, mean WTP is € 9.2 and median WTP € 8.0 (based on 831 observations).
[185] Since monthly electricity costs have been queried in categories covering steps of € 10, the analysis is based on the midpoint estimates of the categories.

Table 55: WTP for hydropower expansion according to individual characteristics

Variable	Subgroup	Mean (in €)	Median (in €)	n	F-test (p-value)
Age	18-29 years	24.9	10.0	215	1.28* (0.081)
	30-59 years	12.6	8.0	535	
	>59 years	11.1	5.0	142	
Gender	Male	18.8	10.0	435	12.19*** (0.001)
	Female	12.0	7.0	457	
Household income	≤ € 5,000	15.0	10.0	863	3.83* (0.051)
	> € 5,000	25.8	15.0	29	
Donator	Donator	21.0	10.0	305	17.30*** (0.000)
	No donator	12.4	6.0	587	
Distance of home to a river	≤ 10 km	14.1	10.0	753	8.82*** (0.003)
	> 10 km	22.1	10.0	139	
Distance of home to hydropower plant	≤ 20 km	14.4	8.0	580	3.15*** (0.000)
	21-50 km	16.2	10.0	205	
	> 50 km	18.8	10.0	107	
Many hydropower plants near home	Yes	25.6	10.0	67	8.94*** (0.003)
	No	14.5	10.0	825	
Electricity payment	Him-/herself	13.1	7.5	544	7.74*** (0.006)
	Someone else	18.7	10.0	348	
Perception of positive effects of hydropower	Index ≤ 2	14.9	10.0	779	1.44 (0.168)
	Index > 2	18.1	5.0	113	
Perception of negative effects of hydropower	Index ≤ 2	17.6	7.0	258	1.64 (0.133)
	Index > 2	14.4	10.0	634	

Significance: ***1 % level **5 % level *10 % level

Source: OWN CALCULATIONS AND REPRESENTATION

Table 55 shows the stated WTP in relation to selected individual characteristics. As expected, the analysis of variance (F-test)[186] reveals that differences in the WTP for hydropower expansion can be explained by individual as well as household characteristics. For instance, mean WTP for additional green electricity from hydropower decreases with growing age. Moreover, female respondents exhibit on average a significantly lower WTP than males. In line with economic theory, mean WTP is significantly higher for respondents with a high household income whereas this relationship only comes into effect when exceeding a relatively high income threshold level of € 5,000. WTP may also be explained by environmental awareness and respondents' present experience with the hydropower tech-

---

[186] More on the F-test for equal variances can be found in PARK (2009:18). Alternatively, it is possible to perform a t-test. However, a t-test is only applicable in the two-group case, i.e. when comparing the means of two groups. Moreover, equal population variances between the two groups are required so that the t-test gives reliable results (PARK, 2009:18). For those reasons, F-tests have been performed for each variable. In the two-group case, t-tests have additionally been conducted; the results are the same as for the F-test.

nology. Accordingly, people giving regular donations to environmental organisations exhibit on average a higher WTP than people who do not donate. In addition, mean WTP increases with distance between a respondent's home and the next hydropower station indicating a significant distance decay effect.[187] Another important result refers to the payment of the electricity bill. Respondents whose electricity bill is paid by another household member stated on average significantly higher WTP values than the survey participants paying the electricity bill themselves. By contrast, people's perception of the positive and negative impacts associated with increased hydropower use does not contribute to explain stated WTP for increased hydropower use.

8.2 Determinants of stated WTP

Generally, WTP for green electricity from increased hydropower use can also be analysed by means of an econometric model. The problem, however, is that data on respondents' WTP are censored. Thus, the dependent variable, in our case WTP, is observed only over some interval of its support. As we have previously seen, a significant proportion of respondents have zero WTP, the rest a positive level of WTP. Accordingly, our sample is a mixture of observations with zero and positive values. More formally, the dataset consists of (y_i, x_i), $i=1,...,n$ where x_i is the set of explanatory variables and fully observed while the dependent variable y_i is not always observed. As mentioned above, some y_i are zero. These zeros can be interpreted as censored observations (CAMERON AND TRIVEDI, 2010:535).

But what is the appropriate estimation procedure in the presence of censored observations? A first option would be to estimate an OLS regression of y_i on x_i using all observations including the censored ones. Although this might be a good approximation, OLS would possibly result in negative fitted values, which leads to negative predictions for y_i representing an invalid result (WOOLDRIDGE, 2000:540).[188] Hence, using a linear model (OLS) for

---

[187] Since a hydropower plant may represent a "bad", WTP can be expected to rise with increasing distance.
[188] This problem is analogous to the prediction of probabilities greater than 1 or smaller than 0 in a Linear Probability Model for binary outcomes (see section 7.2).

censored data produces inconsistent estimates (LONG, 1997:189). Consequently, in case of censoring we have to draw on alternative estimation techniques. One approach is to estimate a *Tobit* model, also referred to as *censored regression model*. The *Tobit* model can basically be defined as a latent variable model. Following this, it is assumed that individuals have a latent (unobserved) demand for green electricity from hydropower, denoted by y^*, that is not expressed as WTP until some threshold, denoted by L, is passed. That is:

$$y = \begin{cases} y^* & \text{if } y^* < L \\ L & \text{if } y^* \leq L \end{cases} \tag{8.1}$$

In this case zero WTP can be interpreted as a left-censored variable that equals zero when $y^* \leq L$ (CAMERON AND TRIVEDI, 2010:535f; LONG, 1997:188). The regression of interest is specified as a latent variable model given by equation 8.2 with x_i denoting the vector of exogenous and fully observed explanatory variables. The error term ε_i is assumed to be normally distributed with zero mean and variance σ^2.

$$y_i^* = \beta_0 + \vec{\beta}_i \vec{x}_i + \varepsilon_i \qquad i = 1,...,n$$
$$\text{with } \varepsilon_i \sim N(0, \sigma^2) \tag{8.2}$$

If y_i^* were observed, the model could easily be estimated by OLS. However, according to equation 8.1 this is not the case (CAMERON AND TRIVEDI, 2010:536; GREENE, 2002:764). Given $L=0$, $y=WTP$ and $y^*=WTP^*$, the model can be written as:

$$WTP_i = \begin{cases} WTP_i^* = \beta_0 + \vec{\beta}_i \vec{x}_i + \varepsilon_i & \text{if } WTP_i^* > 0 \\ 0 & \text{if } WTP_i^* \leq 0 \end{cases} \tag{8.3}$$
$$\text{with } \varepsilon_i \sim N(0, \sigma^2)$$

The variables contained in the set of regressors, denoted by x_i, are described in detail in Table 56. Generally, WTP is assumed to depend on socio-demographic characteristics like age, gender or income, as well as on respondent's individual attitude towards hydropower and their experience with the technology. That stated WTP varies with socio-demographic and other attitudinal characteristics has already been shown in the previous section in Table 55. Now, we try to confirm these correlations with an econometric model.

Table 56: Description of the variables used in the Tobit and Hurdle model

Variable	Type	Description	Relative frequency
Age	Continuous variable	Age of the respondent in years	Mean = 42.3
Gender	Dummy variable	1 = Female 0 = Male	1 = 51.2 % 0 = 48.8 %
Income	Dummy variable	1 = Income > € 5,000 0 = Income ≤ € 5,000	1 = 3.3 % 0 = 96.7 %
Electricity payment	Dummy variable	1 = Electricity bill paid by respondent 0 = Otherwise	1 = 61.0 % 0 = 39.0 %
Electricity bill	Continuous variable	Amount of respondents' current electricity bill in €	Mean = 69.7
Donator	Dummy variable	1 = Respondent gives regular donations to environmental organisations. 0 = Otherwise	1 = 34.2 % 0 = 65.8 %
Importance renewables	Dummy variable	1 = Respondent appreciates electricity from renewable sources. 0 = Otherwise	1 = 86.9 % 0 = 13.1 %
Distance river	Dummy variable	1 = Respondent lives at a distance of maximum 10 km to a river. 0 = Otherwise	1 = 84.4 % 0 = 15.6 %
Many plants	Dummy variable	1 = Many hydropower plants in respondents' surrounding area. 0 = Otherwise	1 = 7.5 % 0 = 92.5 %
Distance plant	Continuous variable	Stated distance in km between respondents' home and the next hydropower station	Mean = 27.9
Recreation impact	Dummy variable	1 = New hydropower plant improves recreational activities. 0 = Otherwise	1 = 34.5 % 0 = 65.5 %
Demand effect	Dummy variable	1 = Increased hydropower use is considered to be important for satisfying increasing electricity demand. 0 = Otherwise	1 = 95.0 % 0 = 5.0 %
Positive impacts index	Continuous variable	Index of people's agreement to the positive impacts of hydropower (between 1 and 4)	Mean = 1.5
Negative impacts index	Continuous variable	Index of people's agreement to the negative impacts of hydropower (between 1 and 4)	Mean = 2.7

Source: OWN CALCULATIONS AND REPRESENTATION

The results of the *Tobit* model are shown in the first two columns of

Table 57.[189] The estimation is based on 858 observations[190], of that, 798 uncensored observations and 60 left-censored observations at WTP ≤ 0. The model is highly significant as shown by the χ^2 statistic and the associated p-value. The results are largely consistent with the findings reported in Table 55. It follows from the estimation results of the *Tobit* model that age and gender are negatively related to stated WTP, while household income is positively related to the WTP for green electricity from increased hydropower use. Moreover, the circumstance that the electricity bill is paid by the respondent him- or herself negatively affects WTP. Generally, environmental awareness can be considered by environmental organisation participation (KOUNDOURI ET AL., 2009:5). For that reason, the variable "donator" was included in the *Tobit* regression and it was found to have a statistically significant positive influence on the WTP for hydropower expansion. In addition, households living at a distance of maximum 10 km to a river exhibit on average a lower WTP compared to those living further away. A similar correlation was found for the actual distance between respondent's home and the nearest hydropower plant: WTP increases with growing distance from the hydropower station. By contrast, WTP is influenced positively by the fact that a respondent has many hydropower plants in his or her neighbourhood. Additionally, people holding the opinion that hydropower plants improve recreational activities are willing to pay more. The same applies to respondents who think that the intensified use of hydropower will help to satisfy the growing demand for electricity in the future.

A problem arises when it comes to the interpretation of the magnitude of the *Tobit* estimates. This is because β_i measures the partial effect of x_i on the conditional mean of the unobserved latent variable y^* which corresponds to WTP^* (WOOLDRIDGE, 2000:541). This is given by equation 8.4.

$$\frac{\partial E(WTP^* \mid \bar{x}_i)}{\partial x_i} = \beta_i \quad (8.4)$$

---

[189] Intercooled Stata 9.2 econometric software was used to estimate the models.
[190] The lower number of observations compared to the full sample with n=892 is attributable to missing values for the variables „electricity bill" and „distance plant".

However, the variable we want to explain is the observed outcome y, i.e. *WTP*. Hence, we have to focus on the partial effect of x_i on the conditional expectation $E(WTP|WTP>0,x_i)$, so called because it is conditional on $WTP>0$. That is:

$$\frac{\partial E(WTP|WTP>0,\bar{x}_i)}{\partial x_i} = \beta_i \{1 - \omega\lambda(\omega) - \lambda(\omega)^2\} \quad (8.5)$$

$$\text{with} \quad \omega = \bar{x}_i \bar{\beta}_i / \sigma \quad \text{and} \quad \lambda(\omega) = \phi(\omega)/\Phi(\omega)$$

According to equation 8.5, the partial effect of x_i on $E(WTP|WTP>0,x_i)$ is not simply determined by β_i, but by an additional adjustment factor given by the term in brackets, $\{\cdot\}$. This adjustment factor depends on a linear function of x_i as shown in equation 8.6 and is strictly between zero and one.

$$\omega = \frac{\bar{x}_i \bar{\beta}_i}{\sigma} = \frac{\beta_0 + \beta_1 x_1 + ... + \beta_k x_k}{\sigma} \quad (8.6)$$

The partial or marginal effects can be estimated by plugging in the maximum likelihood estimators of β_i and σ; for the x_i usually mean values are inserted (WOOLDRIDGE, 2000:542; CAMERON AND TRIVEDI, 2010:541). The outcomes of this calculation procedure are given in column 2 of Table 57.

The marginal effect reflects the effect of a one unit change in x_i on the left-truncated conditional mean of WTP given by $E(WTP|WTP>0, x_i)$. In case x_i is a binary variable, the marginal effect is for a discrete change of the dummy variable from 0 to 1 (WOOLDRIDGE, 2000:543). For instance, when a respondent regularly donates to environmental organisations WTP is € 4.7 higher than for non-donators. Or another example, when respondent's household income is higher than € 5,000, WTP increases by € 6.4. The marginal effect of a continuous variable can be illustrated by looking at the variable "distance plant". WTP increases by € 0.03 with every additional km, starting from the mean distance of 27.9 km.

Table 57: Results of the Tobit and Hurdle model

Variable	Tobit model Coefficients	Tobit model Marginal effects	Probit model for P(WTP>0)	Truncated model for WTP>0
Constant	13.249 (0.185)	-	2.114*** (0.002)	2.171*** (0.000)
Age	-0.277*** (0.000)	-0.133*** (0.000)	-0.021*** (0.000)	-0.010*** (0.000)
Gender	-5.434*** (0.010)	-2.608*** (0.010)	0.136 (0.368)	-0.243*** (0.000)
Income	12.001** (0.032)	6.400* (0.052)	0.396 (0.380)	0.485*** (0.006)
Electricity payment	-3.848* (0.084)	-1.861* (0.086)	-0.075 (0.651)	-0.129* (0.066)
Electricity bill	0.008 (0.699)	0.004 (0.698)	-0.003*** (0.001)	0.002*** (0.001)
Donator	9.566*** (0.000)	4.728*** (0.000)	0.602*** (0.002)	0.341*** (0.000)
Importance renewables	-1.509 (0.631)	-0.731 (0.634)	0.394** (0.032)	0.066 (0.513)
Distance river	-7.303** (0.012)	-3.680** (0.017)	-0.137 (0.534)	-0.166* (0.070)
Many plants	7.765** (0.044)	3.966* (0.058)	-0.074 (0.805)	0.223* (0.066)
Distance plant	0.053*** (0.005)	0.026*** (0.005)	0.001 (0.664)	0.002*** (0.001)
Recreation impact	4.080* (0.076)	1.981* (0.079)	0.298 (0.103)	0.169** (0.019)
Demand effect	11.583** (0.034)	5.025** (0.018)	0.160 (0.603)	0.318* (0.076)
Positive impacts index	2.570 (0.239)	1.233 (0.239)	-0.385** (0.011)	0.024 (0.726)
Negative impacts index	0.867 (0.531)	0.416 (0.531)	0.219** (0.024)	-0.011 (0.806)
Sigma (σ)	29.136***	-	-	-
Observations	858	858	858	798
Log likelihood	-3,871.731	-3,871.731	-182.390	-1,036.967
LR χ^2 (14 d.f.)	92.400	92.400	70.150	8.86 (F)
Prob. > χ^2	0.000	0.000	0.000	0.000 (Prob. > F)
(Pseudo) R^2	0.012	0.012	0.161	0.137

p-values in parentheses
Significance: ***1 % level   **5 % level   *10 % level

Source: OWN CALCULATIONS AND REPRESENTATION

Generally, *Tobit* models are based on the strong assumption that the uncensored (positive) observations and the censored zeros are generated by the same probability mechanism. However, it would sometimes be more flexible to allow for the possibility that the zero and positive values are generated by different mechanisms (CAMERON AND TRIVEDI, 2010:553). According to that, decisions on whether and how much to contribute were modelled separately. This is usually known as the two-part or *Hurdle* model. The first part of the *Hurdle* model includes the estimation of a binary outcome model (typically *Probit*) which tries to explain the probability of willing to pay, that is *P(WTP>0)*. The second part of the *Hurdle* model uses a linear regression for the truncated sample in order to estimate *E(ln WTP|WTP>0)*, i.e. the amount people are willing to pay (CAMERON AND TRIVEDI, 2010:553f; WOOLDRIDGE, 2000:546).[191] The outcomes of this two-step procedure are shown in the last two columns of Table 57. All relevant explanatory variables are simultaneously included in the model. The estimation results of the Probit model show that age and the amount of the current electricity bill have a negative influence on the decision whether to be willing to pay, while environmental behaviour (donator) and people's appreciation of electricity from renewable sources (importance renewables) are positively associated with people's decision whether to pay for hydropower expansion or not. Additionally, people's perceptions of the positive and negative impacts associated with an intensified use of hydropower play an important role for explaining *P(WTP>0)*.[192] Accordingly, people who think that hydropower has adverse effects on wildlife and landscape are less likely to be willing to pay for hydropower expansion. Conversely, people agreeing that the increased use of hydropower will contribute to reduce CO_2 emissions or the dependency on fossil fuel imports are more likely to be willing to pay.

The estimated truncated model using the logarithmised WTP as the dependent variable suggests that WTP is negatively associated with age and gender where female respondents are willing to pay less than males. However, gender did not show up to be statistically significant in the *Probit* model. The same applies to household income. While household income did not prove to have a significant influence on the decision whether to be willing to pay, it is positively related to the amount of the contribution for

---

[191] In the *Hurdle* model, the two parts are assumed to be independent.
[192] A detailed description on how the indices relating to people's agreement to the positive and negative impacts of hydropower were generated has already been elucidated in the previous chapter.

green electricity from increased hydropower use. Furthermore, households with a higher electricity bill are willing to pay more for hydropower expansion while the electricity bill was negatively related to the decision whether to pay. In contrast to the *Probit* model, WTP additionally depends on the fact who pays the electricity bill. Thus, respondents who have to pay the electricity bill themselves have on average a lower WTP. Environmental behaviour as measured by regular donations to environmental organisations has a positive influence in both, the *Probit* and the truncated model. On the contrary, the appreciation of electricity from renewable energy sources, as well as people's perception of the positive and negative hydropower impacts do not play a role in explaining the amount of stated WTP for hydropower expansion. Instead, the amount people are willing to pay depends, as opposed to the *Probit* model, on several variables describing people's experience with the hydropower technology. These measures include, for instance, the distance between a respondents' home and the next hydropower station or the nearest river, the number of hydropower plants in people's neighbourhood or the perceived impact of hydropower on recreational activities.[193]

Previous applications have shown that the two-part or *Hurdle* model can provide a better fit than a *Tobit* model by relaxing the *Tobit* model assumptions (CAMERON AND TRIVEDI, 2010:553; ZORIC AND HROVATIN, 2012:186). To test whether the *Hurdle* model is superior to the *Tobit* model, a likelihood ratio test (LR-test) has been carried out. The log likelihood (LL) of the *Hurdle* model is calculated as the sum of the LL of the *Probit* and the LL of the truncated regression model, given the assumption that the two parts are independent (CAMERON AND TRIVEDI, 2010:555). As shown in Table 58, the obtained significant value of the LR-test implies that the more general *Hurdle* model is preferable over the *Tobit* model.

---

[193] The results of the *Tobit* and *Hurdle* models are generally in line with previous research as for instance ZORIC AND HROVATIN (2012). For further details please refer to the literature review in chapter 5.

Table 58: Likelihood ratio test comparing the Tobit and Hurdle model

Statistical measure	LR test
LL Tobit model	-3,871.731
LL Probit model (part 1)	-182.309
LL Truncated regression model (part 2)	-1,036.967
LL Hurdle model (= LL Probit + LL Truncated)	-1,219.357
LR statistic	5,304.749
Degrees of freedom (df)	14
p-value	0.000

Source: OWN CALCULATIONS AND REPRESENTATION

9 Preferences for the attributes of hydropower expansion

In the previous chapters people's attitude towards hydropower, as well as their overall WTP for green electricity from increased hydropower use have been analysed. However, what we are especially interested in are the multiple impacts associated with an expansion of hydropower. For that reason, the choice experiments presented in section 6.2 have been conducted. In this chapter, the results of the choice experiment (CE) on hydropower expansion are reported. But first, the econometric framework behind the estimated choice models presented in the following chapters is elucidated.

9.1 The econometric framework

As previously mentioned, choice experiments belong to the family of stated preference techniques and contain elements of the traditional microeconomic theory of consumer behaviour. An essential point refers to Lancaster's characteristics theory of value which states that consumers derive utility from the properties or characteristics of a good and not from the good per se (LANCASTER, 1966:134). Lancaster's theory implies that the value of a good, service or policy can be expressed by its characteristics or attributes (LOUVIERE ET AL., 2000:2; RYAN ET AL., 2001:55). Thus, the value of a hydropower expansion policy can be expressed by its multiple impacts like the obtainable reduction of CO_2 emissions or the impact on landscape and biodiversity.[194]

Generally, people derive utility from the alternatives (i.e. the hydropower expansion strategies) in a choice set, denoted by C_n. C_n is assumed to represent a subset of all possible alternatives C and is determined by the constraints faced by an individual, as for instance income and time budgets or other external restrictions. This is shown in equation 9.1.

---

[194] A detailed description of the attributes used in the different choice experiments is given in section 6.2.

$$u = U(C_n)$$
$$C_n \subseteq C \qquad (9.1)$$

Given the assumption that individual preferences are stable, transitive and monotonic[195] over the alternatives that determine a unique preference ranking, the utility of decision maker n for choice alternative i can be defined as follows (BEN-AKIVA AND LERMAN, 2000:47):

$$U_{in} \quad i \in C_n \qquad (9.2)$$

Using Lancaster's approach, utility of individual n for alternative i is determined by the characteristics or attributes of the alternative. This is given by

$$U_{in} = U(\vec{z}_{in}) \qquad (9.3)$$

where z_{in} is a vector of the attribute values for alternative i. In addition to the attribute levels, utility for alternative i may be influenced by socio-economic characteristics that usually explain the variability of tastes across the portion of the population to which the choice model applies. Including a vector S_n of socio-economic characteristics, the utility function can be rewritten as (BEN-AKIVA AND LERMAN, 2000:48):

$$U_{in} = U(\vec{z}_{in}, \vec{S}_n) \qquad (9.4)$$

Moreover, choice theory is based on the assumption that individuals are acting rationally, meaning that they compare alternatives and choose the one which gives them the highest level of utility. More precisely, individuals act as if they are maximizing utility (HENSHER ET AL., 2005:80). For instance, if two alternatives, say i and j, are at choice, utility maximizing be-

---

[195] Consumer theory generally implies three basic assumptions for preferences. First, preferences are assumed to be complete, meaning that consumers have the ability to compare consumption bundles and rank them. Second, transitivity means that if alternative A is preferred over alternative B, and B is preferred over C, then alternative A is also preferred over alternative C. Finally, monotonic preferences imply that larger bundles are always preferred to smaller bundles of the same good (PINDYCK AND RUBINFELD, 2003:107).

haviour implies that individual n will choose alternative i only if its utility is higher than the utility of alternative j (LOUVIERE ET AL., 2000:40). This fact is represented by equation 9.5:

$$U_{in} > U_{jn} \quad \forall\, i \neq j \in C_n \tag{9.5}$$

The problem is, however, "that utility is a latent construct that exists (if at all) in the mind of the consumer, but cannot be observed directly by the researcher" (BENNETT AND BLAMEY, 2001:15). Instead, it is possible to explain a significant proportion of the unobservable consumer utility, but some part of the utility will always remain unobserved. That is:

$$U_{in} = V_{in} + \varepsilon_{in} \quad \forall\, i \in C_n \tag{9.6}$$

V_{in} represents the systematic or explainable component of the utility held by consumer n for choice alternative i and ε_{in} is the random or unexplainable component of latent utility (BENNETT AND BLAMEY, 2001:15). Given equation 9.5, the probability that individual n will choose alternative i over alternative j is given by equation 9.7 (TRAIN, 2003:19; HENSHER ET AL., 2005:82f). The model of this equation is called a random utility model (LOUVIERE ET AL., 2000:40).

$$\begin{aligned} P_{in} &= P(U_{in} > U_{jn}) & \forall\, i \neq j \in C_n \\ P_{in} &= P(V_{in} + \varepsilon_{in} > V_{jn} + \varepsilon_{jn}) & \forall\, i \neq j \in C_n \\ P_{in} &= P(\varepsilon_{jn} - \varepsilon_{in} < V_{in} - V_{jn}) & \forall\, i \neq j \in C_n \end{aligned} \tag{9.7}$$

Equation 9.7 in principle states that the probability of consumer n choosing alternative i is equal to the probability that the difference between the unobserved components of utility of alternative j compared to i is smaller than the difference between the systematic components of utility associated with alternative i compared to alternative j. However, the random parts of utility, ε_{in} and ε_{jn}, are unobserved and thus have to be treated as a random piece of information (HENSHER ET AL., 2005:83). Since we have no idea what numerical value to assign to ε_{in} and ε_{jn}, we have to make assumptions about the distribution of these random components of utility in order to calculate the choice probabilities. Usually, the random part of utility is assumed to be independently and identically distributed (IID) with an extreme value type 1 (EV1) distribution (HENSHER ET AL., 2005:84; LOUVIERE ET AL., 2000:45).[196] More precisely, IID means that the unobserved components of utility have no cross-correlated terms (independent) and exactly the same distributions (identically distributed). This can easily be illustrated by an example with four alternatives. In this case, the full variance-covariance matrix, also referred to as covariance matrix, is a 4x4 matrix as shown in 9.8.

$$\begin{bmatrix} \sigma_{11}^2 & \sigma_{12} & \sigma_{13} & \sigma_{14} \\ \sigma_{21} & \sigma_{22}^2 & \sigma_{23} & \sigma_{24} \\ \sigma_{31} & \sigma_{32} & \sigma_{33}^2 & \sigma_{34} \\ \sigma_{41} & \sigma_{42} & \sigma_{43} & \sigma_{44}^2 \end{bmatrix} \quad (9.8)$$

By imposing the IID assumption, the covariance matrix simplifies to the matrix shown in 9.9. First, all covariances or cross-correlated terms are set to zero because, under IID, the alternatives are assumed to be independent. Second, the variances on the diagonal are identical (i.e. without subscripts) and equal to σ^2. A further simplification is to normalize the variances to 1, known as the "constant variance assumption" which is equivalent to IID (HENSHER ET AL., 2005:77).

$$\begin{bmatrix} \sigma^2 & 0 & 0 & 0 \\ 0 & \sigma^2 & 0 & 0 \\ 0 & 0 & \sigma^2 & 0 \\ 0 & 0 & 0 & \sigma^2 \end{bmatrix} = \begin{bmatrix} 1 & 0 & 0 & 0 \\ 0 & 1 & 0 & 0 \\ 0 & 0 & 1 & 0 \\ 0 & 0 & 0 & 1 \end{bmatrix} \quad (9.9)$$

Generally, different assumptions about the distribution of the random components of utility lead to different choice models (TRAIN, 2003:19f). The IID EV1 distribution is associated with the popular binary or multinomial logit (MNL) models (BENNETT AND BLAMEY, 2001:16). If IID is violat-

---

[196] This distribution is also referred to as Weibull, Gumbel or douple-exponential distribution.

ed[197], the MNL model is not applicable and other, more complex choice models are required. In the mixed logit (MXL) model, also referred to as random parameter logit (RPL) model, for instance, it is assumed that the unobserved portion of utility consists of two parts. One part is IID with extreme value distribution as in the case of MNL models. The other part of unobserved utility follows a distribution, which is specified by the researcher (TRAIN, 2003:20).

Another assumption that is closely related to IID is the independence from irrelevant alternatives (IIA) property. "This states that the ratio of the probabilities of choosing one alternative over another (given that both alternatives have a non-zero probability of choice) is unaffected by the presence or absence of any additional alternatives in the choice set" (LOUVIERE ET AL., 2000:44). The IIA property, in turn, implies that the unobserved parts of the utility function (the $\varepsilon_j s$) are independently and identically distributed (LOUVIERE ET AL., 2000:45). As a consequence, violation of IIA requires more complex statistical models like nested or MXL models that relax some of the assumptions regarding the unobserved part of utility.[198]

A general assumption that is made in order to estimate the explainable or systematic component of utility, V_{in}, is the linearity in parameters (BEN-AKIVA AND LERMAN, 2000:108). Hence, the indirect utility function V_{in} may be parameterised as follows:

$$V_{in} = \beta_0 + \beta_1 X_{in1} + ... + \beta_k X_{ink} = \beta_0 + \sum_{k=1}^{K} \beta_k X_{ink} \qquad (9.10)$$

In equation 9.10, β_0 represents the intercept term of the equation and X_{ink} the vector of $k=1,...,K$ attributes that pertain to the choice options. In addition, indirect utility may depend on socio-economic characteristics, as well

---

[197] The assumption of IID may be violated if the sampled individuals are confronted with a sequence of choices. In this case, responses may be correlated across observations representing a violation of the independence of observations assumption in the classical MNL model. This correlation can have various reasons as for instance learning effects that appear due to the sequencing of offered choice situations (HENSHER ET AL., 2005:617).
[198] The most widely used test for violation of the IIA assumption is the so-called Hausman test, developed by Hausman and McFadden in 1984 (BEN-AKIVA AND LERMAN, 2000:183f; LOUVIERE ET AL., 2000:160ff).

as possible combinations between choice option attributes and individual characteristics. This is given by:

$$V_{in} = \beta_0 + \sum_{k=1}^{K} \beta_k X_{ink} + \sum_{p=1}^{P} \theta_p Z_{inp} + \sum_{k,p=1}^{K,P} \phi_{kp} X_{ink} Z_{inp} \qquad (9.11)$$

where Z_{inp} is a vector of $p=1,...,P$ individual characteristics and $X_{ink}Z_{inp}$ is a matrix of interactions between choice attributes and individual characteristics (BENNETT AND BLAMEY, 2001:16f).

In the classical MNL model, each parameter in the indirect utility specification V_{in} is assumed to be a single fixed estimate, i.e. a fixed parameter (HENSHER ET AL., 2005:76). However, this may often be not appropriate since individuals have different tastes. Consequently, one of the main disadvantages of MNL is the inability to capture preference heterogeneity not embodied in the individual characteristics of the respondent (GREENE AND HENSHER, 2005:2; HENSHER AND GREENE, 2002:5).[199] In the presence of preference heterogeneity, therefore, more complex choice models are required. Such a model would be the previously mentioned mixed logit (MXL) model. In the MXL model parameters are not fixed but random, meaning that they have a mean and standard deviation (HENSHER ET AL., 2005:76). Formally this can be depicted in the following way:

$$\beta_{ink} = \beta_k + \sigma_k \upsilon_{ink} \qquad (9.12)$$

Equation 9.12 implies that the parameter β_k for choice alternative i, varies across individuals, denoted by the subscript n. The parameter is divided into two parts: β_k represents the population mean and υ_{ink} the individual specific heterogeneity (with zero mean and a standard deviation of one). σ_k is the standard deviation of the distribution β_{ink} around β_k.[200] The compo-

---

[199] Another disadvantage of MNL models is certainly the fact that they are based on the IIA and in further consequence the IID property.
[200] By contrast, with fixed parameters, as in the case of MNL models, the standard deviation is assumed to be zero so that all the behavioural information is captured by the mean (HENSHER AND GREENE, 2002:633).

nents estimated by the analyst are the mean β_k and the standard deviation σ_k (BEVILLE AND KERR, 2009:7).

In order to get a better understanding of the sources of preference heterogeneity within a sampled population the MXL model can be extended to allow for variance heterogeneity (GREENE ET AL., 2005:2). With variance heterogeneity in the random parameters, the σ_k in equation 9.12 becomes a heteroscedastic term denoted by σ_{ink} (BEVILLE AND KERR, 2009:7). This heteroscedastic standard deviation of a random parameter can simply be treated as an additional error component. Therefore such models are called error component (EC) models (HENSHER AND GREENE, 2002:5; TRAIN, 2003:143).

As in the case of the previously presented *Logit* and *Tobit* models, MNL, MXL and EC models are estimated by maximum likelihood. The maximum likelihood estimation is an iterative search procedure, searching for a single value of the parameter vector β_k that will maximize the likelihood function L (HENSHER ET AL., 2005:318).

9.2 Model results – hydropower

Based on the dataset presented in chapter 6, several econometric models were estimated in order to explain people's preferences for increased hydropower use in Austria. First, different models containing attributes only have been estimated in order to find the right model specification. A detailed description of the attributes and how they were coded can be found in Table 59. A cardinal-linear coding was used for the attributes jobs, CO_2, distance and cost, while nature was dummy coded with "small impact" as the baseline category.

Table 59: Description of the variables used in the choice model on hydropower

Variable	Description	Coding/relative frequency
Attributes		
Jobs	Number of jobs created by an expansion of hydropower in the residential area of the respondent.	10, 50, 100, 500 jobs
CO_2	Reduction of CO_2 emissions in the electricity sector attainable by the intensified use of hydropower.	-10 %, -20 %, -40 %, -60 %
Nature	Impact of new hydropower plants on the landscape and natural environment.	1 = strong impact 0 = small impact
Distance	Distance of the next planned hydropower station to respondent's home.	2, 4, 8, 20 km
Cost	Increase in respondent's monthly electricity bill.	€ 3, 6, 9, 12, 15, 18
Socio-economic characteristics & other variables		
Gender	1 = Female 0 = Male	1 = 51.2 % 0 = 48.8 %
Age	Age of the respondent in years.	Mean = 42.3
Education	1 = Tertiary level 0 = Below tertiary level	1 = 13.7 % 0 = 86.3 %
Electricity payment *(interacted with cost attribute)*	1 = Electricity bill is paid by another household member. 0 = Electricity bill is paid by the respondent him-/herself.	1 = 25.3 % 0 = 74.7 %
Renewables *(interacted with CO_2 attribute)*	1 = Respondent appreciates electricity from renewable sources. 0 = Otherwise	1 = 86.9 % 0 = 13.1 %
Distance river *(interacted with Nature attribute)*	1 = Respondent lives at a distance of maximum 10 km to a river. 0 = Otherwise	1 = 84.4 % 0 = 15.6 %
Many plants	1 = Many hydropower plants in respondent's surrounding area. 0 = Otherwise	1 = 7.5 % 0 = 92.5 %
Bad information	1 = Respondent feels badly informed about hydropower use in Austria. 0 = Respondent feels well informed.	1 = 43.0 % 0 = 57.0 %

Source: OWN CALCULATIONS AND REPRESENTATION

The results of the attributes only models are given in Table 60. The estimations are based on 5,352 observations, that is, each of the 892 respondents answering six choice tasks. The simplest and most widely used econometric application, the MNL model, has been estimated first. As can be seen, the estimated parameters are statistically significant at least at the 10 % level and have the expected signs. However, as previously mentioned, a standard MNL model may be inappropriate in the presence of taste differences. Additionally, repeated choices for each individual may result in a violation of the IID assumption making the MNL model an improper approach. For those reasons, a MXL model has been estimated in the second step, taking all non monetary attributes as random parameters. As mentioned above, random parameters are differentiated in two parts, a mean and a standard deviation (see equation 10.12).[201] Accordingly, the means are represented by the estimated parameters in the MXL model. The standard deviations are shown separately in the lower part of Table 60. The random parameter standard deviations of the non monetary attributes are all highly significant indicating the presence of preference heterogeneity in the sampled population which is often referred to as unobserved heterogeneity (HENSHER AND GREENE, 2002:5). A further extension of the MXL model is to allow for variance heterogeneity which is treated as an additional error component. As can be seen from Table 60, in the EC model an additional standard deviation for the random effect or error component was estimated which showed up to be highly significant at the 1 % level. Moreover, the random parameter standard deviations, as well as the attribute coefficients are statistically significant in the EC model specification. Comparing the estimated models using attributes only, we can conclude that the model fit has steadily improved when going from a simple MNL to a complex EC model. First, this can be shown by the, in absolute terms, decreasing log likelihood values, as well as the increasing Pseudo-R^2 measures. Second, decreasing information criteria according to Akaike (AIC) and Schwarz (BIC) confirm the improving model fit.[202] According to the results of the model comparison, the EC model appeared to be the most appropriate specification, meaning that it can capture people's preferences best as compared to the standard MNL and the more advanced MXL model.

---

[201] Cardinal-linear coded variables were assumed to be normally distributed, while dummy coded attributes were specified to be uniformly distributed. More on random parameter distributions can be found in HENSHER AND GREENE (2002:7).

[202] Generally speaking, the lower AIC and BIC are, the better the model fit. More on calculation and interpretation of information criteria can be found in section 7.2.2 of this work. For further details reference to econometric literature like LONG (1997) or WOOLDRIDGE (2000) is made.

Table 60: Model estimates – hydropower

Variable	Attributes only MNL	Attributes only MXL	Attributes only EC	Extended model EC
Dependent variable	Choice: Alternative A, B or none of the two			
ASC	1.257*** (0.000)	1.989*** (0.000)	2.727*** (0.000)	4.278*** (0.000)
Jobs	0.0002* (0.064)	0.0003* (0.076)	0.0003** (0.026)	0.0003** (0.028)
CO_2 reduction	0.011*** (0.000)	0.016*** (0.000)	0.016*** (0.000)	0.007* (0.057)
Nature (strong impact)	-1.027*** (0.000)	-1.911*** (0.000)	-1.555*** (0.000)	-1.155*** (0.000)
Distance	0.007*** (0.010)	0.003 (0.471)	0.007* (0.051)	0.007* (0.054)
Cost	-0.087*** (0.000)	-0.131*** (0.000)	-0.124*** (0.000)	-0.129*** (0.000)
Electricity payment*Cost				0.027*** (0.001)
Renewables*CO_2				0.009** (0.021)
Distance river*Nature				-0.436*** (0.007)
Gender				-0.136 (0.557)
Age				-0.032*** (0.000)
Education				0.642* (0.058)
Many plants				0.754* (0.078)
Bad information				-0.596*** (0.010)
Std. Dev. Jobs		0.002*** (0.000)	0.0009** (0.031)	0.0007 (0.132)
Std. Dev. CO_2		0.040*** (0.000)	0.022*** (0.000)	0.021*** (0.000)
Std. Dev. Nature		2.868*** (0.000)	2.177*** (0.000)	2.139*** (0.000)
Std. Dev. Distance		0.077*** (0.000)	0.033*** (0.000)	0.031*** (0.001)
Std. Dev. Random effect (error component)			2.840*** (0.000)	2.758*** (0.000)
Log likelihood	-5,140.288	-4,626.146	-4,331.591	-4,303.841
McFadden Pseudo R^2	0.111	0.213	0.263	0.267
χ^2 (p-value)	-	2,507.3 (0.000)	3,096.4 (0.000)	3,138.7 (0.000)
AIC	1.923	1.732	1.623	1.617
BIC	1.931	1.745	1.636	1.641
Number of respondents	892	892	892	891
Number of observations	5,352	5,352	5,352	5,346

p-values in parentheses
Significance: *** 1 % level   ** 5 % level   * 10 % level

Source: OWN CALCULATIONS AND REPRESENTATION

Table 61: Hausman tests for IIA – hydropower

Excluded...	Chi-squared	p-value
...alternative A	6.316	0.389
...alternative B	7.680	0.263

Source: OWN CALCULATIONS AND REPRESENTATION

Generally, it is recommended to undertake a test for violation of IIA before proceeding to more advanced choice models since splendid simplicity is usually superior to a more complex estimation approach. In general, the Hausman test procedure involves the estimation a standard MNL model with all alternatives first. Subsequently, the model is estimated with a restricted set of alternatives. Under the null hypothesis the IIA property is fulfilled; the alternative hypothesis implies violation of IIA (LOUVIERE ET AL., 2000:160f). According to that, once alternative A, and once alternative B have been excluded from the set of choices in the present application. As can be seen from Table 61, the null hypothesis under which IIA holds cannot be rejected on a usual significance level. Despite this result, it seemed to be superior to stick to the EC model since preference heterogeneity is given in the sampled population, and this unobserved heterogeneity cannot be captured by a standard MNL model.

As we have seen above, unobserved taste heterogeneity can be captured by the derived standard deviations of the distribution of random parameters. However, a systematic part of preference heterogeneity may be identified in household characteristics (DIMITROPOULOS AND KONTOLEON, 2009:1849). Hence, it can be assumed that, in addition to the attributes of the experiment, household characteristics and interactions between these characteristics and attributes may have an influence on choice.[203] This is why in a set of models not presented here a variety of variables including socio-economic characteristics like income or educational level as well as interaction terms between these characteristics and choice attributes were included in the model specification. However, not all of these variables showed up to be statistically significant in the estimated models. The result of this iterative estimation procedure is the statistically best fit (SBF) model

---

[203] This was previously shown in equation 10.11.

which has the following indirect utility form (equation 9.13). The variables used in the final model specification are explained in detail in Table 59.

$$V = \beta_0 + \beta_1 \text{Jobs} + \beta_2 \text{CO}_2 + \beta_3 \text{Nature} + \beta_4 \text{Distance} + \beta_5 \text{Cost}$$
$$+ \beta_6 \text{Electricity payment} * \text{Cost} + \beta_7 \text{Renewables} * \text{CO}_2$$
$$+ \beta_8 \text{Distance river} * \text{Nature} + \beta_9 \text{Gender} + \beta_{10} \text{Age}$$
$$+ \beta_{11} \text{Education} + \beta_{12} \text{Many plants} + \beta_{13} \text{Bad information}$$
(9.13)

The results of equation 10.13 are shown in the last column of Table 60.[204] First, the SBF model is highly significant as shown by the χ^2 statistic calculated for the included variables. Furthermore, the inclusion of interaction terms and socio-demographic characteristics improves the model fit as apposed to an EC model using attributes only. On the one hand, this can be shown by the increasing Pseudo-R^2 and decreasing information criteria, though BIC has slightly risen. On the other hand, a likelihood ratio test (LR test) comparing the attributes only and extended EC models has been performed. Generally, the LR test is used to test the contribution of particular (sub)sets of variables. Under the null hypothesis the particular subset of βs are equal to zero (LOUVIERE ET AL., 2000:53). The outcome of the LR test is given in Table 62.[205] The null hypothesis that the additionally included variables (i.e. socio-demographics and interactions) are jointly equal to zero can clearly be rejected at the 1 % significance level. Thus, based on the LR test, the extended EC model can assumed to be superior to the model specification using attributes only.

Table 62: Result of the likelihood ratio test – hydropower

Comparison of...	LR statistic	p-value
...attributes only and extended EC models	63.350	0.000

Source: OWN CALCULATIONS AND REPRESENTATION

---

[204] NLOGIT 4.0 econometric software was used to estimate the models. The number of draws required to secure stable parameter estimates was set to 100 which usually appears to be a good number (HENSHER ET AL., 2005:615f). Moreover, the panel data structure was accounted for in each of the estimated models.
[205] The way in which the test statistic is calculated has already been elucidated in section 7.2.2 for the binary *Logit* model. The calculation for the choice model at hand is analogous.

Looking at the extended EC model, it can be seen that the coefficients of the five choice attributes, the interaction terms and socio-economic characteristics are all statistically significant and have the expected sign, except for gender which is statistically not significant. In addition, the derived standard deviations of random parameter distributions and the error component are highly significant, except for jobs, justifying the application of a complex EC model.

The alternative specific constant (ASC) can be interpreted similarly to the constant in a regression model. It represents on average the effect on utility of all factors that are not included in the estimated model (HENSHER ET AL., 2005:76; TRAIN, 2003:24). In line with this, the positive ASC indicates that the respondents have some inherent propensity to choose for one of the hydropower expansion scenarios over the opt-out (denoted as none of the two alternatives) for reasons that are not captured in the estimated model. Thus, we can conclude that respondents are in principal for an intensified use of hydropower in Austria. However, respondents' overall utility depends – in addition to the ASC – on the multiple impacts of hydropower that are represented by the attributes of the experiment.

The attribute jobs has a positive sign which implies that respondents have preferences for alternatives where more jobs can be generated by increased hydropower use. A statistically significant positive effect of employment creation has also been found in previous studies examining public preferences for renewable energy investments in general (see for instance LONGO ET AL., 2008 or KU AND YOO, 2010).

The CO_2 attribute exhibits a positive sign too, meaning that alternatives with a higher level of CO_2 reduction are preferred. This outcome is in line with previous econometric results from LONGO ET AL. (2008), FIMERELI ET AL. (2008) or KU AND YOO (2010). In the present analysis, the positive effect of CO_2 emission reduction is additionally amplified if respondents appreciate electricity from renewable energy sources. This can be illustrated by looking at the partial derivative of indirect utility with respect to CO_2:

$$\frac{\partial V}{\partial CO_2} = \beta_2 + \beta_7 \text{Renewables} = 0.007 + 0.009 * \text{Renewables}$$
$$\text{if Renewables} = 1 \quad \rightarrow \quad \frac{\partial V}{\partial CO_2} \uparrow$$
(9.14)

In contrast to the previous results, alternatives with a strong impact on landscape and natural environment are less preferred compared to those with only a small impact. This relationship is captured by the negative sign of the coefficient on the attribute nature. Landscape and wildlife impacts have also been investigated in previous studies with the result that landscape and biodiversity improvements are valued positively (see for instance BERGMANN ET AL., 2004; FIMERELI ET AL., 2008 or KU AND YOO, 2010). In the current application, the effect of the strong nature impact is enhanced if the respondent lives near to a river. More precisely, if the respondent lives at a distance of less than 10 kilometres to a river, a strong impact on landscape and nature is valued more negatively as compared to a situation where respondent's residence is further away from a river. This can easily be shown by equation 9.15.

$$\frac{\partial V}{\partial Nature} = \beta_3 + \beta_8 \text{Distance river} = -1.155 - 0.436 * \text{Distance river}$$
$$\text{if Distance river} = 1 \quad \rightarrow \quad \frac{\partial V}{\partial Nature} \downarrow$$
(9.15)

In addition, a statistically significant distance decay effect was found, meaning that respondents prefer alternatives where new hydropower stations are built further away from their home. This result provides confirmation of the "Not in my backyard" (NIMBY) theory, which has already been mentioned several times in the existing literature (see for instance FIMERELI ET AL., 2008 or MEYERHOFF ET AL., 2010). Thus, people are in general for an expansion of hydropower capacities, but not close to their home.

The cost attribute is negative and highly significant reflecting standard economic theory. It indicates that respondents prefer lower electricity bills, i.e. electricity from hydropower must be provided at a low cost. However, if the electricity bill is not paid by the respondent but instead, by another household member, the negative effect of cost diminishes, suggesting lower price sensitivity.[206] That is:

$$\frac{\partial V}{\partial Cost} = \beta_5 + \beta_6 \text{Electricity payment} = -0.129 + 0.027 * \text{Electricity payment}$$
$$\text{if Electricity payment} = 1 \quad \rightarrow \quad \frac{\partial V}{\partial Cost} \uparrow \qquad (9.16)$$

Another important result of the EC model refers to the impact of people's experience with the hydropower technology in every day life. More specifically, the existence of many hydropower plants near the residence of the respondent increases the acceptance of hydropower expansion. This relationship is captured by the positive sign of the estimated parameter of the variable "many plants". Furthermore, the individual information status seems to create a significant effect on the general acceptance of increased hydropower use. People feeling badly informed about hydropower in general rather tend to choose the opt-out alternative over the hydropower expansion scenarios.

Regarding socio-economic characteristics, it was found that elderly people are less willing to choose one of the hydropower expansion options that are associated with additional costs; a result that is consistent with the outcome of the *Tobit* and *Hurdle* models presented in chapter 8 of this work. Gender-related preference differences do not exist as indicated by the statistical insignificance of the gender parameter. Finally, higher educated people are more likely to vote respectively pay for an expansion of hydropower their less educated counterparts.[207]

---

[206] A similar effect was found in the *Tobit* and *Hurdle* models presented in chapter 8. In these models, people paying household's electricity bill themselves were found to exhibit on average a lower WTP than respondents' whose electricity bill is paid by another household member.

[207] The educational level may represent a proxy variable for income, since higher education is normally associated with higher incomes. And income is usually a strong determinant of stated choices.

9.3 Willingness to pay – hydropower

The estimated model presented in the previous section enabled to find out which attributes significantly influence utility and as a consequence respondents' choice. Due to the linear specification of the utility function, the estimated coefficients (i.e. the βs) exhibit an additional interpretation. In particular, it is possible to give an indication of how much of one attribute respondents are willing to abandon in exchange for one unit of another attribute. Hence, the estimated coefficients can be used to estimate the rate at which respondents are willing to trade-off one attribute for another (BENNETT AND BLAMEY, 2001:63). This relationship is usually referred to as "marginal rate of substitution" (MRS). The MRS between two attributes, say 1 and 2, can be calculated as the ratio between the marginal utilities (MU) of these two attributes. MU is represented by the partial derivative of the utility function with respect to the attribute of interest (MAS-COLELL ET AL., 1995:54). This is shown in equation 9.17.

$$MRS_{12} = \frac{MU_2}{MU_1} = \frac{\partial V / \partial X_2}{\partial V / \partial X_1} = -\frac{\beta_2}{\beta_1} \qquad (9.17)$$

If one of the attributes is measured in monetary units (e.g. electricity price increase) the MRS will correspond to the marginal willingness to pay (MWTP) of the consumer. In that case the MRS, or more precisely the MWTP, measures how many monetary units an individual is willing to give away in exchange for one unit of the non-monetary attribute. Usually, in case of a standard MNL application, MWTP is calculated dividing the coefficient of the attribute of interest by the coefficient of the monetary attribute (see equation 9.18). This ratio is also referred to as "part-worth" or "implicit price" (BENNETT AND BLAMEY, 2001:63).

$$MWTP = -\frac{\beta_{attribute}}{\beta_{monetary}} \qquad (9.18)$$

However, with random parameters this simple approach is not appropriate, since unobserved preference heterogeneity (i.e. taste differences) must be taken into account when calculating MWTP. Therefore, MWTP has been simulated for each respondent $n=1,...,N$ and each choice attribute $k=1,...,K$ using the conditional and constrained parameter estimates for β_{kn} which ensures that the simulations yield plausible values (HENSHER ET AL., 2005:691f).[208] That is:

$$MWTP_n = -\frac{\beta_{kn}}{\beta_{monetary}} \quad \forall\, n = 1,...,N \text{ and } k = 1,...,K \qquad (9.19)$$

Then the means, standard deviations and confidence intervals were taken from these simulations. Confidence intervals have been calculated according to equation 9.20 where $\bar{x}$ is the mean, σ the standard deviation and $u_{1-\alpha/2}$ the $1-\alpha/2$ quantile of the standard normal distribution (HARTUNG ET AL., 1991:130).

$$Confidence_{upper,lower} = \bar{x} \pm \frac{\sigma}{\sqrt{n}} u_{1-\frac{\alpha}{2}} \qquad (9.20)$$

By calculating MWTP, the estimated parameters of the econometric model are given more meaningfulness since it is not possible to gain information on the relative values of attributes simply through the comparison of the estimated coefficients. It would be an incorrect approach to compare coefficients and conclude that these represent the contribution to indirect utility. This is because the estimated coefficients are confounded with a "scale parameter" that is dependent on the variance of the error term. Through the calculation of part-worth, i.e. the division of the β coefficients, the scale parameter is cancelled out eliminating the confounding effect of the error variance. Nevertheless, the comparison of MWTP requires full recognition of the differing units of the attributes. Thus, one should take care when comparing, for instance, the MWTP for one additional job and a one percent reduction of CO_2 emissions. "A comparison of the implicit prices of attributes affords some understanding of the relative importance that respondents hold for them" (BENNETT AND BLAMEY, 2001:64). This is why

---

[208] In particular, for each of the four choice attributes 892 measures of MWTP have been calculated since the number of respondents in the sample was 892.

the importance of the attributes was elicited directly within the debriefing section of the questionnaire. Particularly, respondents were asked to indicate for each attribute their perceived importance ranging from (1) "very important" to (4) "totally unimportant". Analogous to the recreation index presented in section 7.1, an index was calculated by taking the means of the indicated importance for each attribute. This is given by equation 9.21.

$$Index_k = \frac{\sum_{i=1}^{n} importance_i}{n} \quad for\ all\ attributes\ k = 1,......5 \qquad (9.21)$$

The lower the index, the more important is the attribute. The outcomes are shown in Table 63 and will later be compared with the calculated values of MWTP.

Table 63: Importance of the attributes – hydropower

Attribute	Importance index
CO_2 reduction	1.61
Impact on nature and landscape	1.68
Increase in monthly electricity bill	1.88
Job creation	1.97
Distance to home	2.66
Observations	n=892

Source: OWN CALCULATIONS AND REPRESENTATION

The outcomes of part-worth calculation are given in Table 64. The estimated measures of MWTP are based on a "ceteris paribus" assumption, that is, MWTP is calculated for a change in the attribute of concern, given that all other parameters are held constant (BENNETT AND BLAMEY, 2001:63).

Table 64: Estimates of marginal WTP – hydropower

Variable	Measurement	MWTP
Hydropower	effect of the ASC	€ 33.075 [31.721, 34.428]
Jobs	per 100 jobs	€ 0.229 [0.215, 0.244]
CO_2 reduction	per 10 % reduction	€ 1.324 [1.245, 1.403]
Impact on nature and landscape	from small to strong impact	€ -13.508 [-13.907, -13.110]
Distance	per 5 km	€ 0.325 [0.304, 0.346]

95 % confidence intervals in parentheses

Source: OWN CALCULATIONS AND REPRESENTATION

First, people are on average willing to pay € 33.1 in addition to their monthly electricity bill for the expansion of hydropower, independent from the attribute levels. This value reflects the effect of the positive ASC representing the inherent propensity of respondents to vote in favour of hydropower expansion. Here, WTP was simply calculated dividing the ASC by the coefficient of the monetary attribute, since neither ASC nor the cost attribute were treated as random variables in the econometric model.[209]

Regarding the attributes, it can be shown that people exhibit a positive WTP for job creation. In particular, respondents are willing to pay € 0.2 per month for 100 additional jobs in their residential area. The reduction of climate-damaging emissions is positively valued as well. Accordingly, for a 10 % reduction of CO_2 emissions people are willing to pay € 1.3 on top of their monthly electricity bill.

The implicit price for the nature attribute is negative, reflecting the fact that people do not desire alternatives with a strong environmental impact. Generally, negative values of MWTP imply a disutility. More precisely, a negative MWTP can be interpreted as a reduction of individual utility resulting from the environmental changes (MEYERHOFF ET AL., 2008:14). In monetary terms, the disutility associated with a strong environmental impact as

---

[209] The standard error which is required for the calculation of the 95 % confidence interval was estimated using the delta method. For further details on the delta method see OEHLERT (1992).

compared to a small impact is estimated at € 13.5. Or differently interpreted, people wish to be compensated for the loss of nature and landscape when new hydropower plants are built.

Finally, people are willing to pay € 0.3 on top of their monthly electricity bill if a new hydropower station is not built in their "backyard" but 5 km further away. This result provides confirmation of the famous NIMBY theory. Since hydropower plants principally represent a "bad", MWTP rises with increasing distance from home.

The results of MWTP basically confirm the findings relating to the importance of attributes shown in Table 63. The attribute referring to CO_2 emission reduction is seen as the most important attribute by respondents, followed by the impact on nature and landscape. However, this result is not clearly reflected when looking at implicit prices only. From MWTP we would conclude that the environmental impact represents the most important attribute, followed by the reduction of CO_2 emissions. The attributes jobs and distance are less important for respondents, an outcome that is also reflected by the relative low values of MWTP.

9.4 Welfare analysis – hydropower

Implicit prices or MWTP represent important information for policy makers and management regimes. However, they cannot be interpreted as valid welfare measures to be used in cost-benefit-analysis (CBA). Implicit prices correspond to the marginal rates of substitution within an alternative or indirect utility function respectively. Welfare measures take into account not just a single attribute but the simultaneous change in all the attributes (BENNETT AND BLAMEY, 2001:64). For that reason, a number of policy scenarios represented by different combinations of attribute levels, has been evaluated and their welfare implications estimated. Generally, the assessment of economic welfare (EWF) involves the investigation of utility differences associated with a baseline scenario and some other alternative. Accordingly, EWF can be written as follows (BENNETT AND BLAMEY, 2001:66).

$$EWF = -\frac{1}{\beta_{monetary}}(V_1 - V_0) \qquad (9.22)$$

The utility difference is divided through the coefficient of the monetary attribute ($\beta_{monetary}$) which can be interpreted as the marginal utility of income. V_0 is the utility of the reference scenario which can be assumed to equal 0, while V_1 is the observed utility associated with a linear combination of attribute levels. Different measures of V_1 have been estimated on the basis of the statistically best fit model presented above. In doing so, the values of the attributes associated with the policy scenario to be evaluated, socio-economic variables and interaction terms were substituted into the estimation equation.[210] In this calculation, the ASC was included, because it captures systematic but unobserved information of not choosing the opt-out alternative (BENNETT AND BLAMEY, 2001:66).[211] This is given by equation 9.23.

$$V_{n1} = ASC + \sum_{k=1}^{K}\beta_{nk}X_{nk} + \sum_{k,p=1}^{K,P}\phi_{kp}X_{nk}Z_{np} + \sum_{p=1}^{P}\theta_{p}Z_{np} \qquad \forall n = 1,...,N \qquad (9.23)$$

Due to the fact that the estimated attribute parameters vary across individuals (random parameters), EWF has been simulated for each respondent analogous to the calculation of MWTP.[212] For socio-economic variables usually sample means are substituted into the equation. However, in the present case, the full set of information has been used computing individual specific welfare measures. The same applies to interactions between choice attributes and socio-economic characteristics. Finally, the means, standard deviations and confidence intervals have been calculated on the basis of the simulations. The outcomes for seven different policy scenarios are shown in Table 65.

---

[210] The monetary attribute is set at zero, because the coefficient of the monetary attribute is used to translate indirect utility into a monetary value (see equation 10.21).
[211] However, there is a lot of discussion whether the ASC should be included or not since it may represent a yea-saying problematic (KATARIA, 2009:74).
[212] This is indicated by the subscript n in equation 10.22. So, indirect utility of a specific policy scenario has been calculated for each respondent $n=1,...,892$.

Scenario (1) represents the "worst case scenario" and yields an EWF of € 12.2 per household and month. In contrast, the "best case" is represented by scenario (4). Here, EWF is almost three times higher as in the worst case, namely € 33.2 per household and month. This increase in welfare is due to an increased number of generated jobs, a higher CO_2 reduction level, a reduced environmental impact and an increased distance between respondents' home and the next hydropower station. The welfare gain of € 6.3, when going from scenario (1) to (2) is fully attributable to the increased reduction of CO_2 emissions. Comparing scenario (3) with scenario (4), it can be seen that the creation of 400 additional jobs is valued with € 1.0 per household and month. A major result of welfare analysis is given by the comparison of scenarios (5) and (6). As can be seen from Table 65, everything is held constant except for the impact on nature and landscape which changes from small to strong impact. This change causes a substantial decrease in welfare by € 12.5 per household and month. Finally, the effect of distance on economic welfare can be shown by looking at scenarios (5) and (7). Reducing the distance of the next hydropower plant from 20 km to 2 km causes a household's welfare loss of approximately € 1.0 monthly.

Table 65: EWF for different scenarios (per household/month) – hydropower

No.	Jobs	CO_2 reduction	Nature/landscape	Distance	Economic welfare
(1)	10	-10 %	Strong impact	2 km	€ 12.160 [11.568, 12.752]
(2)	10	-60 %	Strong impact	2 km	€ 18.466 [17.772, 19.160]
(3)	100	-40 %	Small impact	8 km	€ 28.984 [28.482, 29.486]
(4)	500	-40 %	Small impact	8 km	€ 29.941 [29.432, 30.450]
(5)	500	-60 %	Small impact	20 km	€ 33.151 [32.561, 33.742]
(6)	500	-60 %	Strong impact	20 km	€ 20.670 [19.963, 21.377]
(7)	500	-60 %	Small impact	2 km	€ 32.119 [31.537, 32.700]

95 % confidence intervals in parentheses

Source: OWN CALCULATIONS AND REPRESENTATION

One problem of non-market valuation is that outcomes of stated preference studies are based on the analysis of individual behaviour (BATEMAN ET AL., 2006:1). Hence, the welfare measures presented above describe the mean of the respondents included in the sample of the study. However, the mean of the sample may not be policy relevant, but rather the mean of the relevant population. For that reason, the estimated measures of economic welfare have to be aggregated from the sample to the population. Usually, this can be done by simply multiplying the estimated EWF by the number of people or households in the population, denoted by N in equation 10.23 (PEARCE ET AL., 2002:89f). This is a valid approach as long as a representative sample was drawn from the entire population. Representativeness refers to the representation of population socio-economic and demographic characteristics within a sample (BATEMAN ET AL., 2006:3). For the current sample, representativeness is in principle given with respect to the most important socio-demographics like age, gender or income (see section 6.9). Accordingly, equation 9.24 should give reliable estimates of aggregate values.

$$Aggregate\ EWF = N \cdot EWF \qquad (9.24)$$

However, the aggregation of welfare measures is associated with two further issues. First, the estimated values of EWF are on a monthly basis. Hence, the welfare measures presented in Table 65 must be multiplied by 12 in order to arrive at a yearly measure of total EWF. The second problem, being of greater importance, refers to the treatment of non-responses. As previously shown, the response rate of the survey was 18.5 %. For this part of the population, a positive willingness to pay can be assumed with certainty. However, we do not have any reliable information about the population that did not response to the survey.[213] Generally, there are two possibilities. First, non-responses can be treated as zero-bids or second, it can be assumed that they behave like the respondents in the sample which can usually be assumed in case of representativeness. In order to capture both

---

[213] There may be a significant proportion of the population that is likely to be unwilling to pay anything for hydropower expansion (CARSON, 1999:14).

opportunities, a range of aggregated welfare measures was calculated. The lower level corresponds to a conservative estimate assuming that non-responses have a zero WTP. In contrast, the upper threshold value anticipates sample behaviour to total population. Accordingly, the range of aggregated welfare measures is given by equation 9.25, where N is the number of households in the area of investigation, i.e. in the provinces of Carinthia, Salzburg, Styria, Tyrol, Vorarlberg and Vienna.[214] The results of these estimations are given in Table 66.

$$Welfare_{lower} = EWF * 12 * N * response\,rate$$
$$Welfare_{upper} = EWF * 12 * N \quad\quad\quad\quad (9.25)$$

Table 66: Aggregation of economic welfare (in million € per year) – hydropower

No.	Jobs	CO_2 reduction	Nature/landscape	Distance	Welfare lower level	Welfare upper level
(1)	10	-10 %	Strong impact	2 km	€ 60.9 mill.	€ 329.3 mill.
(2)	10	-60 %	Strong impact	2 km	€ 92.5 mill.	€ 500.1 mill.
(3)	100	-40 %	Small impact	8 km	€ 145.2 mill.	€ 785.0 mill.
(4)	500	-40 %	Small impact	8 km	€ 150.0 mill.	€ 810.9 mill.
(5)	500	-60 %	Small impact	20 km	€ 166.1 mill.	€ 897.9 mill.
(6)	500	-60 %	Strong impact	20 km	€ 103.6 mill.	€ 559.8 mill.
(7)	500	-60 %	Small impact	2 km	€ 160.9 mill.	€ 869.9 mill.

Source: OWN CALCULATIONS AND REPRESENTATION

The "worst case scenario" (1) is associated with an aggregated economic welfare of € 329.3 million referring to the upper threshold value. Assuming that non-responses have a zero WTP, the surplus goes down to € 60.9 million. The "true" welfare measure can be expected somewhere between these lower and upper threshold values. The following discussion refers to the upper welfare levels which appeared to be a reliable estimate due to the representativeness of the sample. The highest economic surplus of € 897.9 million can be attained with scenario (5), which represents the best case of hydropower expansion. A higher CO_2 reduction level (-60 % compared to -10 %) is associated with a welfare gain of € 170.8 million, as can be seen from the comparison of scenarios (1) and (2). Comparing the policy scenar-

---

[214] According to STATISTIK AUSTRIA (2011a:70), the number of households in the area of investigation is 2,257,000.

ios (3) and (4), it can be seen that 400 additional jobs are worth € 25.9 million per year. The greatest welfare loss is caused by a strong environmental impact amounting to € 338.1 million per year when going from scenario (5) to (6). This result has an alternative interpretation as well: holding the environmental impact as small as possible when building new hydropower plants, causes a welfare gain of € 338.1 million per year. Finally, the positioning of new hydropower plants far from residential areas is associated with an increase of EWF. In particular, an additional distance of 18 km is worth € 28.0 million per year. This can be obtained by a comparison of scenarios (5) and (7).

To conclude, an expansion of hydropower is in general associated with positive welfare effects, even in the "worst case scenario" with only few generated jobs, a low CO_2 reduction level, a strong environmental impact and a close distance of hydropower stations to residential areas. Based on that, an increase of generated jobs, CO_2 reduction or distance leads to a significant rise of total EWF. The effect of differing environmental impacts must be particularly highlighted. A hydropower expansion policy causing a strong impact on landscape and natural environment is associated with a huge welfare loss as compared to a strategy where the environmental impacts are held small. This result illustrates how important it is to hold the environmental impact as small as possible when new hydropower stations are built.

9.5 The impact of current electricity bill on WTP

An issue that is often criticised in stated preference studies, or more precisely, choice experiments is the hypothetical bias. The problem of hypothetical bias refers to the fact that respondents may state high WTP values knowing that there will not be any real consequences from their indicated answers or choices (ALRIKSSON AND ÖBERG, 2008:250). In the current CE, WTP is elicited indirectly through the inclusion of a monetary attribute, which was defined as an increase in monthly electricity bill. The hypothetical character of the decision situation may be alleviated by making people aware of the amount they are currently paying for their electricity bill. If people have in mind what they are already paying for electricity, they won't state their choices as recklessly as in a situation where they don't have in mind the amount of their current electricity bill. Thus, the hypothesis is that people being aware of the current electricity bill are suspected to exhibit a lower WTP for the multiple impacts of future hydropower investments. In order to test this hypothesis, the sample was divided into two subsamples. One half of the respondents (in total 446) were questioned about the current electricity bill prior to the CE, the other half (i.e. the other 446 respondents) subsequently to the CE.[215] The question of interest is whether people value hydropower expansion differently depending on their awareness of current electricity bill. To answer this question, a comparison of the two datasets mentioned above is required. However, comparing two or more datasets with each other is not as trivial as one might think. The problem is that the true parameters of a choice model are confounded with the scale factor (λ). Hence, the estimated coefficients represent in fact a multiplicative form of the scale factor and the true underlying parameter (ADAMOWICZ ET AL., 1998a:16). That is:

$$\text{estimated } \beta = \lambda\beta \qquad (9.26)$$

The scale parameter λ is inversely related to the variance of the error terms (σ^2). This is illustrated by equation 9.27 and means that the higher the scale parameter, the smaller the error variance. Therefore, high-fit models have larger scales (ADAMOWICZ ET AL., 1998a:17; ALPIZAR ET AL., 2001:11; HENSHER ET AL., 2005:488).

$$\sigma^2 = \frac{\pi^2}{6\lambda^2} \qquad (9.27)$$

Generally, it is not possible to identify the scale parameter within a dataset. For the interpretation of the signs and relative sizes of the estimated parameters the unknown scale factor is irrelevant, since all parameters within an

---

[215] The questions on electricity bill include a question on the amount of the monthly electricity bill, a question on how accurately people know about the amount they are currently paying, and finally, a question on who pays the bill in the respondent's household. For further details, please refer to the full questionnaire to be found in the appendix of this work.

estimated model have the same scale. However, when comparing two or more estimated models from different data sources the confounding of scale and preference parameters turns out to be a problem (ALPIZAR ET AL., 2001:11). "One should never directly compare the coefficients from different choice models and conclude that one is larger than another" (ADAMOWICZ ET AL., 1998a:16). Hence, in order to make a clear statement about the possible presence of preference differences between the sample asking about current electricity bill prior to the CE and the one asking subsequently to the CE, it is necessary to identify the scale factor. SWAIT AND LOUVIERE (1993) provide a method to estimate the ratio of scale parameters for two different datasets. Accordingly, a sequential testing procedure is applied in line with the approach described in SWEAT AND LOUVIERE (1993). In the following, the sample with the questions about the electricity bill before the CE is referred to as sample A; the other sample is denoted sample B.

The first step of the sequential testing procedure involves a test for differences in the preference parameter vector β by allowing for varying scale parameters λ between the two samples A and B. In the second step, a test for scale parameter equality is performed. The latter test can only be conducted if the preference parameters are equal between the two samples, because the confoundedness of preference and scale parameters prevents the attribution of observed differences to parameter vector inequality and scale equality ($\beta_A \neq \beta_B$, $\lambda_A = \lambda_B$) or to both parameter and scale inequality, i.e. $\beta_A \neq \beta_B$, $\lambda_A \neq \lambda_B$ (BROUWER ET AL., 2010:97; SWAIT AND LOUVIERE, 1993:307).

The first step described above is performed by estimating a separate EC model for the two subsamples. This gives efficient estimates for the confounded coefficients, i.e. $\lambda_A \beta_A$ and $\lambda_B \beta_B$, as well as a likelihood value for both samples. The problem is that the scale parameter cannot be identified in any particular dataset, but the ratio of the scale parameter of one dataset relative to another can be identified. This is why the scale parameter of sample A is normalized to $\lambda_A = 1$ which implies that estimates of scale can be interpreted as relative scale parameters to sample A (i.e. λ_B/λ_A). Then, in a second step, a pooled model is estimated across the two samples. In the pooled model preference parameter equality is imposed (i.e. $\beta_A=\beta_B$)[216] while scale parameters are allowed to vary (BROUWER ET AL., 2010:98). What follows is an optimisation exercise searching for the optimal combination of relative scale and pooled preference parameters that maximizes the log likelihood function of the model (SWEAT AND LOUVIERE, 1993:308). Once the best model fit has been identified, a standard likelihood ratio (LR) test using the log likelihood (LL) values of the separately estimated and pooled models can be performed. With this, it is possible to test the difference in preference parameters for the attributes between the two samples.[217] Under the null hypothesis preference parameters are equal between the two samples (BROUWER ET AL., 2010:98). The test statistic is calculated as follows:

$$-2(LL^{pooled} - (LL^A + LL^B)) \quad with \ d.f. \ |\beta|-1 \qquad (9.28)$$

The degrees of freedom required to perform the test are determined by the number of imposed parameter restrictions ($|\beta|$). If the LR test supports the retention of the null hypothesis indicating that preference parameters (confounded with scale) are equal across the two samples, it is then possible to isolate scale factor differences (SWEAT AND LOUVIERE, 1993:305). In order to test for differences in scale factors, it is necessary to estimate an EC model for the pooled sample. However, in contrast to the pooled model in step 1, now preference and scale parameter equality is imposed; this is $\beta_A=\beta_B$ and $\lambda_A=\lambda_B$. Then a LR test can be applied comparing the log likelihood function of the pooled model with imposed scale factor equality and the log likelihood of the pooled model allowing for varying scale parameters (BROUWER ET AL., 2010:98; SWEAT AND LOUVIERE, 1993:309). That is:

$$-2(LL^{equalscale} - (LL^{pooled})) \quad with \ d.f. \ 1 \qquad (9.29)$$

---

[216] Strictly speaking, parameter equality also means that the constants (ASC) are the same, i.e. $ASC_A=ASC_B$.
[217] As we have seen above, the estimated preference parameters β are confounded with the scale factor λ.

Under the null hypothesis scale parameters are assumed to be equal between sample A and B. Hence, a rejection of the null hypothesis would imply that the scale factors of the two samples are different (BROUWER ET AL., 2010:98).

Table 67: Procedural test results for preference and scale parameter equality

Test procedure	Result
STEP 1 – Test for preference parameter equality	
LL sample A (electricity bill before CE)	-2,157.714
LL sample B (electricity bill after CE)	-2,163.392
LL sample A + LL sample B	-4,321.106
LL pooled sample (allowing scale parameters to vary)	-4,331.587
LR-test – test-statistic (11 d.f.)	20.962
LR-test – p-value (11 d.f.)	0.034
Relative scale (λ_B/λ_A)	0.990
Relative variance (σ_B^2/σ_A^2)	1.020
Reject H_0: $\beta_A = \beta_B$?	Yes
STEP 2 – Test for scale parameter equality	
Not performed since H_0 is rejected in step 1	

Source: OWN CALCULATIONS AND REPRESENTATION

The results of the sequential test procedure described above are illustrated in Table 67.[218] As already mentioned before, sample A refers to the sample which included the questions on household's electricity bill prior to the CE. In sample B the said questions were included subsequently to the CE. In the first step, it was tested whether preference parameters β that are confounded with the scale factor λ are equal across the two samples. The LL value for the pooled model with varying scale parameters corresponds to the maximum LL function obtained at the optimum relative scale. In the present sample, the optimum relative scale was found to be 0.99 in the present sample. This is shown in Figure 56. The subsequent LR test revealed that the preference parameters of sample A and B are statistically significant different from each other, i.e. the null hypothesis (H_0) of preference parameter equality can be rejected at the 5 % significance level. The second step of the sequential testing procedure, isolating differences in scale, could

---

[218] NLOGIT 4.0 econometric software was used to perform the test. For simplicity, the test was performed using attributes only EC models.

not be performed due to the rejection of the null hypothesis in step 1. Thus, it can be concluded that the models are different but we cannot attribute this difference to preference or scale parameter inequality. As a consequence, the two samples are considered separately in the following remarks.

The separate models for the two samples are presented in Table 68. For simplicity, the models have been estimated using attributes only. Each model is based on 2,676 observations, that is, 446 respondents answering six choice tasks in total. Both models are highly significant as shown by the χ^2 statistic calculated for the included variables. In addition, the Pseudo-R^2 measure is with a value of approximately 0.26 nearly equal in both samples.

Figure 56: Result of the optimization exercise searching for optimal relative scale

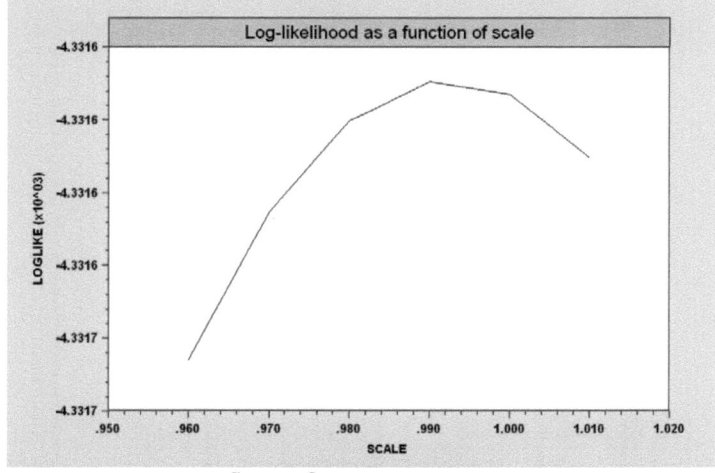

Source: OWN CALCULATIONS

Looking at Table 68, we can first assess the significance of the estimated parameters. In sample B (electricity bill after CE) all coefficients are statistically significant at least at the 10 % level and exhibit the expected signs. Significant parameter and random effect standard deviations justify the complex EC model estimation which enables to capture unobserved preference heterogeneity in the sampled population. In sample A (electricity bill prior to CE), by contrast, not all the parameters showed up to be statistically significant. In particular, the attributes jobs and distance do not have a statistically significant impact on choice. Derived parameter standard deviations are still significant indicating the existence of unobserved taste differences.

Table 68: Separate EC models for sample A and B

Variable	EBILL before CE (sample A)	EBILL after CE (sample B)
ASC	2.522*** (0.000)	2.918*** (0.000)
Jobs	0.0002 (0.417)	0.0004** (0.036)
CO_2 reduction	0.017*** (0.000)	0.015*** (0.000)
Nature (strong impact)	-1.633*** (0.000)	-1.516*** (0.000)
Distance	0.0005 (0.930)	0.010* (0.066)
Cost	-0.121*** (0.000)	-0.130*** (0.000)
Std. Dev. Jobs	0.001** (0.022)	0.001** (0.016)
Std. Dev. CO_2	0.022*** (0.000)	0.024*** (0.000)
Std. Dev. Nature	2.269*** (0.000)	2.201*** (0.000)
Std. Dev. Distance	0.050*** (0.000)	0.025* (0.082)
Std. Dev. Random effect (error component)	3.031*** (0.000)	2.791*** (0.000)
Log likelihood	-2,157.714	-2,163.392
McFadden Pseudo-R^2	0.266	0.264
χ^2 (p-value)	1,564.3 (0.000)	1,553.0 (0.000)
Number of respondents	446	446
Number of observations	2,676	2,676

p-values in parentheses
Significance: ***1 % level **5 % level *10 % level

Source: OWN CALCULATIONS AND REPRESENTATION

Although the sequential testing procedure according to SWEAT AND LOUVIERE (1993) showed that preference parameters differ between the two samples, we cannot distinguish whether these differences are due to differences in the true underlying parameters or scale factor differences. This fact makes the comparison of the two models difficult. For that reason, MWTP measures have been estimated according to the procedure described in section 9.3. By dividing attribute parameters through the coefficient of the monetary attribute the scale factor is cancelled out allowing for the direct comparison of implicit prices (BENNETT AND BLAMEY, 2001:64).

Respondents being asked about their current electricity bill prior to the CE generally exhibit lower values of MWTP as compared to the group of respondents being asked subsequently to the CE. This is shown in Table 69. Overall WTP for the expansion of hydropower is significantly positive in both samples, with sample B (electricity bill after CE) showing a slightly higher value. However, since 95 % confidence intervals overlap we must conclude that the estimated values are statistically not different between the two samples. MWTP for the creation of 100 additional jobs is approximately twice as high in sample B as compared to sample A. The values are significantly different from each other as shown by the non-overlap of confidence intervals.[219] A somewhat controversial result is given for CO_2 reductions. Here, a significantly higher MWTP was found for sample A in comparison to sample B, contradicting a priori expectations. The disutility associated with a strong environmental impact is significantly higher in sample A as compared with sample B. This fact shows up in the lower MWTP, or alternatively, the higher negative MWTP. Statistical significant difference between the two values is given as confidence intervals do not overlap. Finally, MWTP for increasing distance of the next hydropower station to respondent's home is significantly lower in sample A which corresponds to prior expectations.[220]

---

[219] However, the number of generated jobs was found to be a statistically insignificant determinant of choice in the estimated EC model making the interpretation of MWTP doubtful.
[220] As in the case of job creation, the question rises whether the estimated value can be interpreted as MWTP since the attribute parameter on distance was statistically not significant in the estimated EC model.

Table 69: Estimates of MWTP for sample A and B

Variable	Measurement	MWTP EBILL before CE (A)	MWTP EBILL after CE (B)
Hydropower	effect of the ASC	€ 20.880 [17.693, 24.066]	€ 22.496 [19.596, 25.396]
Jobs	per 100 jobs	€ 0.149* [0.136, 0.163]	€ 0.333 [0.303, 0.363]
CO_2 reduction	per 10 % reduction	€ 1.453 [1.331, 1.574]	€ 1.280 [1.171, 1.390]
Impact on nature and landscape	from small to strong impact	€ -13.339 [-13.926, -12.752]	€ -11.389 [-11.898, -10.880]
Distance	per 5 km	€ 0.253* [0.230, 0.277]	€ 0.454 [0.413, 0.495]

95 % confidence intervals in parentheses
*coefficient estimate was statistically not significant

Source: OWN CALCULATIONS AND REPRESENTATION

10 Preferences for a specific hydropower project

In addition to the valuation of hydropower expansion in general, a specific hydropower project was considered in the present work. This regional case study refers to a planned hydropower station in the province of Styria, more precisely, in Graz-Puntigam. The key data of the hydropower project, known as "Murkraftwerk Graz", are given in Table 70.

Table 70: The hydropower project Graz-Puntigam

Installed capacity	16.3 MW
Electric power generation	74 GWh per year
Investment volume	€ 95 million
Start of construction works	autumn 2013
Completion of the project	end of 2015

Source: ENERGIE STEIERMARK (2010c and 2010d, online); OWN REPRESENTATION

The hydropower station is planned to be built within the city limits of Graz along the river Mur[221], in the part of town called Puntigam. The project is being implemented by "Energie Steiermark AG" in collaboration with "Verbund", Austria's leading electricity company and one of Europe's largest hydropower producers. The overall investment volume of the project is € 95 million. Total installed capacity will be 16.3 MW.[222] With this, an electricity amount of 74 GWh per year can be generated. Hence, about 20,000 households can be provided with green electricitiy from the power station (ENERGIE STEIERMARK, 2010a and 2010c, online; DOBROWOLSKI AND SCHLEICH, 2009:10). The construction works are scheduled to start in autumn 2013; the completion and start-up of the power plant is planned for the end of 2015 (ENERGIE-STEIERMARK, 2010d, online).

The design of the hydropower project is shown in Figure 57. On the one hand, the power plant will contribute to the emission-free generation of

---
[221] Graz represents the provincial capital of Styria and is situated about 150 km south-west of Vienna, the capital of Austria. The number of inhabitants amounts to 265,318 (per 1.1.2012). With this, Graz is the second largest city in Austria (LAND STEIERMARK, 2012:83).
[222] With this, the planned hydropower station ranks among the large-scale projects. Smale-scale facilities, by contrast, are defined to have a capacity of less than 10 MW.

electricity from domestic hydropower. Since the hydropower plant is situated in the city of Graz, a high availability of transmission network capacities will be given. At the same time, grid losses can be minimized. Consequently, the hydropower plant is making an important contribution to a sustainable energy supply (PISTECKY, 2010:4). On the other hand, the project is criticised due to the environmental impacts that arise from the power plant.[223]

Figure 57: Design of the hydropower plant in Graz-Puntigam

Source: ENERGIE STEIERMARK (2010e, online)

Before we go deeper into public preferences for the multiple impacts associated with the hydropower project in Graz-Puntigam, people's general attitude and knowledge towards the hydropower project is analysed. The following analyses are based on a representative sample of people living in Graz and its surrounding communities.[224]

---

[223] A detailed description of the environmental impacts that are expected to result from the construction of the new hydropower station can be found in section 6.2.3 of this work.
[224] For a detailed description of the sample see sections 6.7 to 6.9.

10.1 Attitude towards the planned hydropower project

As in the Austrian sample referring to hydropower expansion, the majority of the respondents (82.9 %) living in Graz and surroundings consider the intensified use of renewable energy sources such as hydropower, wind power or photovoltaics in the future as very important. Further 16.1 % of the respondents state that it is rather important to increase the use of electricity from renewable sources. Only a minority of 1.0 % considered the prospective expansion of renewable energy as unimportant.

Furthermore, most respondents have a very positive (43.2 %) or rather positive (52.3 %) attitude towards hydropower utilisation in Austria. The share of people with a negative attitude is considerably low with 3.5 % being rather negative and 1.0 % very negative towards hydropower use (see Figure 58). This result conforms to the Austrian outcome (see section 7.1) concluding that the smaller subset of the Austrian population, i.e. people living in the geographical area of Graz, exhibit similar perceptions of hydropower use in Austria. Regarding people's attitude towards the construction of new hydropower plants along the river Mur[225], a quite different picture is provided. The proportion of people exhibiting a very positive attitude towards hydropower expansion along the Mur amounts to 33.7 %, a significantly lower value as compared to the very positive attitude towards hydropower use in general. A similar result is given for the category "rather positive" whereas the difference is not as large as before (48.2 % versus 52.3 %). In contrast, the share of respondents having a rather negative attitude towards the construction of new hydropower plants along the Mur is with 15.1 % significantly higher as before. The same applies to the category "very negative". In total, 3.0 % of the respondents are very negative towards hydropower expansion along the Mur (see Figure 58). Consequently, people are in general pro hydropower. However, if hydropower plants are to be built along a nearby river people's acceptance will diminish. This provides confirmation of the "Not in my backyard" phenomenon which

---

[225] As an aside, the main part of the respondents (86.9 %) is already in knowledge about the plan to expand hydropower utilisation along the Mur. By contrast, 13.1 % of the respondents have never heard about the fact that new hydropower plants are to be constructed.

was found to be a significant factor in the Austrian sample referring to hydropower expansion in general.

Figure 58: Attitude towards hydropower and its expansion along the Mur

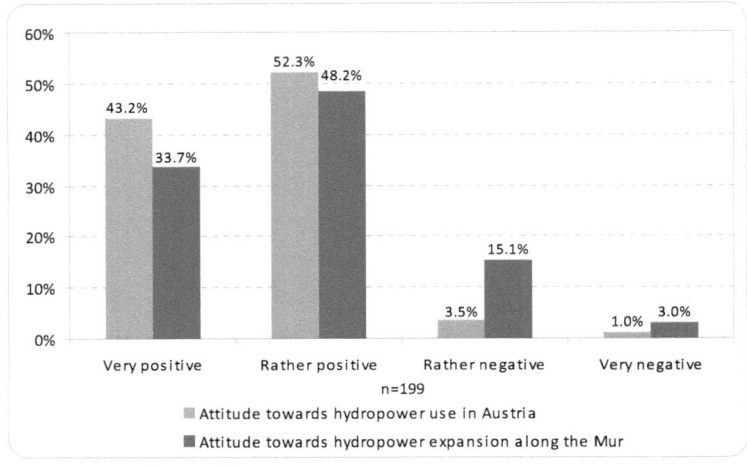

Source: OWN CALCULATIONS AND REPRESENTATION

Since the new hydropower station is expected to create a significant impact on leisure possibilities along the Mur, it appeared to be important to raise the status quo of recreational activities. This is why people were asked to state how frequently they exercise different leisure activities along the river. The outcomes are shown in Table 71. Low recreation index values indicate a high frequency and vice versa.[226] According to that, the most frequently practised recreational activities along the Mur are walking along the riverside, sportive activities and enjoying the landscape. Swimming, boating and fishing can be found at the other end of the scale. Hence, these recreational activities do not play a significant role.

---

[226] For further details on the calculation of the recreation index reference is made to section 7.1.

Table 71: Recreational activities along the Mur

Recreational activity	Index value
Walking along the riverside	1.81
Sportive activities	1.87
Recreating/enjoying the landscape	1.98
Visiting a restaurant/café	2.26
Family trip	2.29
Animal watching	2.46
Picnic on the riverside	2.72
Swimming	2.77
Boating	2.87
Fishing	2.91
Observations	n=199

Source: OWN CALCULATIONS AND REPRESENTATION

In addition, 39.7 % of the respondents think that the planned hydropower station would improve the possibilities for recreation, while 14.6 % hold the opinion that the construction of the hydropower plant would deteriorate recreational activities. A rather large part of the sampled population (45.7 %) is unable to assess the impact of the new hydropower station on leisure opportunities (see Figure 59).

The perceived impact of the hydropower project on recreational activities depends on whether people have knowledge about the plan to build a new hydropower station in Graz-Puntigam. Generally, the degree of recognition of the hydropower project is pretty high. Accordingly, about three quarters (75.4 %) of the respondents explicitly know that there will be built a new hydropower station. Looking at Figure 59, it can be seen that these people intensively think that the new hydropower station will improve the possibilities for recreational use of the Mur (47.3 % compared to 16.3 %). By contrast, within the group of respondents having no knowledge about the planned hydropower station the proportion holding the opinion that the hydropower scheme will limit recreational activities is significantly higher (24.5 % compared to 11.3 %). Finally, the share of "don't knows" is significantly higher in the group of respondents being not aware of the planned hydropower station. The correlation between people's knowledge and their perception of the impact on recreational activities is statistically significant (Cramer's V=0.283; Pearson-χ^2=15.906, p-value=0.000).

Figure 59: Perceived impact of the new hydropower station on recreational activities

	Improvement	Deterioration	Don't know
In total (n=199)	39.7%	14.6%	45.7%
Knowledge plant (n=150)	47.3%	11.3%	41.3%
No knowledge plant (n=49)	16.3%	24.5%	59.2%

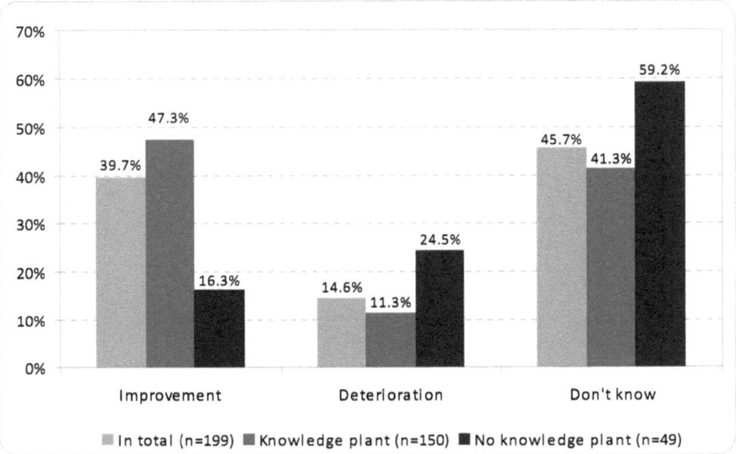

Source: OWN CALCULATIONS AND REPRESENTATION

The mean distance between the location of the hydropower project and respondent's home is 10.8 km (standard deviation: 10.3 km).[227] Median distance amounts to 9.0 km, indicating that 50 % of the respondents live at a distance of less than 9.0 km, the other half at a distance of more than 9 km. From Table 72 it can be seen that a small proportion of the respondents is living in very close proximity (till 2 km) to the project location. Another 30.2 % stated a distance between 2 km and 5 km. The major part (36.2 %) lives at a distance of 5 km to 10 km, and the remaining 30.2 % indicated to live more than 10 km from the regarding location of the hydropower plant.

---

[227] Only people who know that there will be built a new hydropower station in Graz-Puntigam were asked to state the approximate distance between the project location and their home. This is why the following analyses are based on 150 observations instead of 199 in the full sample.

Table 72: Distance between project location and respondents' home

Distance	absolute	in %
Till 2 km	5	3.4 %
More than 2 km to 5 km	45	30.2 %
More than 5 km to 10 km	54	36.2 %
More than 10 km	45	30.2 %
In total	**149**	**100.0 %**

Source: OWN CALCULATIONS AND REPRESENTATION

People who explicitly know that a new hydropower station is planned to be built in Graz-Puntigam were asked about the degree to which they feel affected by the new hydropower plant. A relatively high number of respondents (95 or 63.3 %) reported not to be affected by the new hydropower project. At the same time, 8.7 % of the sample population indicated to feel negatively affected. The share of people feeling positively affected by the hydropower project is 28.0 %. People living further away from the project location (more than 5 km) indicated more often to be not affected than respondents living closer. This is shown in Figure 60. Moreover, respondents that live close to the location of the planned power plant are more likely to feel negatively affected, a result that is to be expected and in line with NIMBY theory. However, a contradictory outcome is obtained for the category "positively affected". Accordingly, people whose residence is located closer to the hydropower project more often indicated to feel positively affected as compared to respondents living further away. This is in contrast to the previous result which indicated that respondents living in close proximity to the project are increasingly affected in a negative way. Nevertheless, this inconsistent result may be attributable to people's intensified awareness of the positive effects (e.g. emission-free generation of electricity, new recreational opportunities) the new hydropower plant will be associated with. The correlation between individual concernment by the planned hydropower project and distance is statistically significant at the 1 % level as shown by the Pearson-χ^2 test (Cramer's V=0.179; Pearson-χ^2 =4.768, p-value=0.000).

Figure 60: Individual concernment of people by the new hydropower plant

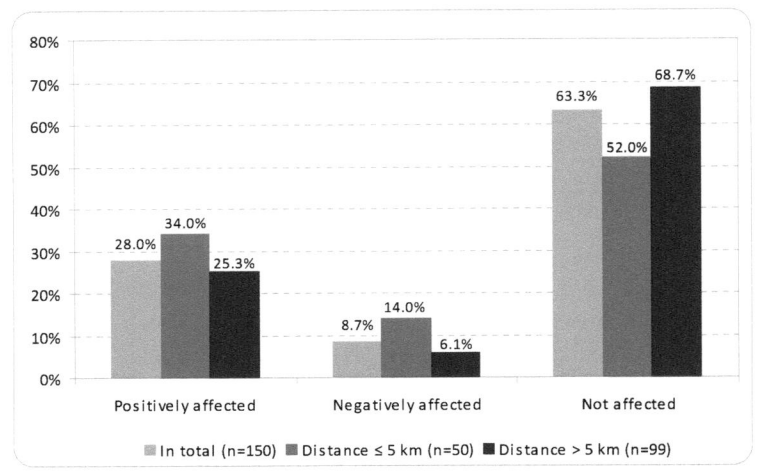

Source: OWN CALCULATIONS AND REPRESENTATION

The positive concernment associated with the new hydropower station can mainly be explained by the possibility to obtain electricity from a highly reliable and renewable energy source. Furthermore, people think that the hydropower project will facilitate new recreational activities. Finally, people are positively affected simply due to the fact that they are generally pro hydropower. Regarding negative concernments, a consistent outcome is given showing that people are in principal against the new hydropower station due to the adverse effects on landscape and wildlife.

10.2 Model results – regional case study

As in the case of overall hydropower expansion (see chapter 9), an econometric model was estimated in order to explore people's preferences for the planned hydropower station in Graz-Puntigam. Different models containing attributes only have been estimated in the first step, so as to find the most appropriate model specification. A detailed description of the attributes and their corresponding coding is given in Table 73. For the attributes households and cost a cardinal-linear coding was used, whereas the household attribute was rescaled by dividing the original values by 1,000. This rescaling prevents a very small coefficient estimate being close to zero. Nature and recreation were coded as dummy variables with "small impact" and "restricted recreational opportunities" as the baseline categories.

Table 73: Description of the variables used in the choice model – regional case study

Variable	Description	Coding/relative frequency
Attributes		
Households	Number of households that can be provided with green electricity from the new hydropower plant (in 1,000).	5, 10, 15 households
Nature	Impact of the new hydropower plant on the landscape and natural environment.	1 = strong impact 0 = small impact
Recreation	Impact of the new hydropower plant on the possibilities for recreation.	1 = extended possibilities 0 = restricted possibilities
Cost	Increase in respondent's monthly electricity bill.	€ 3, 6, 9, 12, 15, 18
Socio-economic characteristics & other variables		
Electricity payment *(interacted with cost attribute)*	1 = Electricity bill is paid by another household member. 0 = Electricity bill is paid by the respondent him-/herself.	1 = 19.6 % 0 = 80.4 %
Donator *(interacted with nature attribute)*	1 = Respondent gives regular donations to environmental organisations. 0 = Otherwise	1 = 33.2 % 0 = 66.8 %
Recreation impact *(interacted with recreation attribute)*	1 = New hydropower plant is expected to improve recreational activities. 0 = Otherwise	1 = 39.7 % 0 = 60.3 %
Children *(interacted with nature attribute)*	1 = Respondent has children living in his or her household. 0 = No children in household.	1 = 29.1 % 0 = 70.9 %
Age	Age of the respondent in years.	Mean = 40.9
Hydropower	1= Hydropower first preferred energy source for future expansion. 0 = Other renewable energy source	1 = 35.2 % 0 = 64.8 %
Bad information	1 = Respondent feels badly informed about hydropower use in Austria. 0 = Respondent feels well informed.	1 = 48.7 % 0 = 51.3 %

Source: OWN CALCULATIONS AND REPRESENTATION

The results of the estimated models are given in Table 75.[228] The estimates are based on 1,194 observations, that is, each of the 199 respondents answering six choice tasks. The simplest and most widely used econometric application, the MNL model, has been estimated first. In this model specification all parameters are highly significant at the 1 % level and exhibit the

---

[228] Again, NLOGIT 4.0 econometric software was used to estimate the models. The panel data structure was accounted for in each of the estimated models. Moreover, models incorporating random parameters (i.e. MXL and EC) were estimated with 100 Halton draws.

expected signs. The problem with a MNL application is, however, the inability to capture unobserved preference heterogeneity. Moreover, the fact that each individual answered six choice tasks may lead to a violation of the IID and in further consequence the IIA assumption making the MNL model an improper approach. The latter problem can be verified by means of a Hausman test. The test is designed so that IIA holds under the null hypothesis. As can be seen from Table 74, the null hypothesis can be rejected at the 10 % significance level in case alternative A is excluded from the set of choices. If the test is performed for alternative B to be excluded from the choice set, the null hypothesis can only be rejected at a marginally statistically significant level.

Table 74: Hausman tests for IIA – regional case study

Excluded...	Chi-squared	p-value
...alternative A	9.728	0.083
...alternative B	9.026	0.108

Source: OWN CALCULATIONS AND REPRESENTATION

Due to this result and the previously mentioned inability of the MNL model to capture unobserved taste differences, a MXL model has been estimated in a second step. Non-monetary attributes were assumed as random parameters. As shown in the lower part of Table 75, the derived random parameter standard deviations are highly significant being indicative for the presence of unobserved preference heterogeneity in the sampled population. A further extension of the MXL model is to allow for variance heterogeneity captured by an additional error component. As can be seen from the EC model, the standard deviation of the random effect (error component) is highly significant at the 1 % level which justifies the complex EC model estimation. Random parameter standard deviations are highly significant as well. The comparison of the estimated attributes only models leads to the conclusion that the EC model is preferable over MNL and MXL. This can be shown by the steadily decreasing log likelihood values, as well as the rising Pseudo-R^2 measures. Additionally, decreasing information criteria (AIC and BIC) provide confirmation of the increasing model fit when going from a MNL to a more complex EC model. Hence, it can be assumed that the EC model captures people's preferences best.

Table 75: Model estimates – regional case study

Variable	Attributes only MNL	Attributes only MXL	Attributes only EC	Extended model EC
Dependent variable	Choice: Alternative A, B or none of the two			
ASC	1.293*** (0.000)	2.210*** (0.000)	2.557*** (0.000)	4.300*** (0.000)
Households	0.033*** (0.003)	0.043** (0.045)	0.054*** (0.001)	0.057*** (0.001)
Nature (strong impact)	-1.492*** (0.000)	-3.203*** (0.000)	-2.807*** (0.000)	-2.161*** (0.000)
Recreation (extended)	0.638*** (0.000)	1.005*** (0.000)	1.056*** (0.000)	0.682*** (0.001)
Cost	-0.143*** (0.000)	-0.222*** (0.000)	-0.239*** (0.000)	-0.255*** (0.000)
Electricity payment*Cost				0.067** (0.016)
Donator*Nature				-1.320*** (0.007)
Children*Nature				-0.791* (0.055)
Impact recreation*Recreation				0.842*** (0.005)
Age				-0.034** (0.018)
Hydropower				0.754* (0.077)
Bad information				-1.163*** (0.005)
Std. Dev. Households		0.185*** (0.000)	0.086*** (0.001)	0.087*** (0.002)
Std. Dev. Nature		4.294*** (0.000)	3.874*** (0.000)	3.666*** (0.000)
Std. Dev. Recreation		1.731*** (0.000)	1.781*** (0.000)	1.693*** (0.000)
Std. Dev. Random effect (error component)			2.263*** (0.000)	2.218*** (0.000)
Log likelihood	-1,052.062	-902.674	-876.465	-855.999
McFadden Pseudo R^2	0.198	0.312	0.332	0.347
Chi-squared (Prob.)	-	818.1 (0.000)	870.6 (0.000)	911.5 (0.000)
AIC	1.771	1.525	1.483	1.461
BIC	1.792	1.559	1.522	1.529
Number of respondents	199	199	199	199
Number of observations	1,194	1,194	1,194	1,194

p-values in parentheses
Significance: *** 1 % level   ** 5 % level   * 10 % level

Source: OWN CALCULATIONS AND REPRESENTATION

In addition to the attributes of the choice experiment, a selection of socio-economic and attitudinal characteristics, as well as plausible interaction terms between attributes and these variables have been included in the model. The "right" selection of socio-demographics and interactions was found within the scope of an iterative estimation procedure. The indirect utility function of the statistically best fit (SBF) model resulting from this procedure is shown below. The model of equation 10.1 can be found in the last column of Table 75.

$$\begin{aligned}V = {}& \beta_0 + \beta_1 \text{Households} + \beta_2 \text{Nature} + \beta_3 \text{Recreation} + \beta_4 \text{Cost} \\ & + \beta_5 \text{Electricity payment} * \text{Cost} + \beta_6 \text{Donator} * \text{Nature} \\ & + \beta_7 \text{Children} * \text{Nature} + \beta_8 \text{Impact recreation} * \text{Recreation} \\ & + \beta_9 \text{Age} + \beta_{10} \text{Hydropower} + \beta_{11} \text{Bad information} \end{aligned} \quad (10.1)$$

The SBF model is highly significant as shown by the χ^2 statistic calculated for the entire set of variables. Another point of interest refers to the question whether the additionally included variables can contribute to improve the model fit. A first indication of improving model fit is given by the slightly rising Pseudo-R^2 and decreasing information criteria, although BIC has slightly risen when going from the attributes only to the extended EC model. Additionally, a likelihood ratio (LR) test has been performed in order to find out whether the subset of socio-demographics and interactions are jointly significant. The outcome of the test is given in Table 76. The null hypothesis that the additionally included variables do not have any explanatory power can clearly be rejected at the 1 % significance level. Accordingly, the extended EC model can be assumed to be superior to the model specification using attributes only.[229]

Table 76: Result of the likelihood ratio test – regional case study

Comparison of...	LR statistic	p-value
...attributes only and extended EC models	40.932	0.000

Source: OWN CALCULATIONS AND REPRESENTATION

Thus, looking at the EC model estimates it can be seen that the coefficients of the four choice attributes, the interaction terms and the remaining variables have the expected signs and are all statistically significant at least at

---

[229] The extended model specification has the further advantage that systematic preference heterogeneity can be explored by the inclusion of socio-economic characteristics and their interactions with the choice attributes.

the 10 % level. The alternative specific constant (ASC) is highly significant and positive indicating that the respondents have some inherent propensity to choose for one of the power plant alternatives over the opt-out (none of the two alternatives) for reasons that are not captured in the estimated model.

The household attribute affects indirect utility positively meaning that respondents prefer alternatives where more households can be supplied with green electricity from the new hydropower station. The impact of the new hydropower plant on recreational opportunities is positive as well. This means that people are more likely to choose an alternative when the possibilities for recreation are extended as compared to an alternative with restricted leisure activities. Furthermore, people holding the opinion that the planned hydropower station would improve leisure opportunities pay increasing attention to the recreation attribute. This relationship is captured by the positive coefficient attached to the interaction term between the attribute recreation and the dummy variable "impact recreation". As can be seen from the partial derivative of indirect utility with respect to recreation (equation 10.2), the marginal effect would increase if people think that recreational activities are to be improved when the new hydropower plant is built, that is, the dummy variable "impact recreation" equals 1.

$$\frac{\partial V}{\partial \text{Recreation}} = \beta_3 + \beta_8 \text{Impact recreation} = 0.638 + 0.842 * \text{Impact recreation}$$

$$\text{if Impact recreation} = 1 \quad \rightarrow \quad \frac{\partial V}{\partial \text{Recreation}} \uparrow \qquad (10.2)$$

In contrast to these positive outcomes, environmental impacts appeared to have a negative effect on choice, providing confirmation of the trade-off between positive consequences and negative environmental side effects. More precisely, alternatives with a strong environmental impact are less preferred as compared to power plant alternatives exhibiting only a small impact. This relationship is captured by the negative sign of the coefficient on the attribute nature. In addition, the effect of the strong nature impact is enhanced if the respondent (or someone else in his or her household) is a donator to environmental organisations; regular donations reflect affinity with environmental issues. Another important result of the model refers to the impact of children on people's perception of a strong environmental impact. Particularly, the strong environmental impact shows a greater impact on choice or utility when children are living in respondent's household. This result implies the presence of bequest values. Consequently, respondents with children are more inclined to preserve a natural river landscape for the sake of future generations (KOUNDOURI ET AL., 2009:1949). The cross effects between nature impact, environmental behaviour and children can easily be shown by equation 10.3.

$$\frac{\partial V}{\partial \text{Nature}} = \beta_2 + \beta_6 \text{Donator} + \beta_7 \text{Children}$$
$$= -1.492 - 1.32 * \text{Donator} - 0.791 * \text{Children}$$
$$\text{if Donator} = 1 \quad \rightarrow \quad \frac{\partial V}{\partial \text{Nature}} \downarrow \qquad (10.3)$$
$$\text{if Children} = 1 \quad \rightarrow \quad \frac{\partial V}{\partial \text{Nature}} \downarrow$$

The negative sign of the cost attribute reflects standard economic theory and indicates that green electricity must be provided at a low cost in order to accept the construction of the new hydropower plant. In simple terms, people prefer cheaper alternatives. However, price sensitivity will diminish if the electricity bill is not paid by the respondent but instead by another household member.[230] This can be shown by the partial derivative of indirect utility with respect to cost.

$$\frac{\partial V}{\partial \text{Cost}} = \beta_5 + \beta_6 \text{Electricity payment} = -0.129 + 0.027 * \text{Electricity payment}$$
$$\text{if Electricity payment} = 1 \quad \rightarrow \quad \frac{\partial V}{\partial \text{Cost}} \uparrow \qquad (10.4)$$

Regarding socio-demographic characteristics, the model outcomes reveal that elder people are less likely to vote for the construction of the new hydropower plant. Instead, they rather tend to choose the opt-out alternative. No other socio-demographic were found to exhibit a statistically significant

---

[230] An analogous result was found in the choice experiment referring to overall hydropower expansion.

impact on choice. However, two additional attitudinal variables appeared to represent significant determinants of people's choice. First, respondents who ranked hydropower first when asked for the two most preferred renewable energy sources for the purpose of future electricity generation are more likely to choose one of the hydropower scenarios over the opt-out. Finally, the level of information has a significant effect on choice as derived from the positive coefficient on "bad information". Specifically, people feeling badly informed about hydropower in general are less likely to accept the new hydropower plant.[231]

10.3 Willingness to pay – regional case study

In order to give the estimated parameters presented above more meaningfulness, implicit prices (i.e. MWTP) have been calculated. However, as shown before, taking the ratio between the coefficient of the attribute of interest and the monetary attribute represents an inappropriate approach in the presence of unobserved preference heterogeneity. Consequently, MWTP for the choice attributes was simulated for each respondent using the conditional constrained parameter estimates.[232] Based on these simulations, means, standard deviations and confidence intervals were calculated. The outcome of this procedure is shown in Table 77.

The estimated values of MWTP are based on a ceteris paribus assumption, meaning that all other effects are held constant except for the attribute for which the implicit price is being calculated. First, people generally exhibit a positive MWTP for the construction of the new hydropower station independent from the attribute levels. This general MWTP, which represents the positive ASC, is € 16.9 per household and month.

Additionally, respondents are willing to pay around € 0.3 on top of their monthly electricity bill for the supply of 1,000 additional households with green electricity from the hydropower plant.

---

[231] Such a result was also found in the Austrian sample referring to hydropower expansion in general, rather than to a specific hydropower project.
[232] More precisely, for each attribute 199 estimates of MWTP have been calculated.

Table 77: Estimates of marginal WTP – regional case study

Variable	Measurement	MWTP
Hydropower plant	effect of the ASC	€ 16.882 [11.544, 22.219]
Households	per 1,000 households	€ 0.258 [0.225, 0.292]
Impact on nature and landscape	from small to strong	€ -9.430 [-10.019, -8.841]
Recreational activities	from restricted to extended recreational activities	€ 3.078 [2.839, 3.318]

95 % confidence intervals in parentheses

Source: OWN CALCULATIONS AND REPRESENTATION

The implicit price for the nature attribute is negative since stated choices are negatively affected by the adverse environmental effects associated with the new hydropower plant. Negative values of MWTP imply a reduction of respondents' utility. According to that, the disutility associated with a strong environmental impact is estimated at € 9.4 per household and month. Conversely, the negative implicit price can be interpreted as a demand for compensation required for the loss of nature and landscape when the new hydropower station is built.

Another important factor for respondents is the creation of leisure activities. Since the survey participants are living near the Mur, recreational activities along the river.[233] Hence, an improvement of the possibilities for recreation is valued positively. More specifically, respondents are willing to pay € 3.1 on top of their monthly electricity bill if the hydropower station opens up new opportunities for leisure activities (such as a cycle paths or canoeing).

Table 78: Importance of the attributes – regional case study

Attribute	Index value
Impact on nature and landscape	1.53
Increase in monthly electricity bill	1.75
Households	1.78
Recreation	2.34

Source: OWN CALCULATIONS AND REPRESENTATION

---

[233] A detailed analysis of practised recreational activities along the Mur has already been elucidated in greater detail in section 10.1 of this chapter.

The results of MWTP in general correspond to the perceived importance of the attributes retrieved from a debriefing question which was asked subsequently to the choice experiment. The index values shown in Table 78 were calculated as before (see section 9.3) and represent the mean of people's importance attached to each of the four choice attributes. Possible answers ranged from (1) "very important" to (4) "totally unimportant". Consequently, lower index values indicate higher importance. According to that, the impact on nature and landscape appeared to be the most important attribute. This is also reflected when looking at implicit prices. Electricity price increases represent an important attribute too, followed by the number of households that can be supplied with electricity from the hydropower station. The attribute describing the impact on recreational activities is on average ranked last. This outcome does not correspond to the result of the choice experiment where people exhibit a higher WTP for recreational activities than for the provision of additional households with electricity.

10.4 Welfare analysis – regional case study

Implicit prices (WTP) for the individual attributes are in fact useful for policy makers. However, these values do not represent valid welfare measures. This is why overall economic welfare (EWF) was estimated for different policy scenarios. Similar to the calculation of implicit prices, the welfare measures were simulated for each respondent based on the statistically best fit model presented above. With this approach, unobserved preference heterogeneity is accounted for. Then means, standard deviations and the corresponding confidence intervals were drawn from the simulations. For further theoretical details on the estimation of welfare measures please refer to section 9.4 of this work.

The outcomes for four different policy scenarios are presented in Table 79.[234] The first scenario corresponds to the worst case, meaning that a small hydropower plant is built with a strong impact on landscape and natural environment and no additional possibilities for recreation. This attribute

---

[234] The hydropower station Graz-Puntigam is expected to provide 20,000 households with green electricity. Therefore, this value was used in the subsequent welfare analysis although it is outside the predetermined range of the attribute levels.

level combination is associated with a very low level of EWF amounting to merely € 0.2 per household and month. Additionally, we cannot conclude that EWF attached to the worst case scenario is significantly positive since the 95 % confidence interval includes the value zero. Improving all attributes leads to a substantial increase of welfare to € 20.0 per household and month. This value is associated with 20,000 households able to be provided with electricity from the hydropower plant, a small environmental impact and the presence of new recreational activities. Starting from this scenario, a deterioration of environmental conditions, that is, a change from small to strong impact is associated with a significant decrease in total EWF. In particular, EWF goes substantially down from € 20.0 in scenario (2) to € 8.0 in scenario (3). The effect of additionally available recreational activities can be shown by the comparison of scenarios (2) and (4). The creation of additional leisure opportunities is associated with an increase of EWF from € 15.8 to € 20.0.

Table 79: EWF for different scenarios (per household/month) – regional case study

	Households	Nature/landscape	Recreation	Economic welfare
(1)	5,000	strong impact	restricted	€ 0.190 [-0.927, 1.306]
(2)	20,000	small impact	extended	€ 19.992 [19.022, 20.963]
(3)	20,000	strong impact	extended	€ 8.039 [6.700, 9.378]
(4)	20,000	small impact	restricted	€ 15.750 [14.972, 16.528]
	95 % confidence intervals in parentheses			

Source: OWN CALCULATIONS AND REPRESENTATION

As we have seen in the previous chapter, what is relevant for policy makers is an aggregated measure of economic welfare. For that reason, the estimated welfare measures per household and month were aggregated across the area of investigation (Graz and surrounding communities) using the number of households living in this area. However, there is a lack of data. As a consequence, the number of households required to aggregate economic welfare had to be calculated manually. First, an estimate of average household size (Hsize) was generated by calculating a population (Pop) weighted average of the household sizes in the districts of "Graz-Stadt" (G) and

"Graz-Umgegung" (GU).[235] This is shown in equation 10.5 and the result is an average household size of 2.26 persons.

$$\phi Hsize = \frac{Hsize_G * Pop_G + Hsize_{GU} * Pop_{GU}}{Pop_{G+GU}} \qquad (10.5)$$

In a second step, the number of inhabitants living in the city of Graz and the directly surrounding communities which is 338,341 (LAND STEIERMARK, 2012b) was divided by the average household size calculated above yielding a number of 149,903 households. This is given in equation 10.6.

$$Households = \frac{Population}{\phi Hsize} \qquad (10.6)$$

Prior to the aggregation of economic welfare with the number of households, the estimated measures were converted to yearly values. Then, as before, a range of total economic welfare was calculated (see Table 80). Within the lower bound non-responses are assumed to have zero WTP, while the upper threshold supposes that people who did not response to the survey have the same preferences as the people in the sample. The following remarks are based on the upper welfare levels since translating sample into population behaviour gives reliable results as long as a representative sample was drawn from the entire population.

Table 80: Aggregation of economic welfare
(in million € per year) – regional case study

	Households	Nature/landscape	Recreation	Welfare lower level	Welfare upper level
(1)	5,000	strong impact	restricted	€ 0.1 mill.	€ 0.3 mill.
(2)	20,000	small impact	extended	€ 6.7 mill.	€ 36.0 mill.
(3)	20,000	strong impact	extended	€ 2.7 mill.	€ 14.5 mill.
(4)	20,000	small impact	restricted	€ 5.2 mill.	€ 28.3 mill.

Source: OWN CALCULATIONS AND REPRESENTATION

First, the worst case scenario is associated with a very low value of total EWF amounting to solely € 0.3 million. Going to the best case (scenario 2)

---
[235] These household sizes were taken from LAND STEIERMARK (2012a, online).

welfare rises substantially to € 36.0 million. A strong environmental impact is associated with a welfare burden of € 21.5 million, as can be seen from the comparison of scenarios (2) and (4). In contrast, the creation of new possibilities for leisure activities is totally worth € 13.8 million.

11 Preferences for different renewable technologies

The final econometric application covered within the scope of this work refers to the expansion of different renewable energy sources. Due to the fact that, beside hydropower, other renewable energy sources may play a role for future electricity generation, hydropower expansion was embedded into a broader strategy for renewable energy. In addition to hydropower, biomass, solar and wind power were used as alternative energy sources between which respondents were asked to choose. These renewable energy sources, in turn, create multiple impacts on employment, CO_2 emission levels, nature and landscape, as well as distance between the location of the renewable energy power plant and respondent's home.[236] The inclusion of alternative renewable energy sources broadens the choice set and makes people aware that alternative sources could also be used to expand renewable energy in Austria. This may significantly influence the value attached to hydropower expansion, an effect that is usually known as framing bias. Furthermore, the consideration of other renewable energy sources enables to study whether people's perception of the multiple impacts differ between renewable technologies.

11.1 Attitude towards renewable energy

Prior to the analysis of public preferences for the multiple impacts of renewable energy investments, people's general attitude towards renewable energy expansion is analysed. The following analyses are based on a representative sample of the Austrian population.[237]

As in the previous analyses, it can be shown that people have in general a positive attitude towards future expansion of renewable energy sources. More specifically, 77.3 % of the respondents think that it is very important to increase energy production from renewable sources such as hydropower, solar or wind power in the future. Further 20.4 % stated that it is rather im-

---

[236] A detailed description of the labelled choice experiment on renewable energy expansion can be found in section 6.2.2 of this work.
[237] For a detailed description of the sample see sections 6.7 to 6.9.

portant. Only a very small proportion of the sampled population (2.3 % in total) perceives the intensified use of renewable energy sources as rather or totally unimportant (see Figure 61).

Regarding people's preferences for different renewable technologies a clear outcome was obtained. As can be seen from Figure 62, solar power (PV) appeared to be the most preferred renewable energy source for future electricity generation chosen by 72.6 % of the respondents. The second popular renewable energy source is hydropower with 35.1 % of the respondents ranking the technology first and 24.1 % second. Wind power was chosen as a preferred energy source by 53.8 % in total. With this, wind power ranks third after solar power and hydropower. Biomass which was chosen by only 13.1 % of the respondents is the least preferred renewable energy source for future electricity generation. This preference order in principle corresponds to the one resulting from the Austrian sample on hydropower expansion (see section 7.1).

Figure 61: Perceived importance of future renewable energy expansion (2)

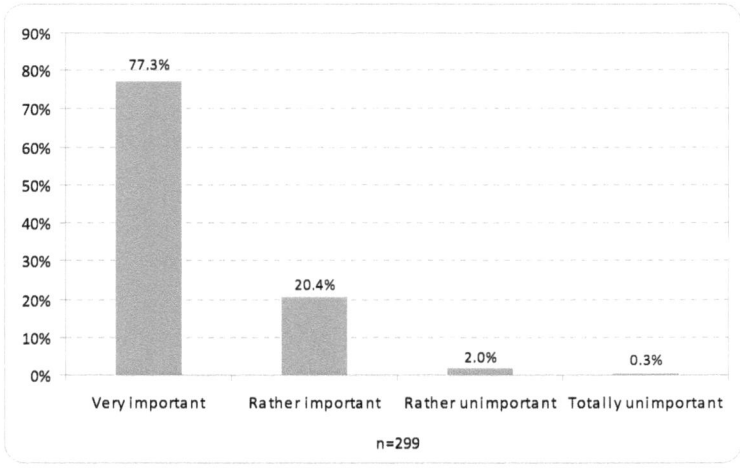

Source: OWN CALCULATIONS AND REPRESENTATION

Figure 62: Preferred renewable energy sources for future expansion in Austria (2)

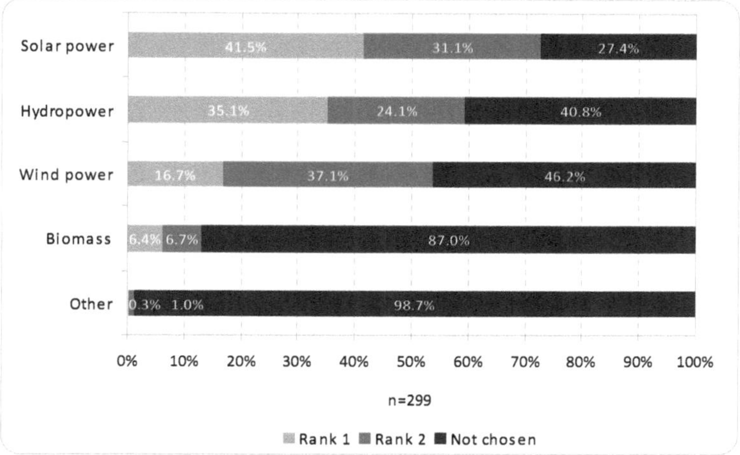

Source	Rank 1	Rank 2	Not chosen
Solar power	41.5%	31.1%	27.4%
Hydropower	35.1%	24.1%	40.8%
Wind power	16.7%	37.1%	46.2%
Biomass	6.4%	6.7%	87.0%
Other	0.3%	1.0%	98.7%

n=299

Source: OWN CALCULATIONS AND REPRESENTATION

Finally, people were confronted with a series of statements referring to the impact of renewable energy investments on the satisfaction of increasing electricity demand, the reduction of CO_2 emissions and the dependence on electricity imports from abroad. The outcome is given in Figure 63. The highest commitment was found for the importance of increased renewable energy use for reducing climate-damaging CO_2 emissions. Here, 72.9 % totally agree and 22.7 % rather agreed whereas only 4.4 % in general disagree to the statement that renewable energy use can contribute to the reduction of CO_2 emissions. A similar result is given with respect to the other two statements. About 70 % of the respondents totally agree that the intensified use of renewable energy sources will help to satisfy growing electricity demand in Austria and reduce the necessity of fossil fuel imports. The share of people who rather disagreed is about one quarter in both cases, while disagreement rates were relatively low.[238]

---

[238] By comparison, the contribution of intensified hydropower use to the issues described above is perceived significantly lower as shown in section 7.1 of this work.

Figure 63: Respondents' agreement to several statements on renewable energy

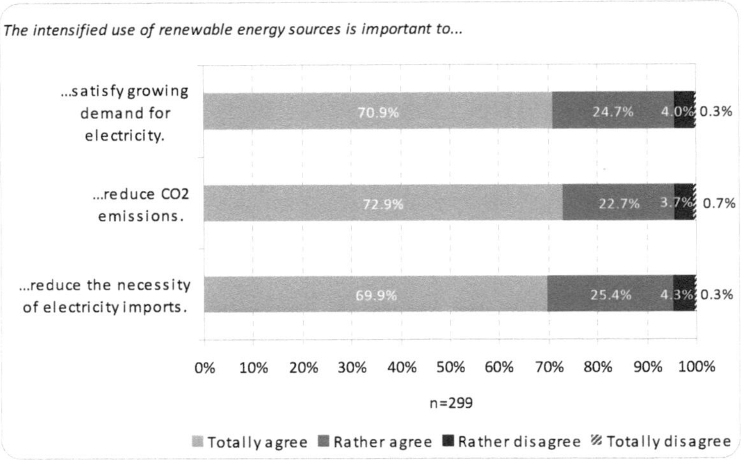

The intensified use of renewable energy sources is important to...

Statement	Totally agree	Rather agree	Rather disagree	Totally disagree
...satisfy growing demand for electricity.	70.9%	24.7%	4.0%	0.3%
...reduce CO2 emissions.	72.9%	22.7%	3.7%	0.7%
...reduce the necessity of electricity imports.	69.9%	25.4%	4.3%	0.3%

n=299

Source: OWN CALCULATIONS AND REPRESENTATION

11.2 Model results – renewable energy

In order to test the impact of framing hydropower demand in the context of demand for other renewable energy sources, a labelled choice experiment (CE) was conducted. The only difference between this labelled CE and the unlabelled hydropower CE presented in chapter 9 was that labels were used to reflect alternative renewable energy sources as substitutes to hydropower.[239] The labelled CE was designed against the background of testing whether the broadening of the choice set significantly affects the value attached to the attributes of hydropower expansion. For that reason, modes using attributes only have been estimated. No interaction terms and socio-demographic characteristics were included in the estimated models.

11.2.1 Generic parameter estimates

A first attempt to quantify people's preferences for different renewable technologies (including hydropower) was made by estimating a choice model with parameters that are generic across alternatives, i.e. technologies. So, for each attribute a single-parameter estimate which is equal for

---

[239] For further details on the attributes and levels used in the CE please refer to section 6.2 of this work.

each technology was calculated (HENSHER ET AL., 2005:311). Differences between renewable technologies are captured by the alternative specific constants (ASCs). Hence, the constant of the estimated model is made specific to each technology or label. This is important since constraining a constant to be equal across two or more alternatives may be an invalid approach when alternatives are labelled (HENSHER ET AL., 2005:310).

The estimated models are shown in Table 82. As in the previous choice models, three different model specifications have been estimated starting with the classical MNL model. According to the model diagnostics, it can be concluded that the EC model is preferable over MNL and MXL. This is because the highest Mc Fadden Pseudo-R^2 is obtained with the EC specification. In addition, information criteria according to Akaike (AIC) and Schwarz (BIC) are lowest in the EC model indicating a better model fit. Consequently, the following remarks refer to the EC model estimates. Based on the χ^2 statistic calculated for the included variables (attributes only), we can infer that the model is highly significant. The included variables are all statistically significant and have the expected signs. Statistical insignificance is given with respect to the random parameter standard deviations of Jobs and Distance. As in the hydropower CE, job creation, CO_2 reductions and distance have a positive influence on consumer utility and in further consequence choice. By contrast, strong environmental impacts and electricity price increases have a negative effect.

Table 81: Model estimates using generic parameters – renewable energy

Variable	MNL model	MXL model	EC Model
Dependent variable:	\multicolumn{3}{c}{Choice: Alternative 1, 2 or none of the two}		
ASC Biomass	1.374*** (0.000)	2.089*** (0.000)	3.096*** (0.000)
ASC Hydropower	1.930*** (0.000)	2.919*** (0.000)	3.834*** (0.000)
ASC Solar power	2.117*** (0.000)	3.259*** (0.000)	4.106*** (0.000)
ASC Wind power	1.735*** (0.000)	2.663*** (0.000)	3.612*** (0.000)
Jobs	0.0005*** (0.003)	0.0005** (0.043)	0.0006*** (0.010)
CO_2 reduction	0.013*** (0.000)	0.017*** (0.000)	0.016*** (0.000)
Nature (strong impact)	-0.755*** (0.000)	-1.257*** (0.000)	-1.008*** (0.000)
Distance	0.009* (0.055)	0.010 (0.259)	0.011* (0.076)
Cost	-0.107*** (0.000)	-0.157*** (0.000)	-0.141*** (0.000)
Std. Dev. Jobs		0.002*** (0.000)	0.0007 (0.376)
Std. Dev. CO_2		0.039*** (0.000)	0.025*** (0.000)
Std. Dev. Nature		2.405*** (0.000)	1.847*** (0.000)
Std. Dev. Distance		0.081*** (0.000)	0.022 (0.262)
Std. Dev. Random effect (error component)			2.846*** (0.000)
Log likelihood	-1,598.348	-1,454.987	-1,390.523
McFadden Pseudo R^2	0.117	0.496	0.518
χ^2 (p-value)	2,525.2 (0.000)	2,864.7 (0.000)	2,993.6 (0.000)
AIC	1.792	1.637	1.566
BIC	1.819	1.676	1.609
Number of respondents	299	299	299
Number of observations	1,794	1,794	1,794

p-values in parentheses
Significance: ***1 % level   **5 % level   *10 % level

Source: OWN CALCULATIONS AND REPRESENTATION

The labels for the energy sources are all highly significant indicating that the type of renewable energy technology indeed affects people's choice for reasons that are not captured in the estimated model, i.e. the attributes. The estimated technology-specific constant is highest for solar power, followed by hydropower and wind power. Biomass is – based on the estimated ASC

– the least preferred energy source.[240] Whether the observed differences between technologies are statistically significant, need to be tested by means of a Wald test. Generally, the Wald test is an opportunity to test for linear restrictions (HENSHER ET AL., 2005:346). In the present case, the null hypothesis (H_0) to be tested is whether the difference between two parameters, say k and j, is equal to zero.[241] "The test is based on how well the corresponding parameter estimates satisfy the null hypothesis" (CAMERON AND TRIVEDI, 2010:404). That is:

$$H_0 : \beta_k - \beta_j = 0 \rightarrow \hat{\beta}_k - \hat{\beta}_j \cong 0 \qquad (11.1)$$

In total, six different pairwise comparisons between technologies must be considered in order to capture all possible combinations of renewable energy sources. This comparison structure is shown in Table 82 and represents the basis for the following wald tests for linear restrictions.

Table 82: Comparison structure between technologies

Comparison between		
Hydropower	←→	Biomass
Hydropower	←→	Solar power
Hydropower	←→	Wind power
Biomass	←→	Solar power
Biomass	←→	Wind power
Solar power	←→	Wind power

Source: OWN REPRESENTATION

According to Table 82, six different linear restrictions have been tested in order to find out whether the labels for the technologies differ between each other. The results of the Wald tests are given in Table 83. As can be seen from the last column, the null hypothesis that technology-specific constants (ASCs) are equal can clearly be rejected for each of the possible linear combinations. Hence, the preference ordering presented above with so-

---

[240] This preference ordering corresponds to the ranking obtained from the direct question asking for the preferred renewable energy sources to be used for future electricity generation (see Figure 62 in the previous section). Hence, preferences for renewable energy expansion seem to be consistent.
[241] Alternatively, any kind of linear restriction can be tested by means of a Wald test.

lar power being the most preferred and biomass the least preferred renewable energy source is indeed statistically significant. From this, we can conclude that respondents have mixed emotions about different renewable energy sources. This is an important result. However, what is of special interest is the difference between MWTP values for the attributes in the original hydropower choice experiment and the more broadened framework considering other renewable technologies.

Table 83: Wald tests comparing alternative specific constants

Linear restriction (H_0)			Wald statistic	p-value
ASC_{hydro}	$- ASC_{bio}$	$= 0$	6.169	0.000
ASC_{hydro}	$- ASC_{solar}$	$= 0$	-2.571	0.010
ASC_{hydro}	$- ASC_{wind}$	$= 0$	1.923	0.054
ASC_{bio}	$- ASC_{solar}$	$= 0$	-8.121	0.000
ASC_{bio}	$- ASC_{wind}$	$= 0$	-4.173	0.000
ASC_{solar}	$- ASC_{wind}$	$= 0$	4.240	0.000

Source: OWN CALCULATIONS AND REPRESENTATION

This difference was analysed by the direct comparison of the MWTP values from the unlabelled hydropower CE presented in section 9.3 and the labelled renewable energy CE of this chapter. As before, MWTP for the attributes of renewable energy expansion has been simulated for each respondent $n=1,...,N$ and each attribute $k=1,...,K$ so as to capture unobserved preference heterogeneity in the sampled population.[242] The outcomes are given in the last column of Table 84 and are set in relation to the MWTP obtained from the previous choice model on hydropower expansion. First, it can be seen that overall WTP for hydropower expansion is significantly lower when considered in combination with other renewable energy sources.[243] Consequently, hydropower is not the highest valued energy source when people are given the option to choose other renewable technologies as substitutes to hydropower. The creation of additional jobs is valued more when referring to the expansion of renewable energy sources in general rather than the expansion of hydropower in particular. The con-

---

[242] Overall WTP for the different renewable energy sources has been calculated by simply dividing the estimate of the corresponding ASC by the cost attribute since neither ASCs nor Cost were treated as random parameters in the estimated model.
[243] Statistical significance between the two values of MWTP is given due to the non-overlap of 95 % confidence intervals.

tribution of renewable energy sources as a whole and hydropower in particular to the reduction of climate-damaging CO_2 emissions is valued equivalently as shown by the overlap of 95 % confidence intervals. By contrast, the negative impact on nature and landscape is considered significantly less of a concern when referring to the overall expansion of renewable energy sources rather than the specific extension of hydropower. In particular, the disutility associated with the environmental side effects of hydropower expansion is nearly twice as high as the disutility related to the environmental impacts of renewable energy in general. Finally, the attribute distance is valued significantly higher in the labelled CE considering different renewable energy sources as compared to the unlabelled CE which simply referred to the expansion of one specific energy source, namely hydropower. Hence, it seems to be more important for people that renewable energy plants such as wind parks, photovoltaic installations or biomass power stations are further away from their residence than hydropower plants.

Table 84: Comparison of MWTP from the unlabelled and labelled CE (with generic parameters)

Variable	Measurement	MWTP unlabelled CE	MWTP labelled CE
Hydropower	effect of the ASC	€ 33.075 [31.721, 34.428]	€ 27.159 [22.947, 31.371]
Biomass	effect of the ASC		€ 21.929 [17.684, 26.175]
Solar power	effect of the ASC		€ 29.086 [24.874, 33.297]
Wind power	effect of the ASC		€ 25.582 [21.336, 29.827]
Jobs	per 100 jobs	€ 0.229 [0.215, 0.244]	€ 0.431 [0.384, 0.478]
CO_2 reduction	per 10 % reduction	€ 1.324 [1.245, 1.403]	€ 1.305 [1.173, 1.436]
Impact on nature and landscape	from small to strong impact	€ -13.508 [-13.907, -13.110]	€ -7.612 [-8.052, -7.173]
Distance	per 5 km	€ 0.325 [0.304, 0.346]	€ 0.483 [0.430, 0.537]
95 % confidence intervals in parentheses			

Source: OWN CALCULATIONS AND REPRESENTATION

11.2.2 Alternative-specific parameter estimates

In the previous section a choice model with generic parameter estimates was presented. Hence, the impacts of renewable energy investments on employment, CO_2 reduction, the environment, distance and price increases were assumed to be equal across technologies. However, this may be an inappropriate assumption since people can be expected to perceive the impacts of various renewable energy sources differently. In order to test this, a second model was estimated using alternative-specific parameters. This means that the parameter estimates are specific to each alternative, i.e. technology (HENSHER ET AL., 2005:311). The results are shown in Table 85.[244] For each technology (biomass, hydropower, solar and wind power) specific parameters have been estimated.

The alternative specific constant (ASC) is highly significant for each technology indicating that respondents have some inherent propensity to choose for the expansion of biomass, hydropower, solar or wind power over the opt-out for reasons that are not captured in the estimated model. Job creation is a significant determinant of choice for the technologies biomass, hydropower and wind power. By contrast, the creation of additional jobs does not influence consumer utility for the expansion of solar power. The reduction of climate-damaging CO_2 emissions creates a significant impact on consumer utility when referring to biomass, hydropower or solar power expansion. The choice of wind power expansion is, however, not affected by the attainable level of CO_2 reductions. A consistent result is given with respect to environmental impacts. Accordingly, a strong impact on landscape and natural environment creates a disutility within the expansion of each of the four technologies. Distance between the next power plant and respondent's home has a significant influence on indirect consumer utility for solar power expansion. For the other technologies, confirmation of the "Not in my backyard" theory could not be found indicating that people do not care whether a biomass plant, hydropower facility or wind park is built

---

[244] An EC model was used to estimate the alternative-specific parameters. Due to the fact that the EC model appeared to be the most appropriate specification in the generic parameter case, it can be assumed that the EC estimation is a valid approach also for a model with alternative-specific parameters.

near their home. Finally, price increases have a negative effect in each of the four utility specifications reflecting standard economic theory. Thus, people are sensitive to price increases meaning that they principally prefer an expansion of renewable energy, but the electricity must be provided at a low cost. This relationship applies to every technology.

Table 85: Model estimates alternative-specific parameters – renewable energy

Variable	Biomass	Hydropower	Solar power	Wind power
Dependent variable	\multicolumn{4}{c}{Choice: Alternative 1, 2 or none of the two}			
ASC	3.254***	3.955***	4.349***	3.806***
	(0.000)	(0.000)	(0.000)	(0.000)
Jobs	0.0012*	0.0014**	-0.0004	0.0011*
	(0.085)	(0.031)	(0.477)	(0.085)
CO_2 reduction	0.030***	0.020***	0.016***	0.006
	(0.000)	(0.002)	(0.006)	(0.338)
Nature (strong impact)	-1.555***	-1.061***	-0.989***	-1.181***
	(0.000)	(0.000)	(0.000)	(0.000)
Distance	-0.001	-0.003	0.031*	0.016
	(0.949)	(0.876)	(0.073)	(0.367)
Cost	-0.192***	-0.166***	-0.159***	-0.142***
	(0.000)	(0.000)	(0.000)	(0.000)
Std. Dev. Jobs	0.0002	0.0019	0.0009	0.0004
	(0.950)	(0.226)	(0.623)	(0.842)
Std. Dev. CO_2	0.020**	0.011	0.025***	0.011
	(0.019)	(0.258)	(0.000)	(0.178)
Std. Dev. Nature	2.454***	2.277***	2.208***	2.005***
	(0.000)	(0.001)	(0.000)	(0.000)
Std. Dev. Distance	0.033	0.065***	0.034*	0.025
	(0.281)	(0.002)	(0.097)	(0.367)
Std. Dev. Random effect (error component)	3.106***	3.106***	3.106***	3.106***
	(0.000)	(0.000)	(0.000)	(0.000)
Log likelihood	\multicolumn{4}{c}{-1,394.102}			
McFadden Pseudo R^2	\multicolumn{4}{c}{0.517}			
χ^2 (p-value)	\multicolumn{4}{c}{2,986.5 (0.000)}			
Number of respondents	\multicolumn{4}{c}{299}			
Number of observations	\multicolumn{4}{c}{1,794}			

p-values in parentheses
Significance: *** 1 % level   ** 5 % level   * 10 % level

Source: OWN CALCULATIONS AND REPRESENTATION

According to the model results presented in Table 85, the statistical significance of parameter estimates was found to differ between technologies. Although this result provides an important piece of information, it would be of greater interest to reveal whether significant effects distinguish in magnitude between technologies. For that purpose, several Wald tests for linear parameter restrictions have been applied. The comparison structure used for the Wald tests corresponds to that shown in Table 82. Accordingly, the derived alternative-specific parameter estimates presented above were compared across all possible combinations of technologies. Under the null hypothesis the corresponding parameters are equal. The results of the implemented Wald tests are given in Table 86.

Table 86: Wald tests comparing technology-specific parameter estimates

Linear restrictions (H_0)			Wald statistics			
	ASC	β Jobs	β CO2	β Nature	β Distance	β Cost
Hydropower – Biomass = 0	1.293 (0.196)	0.209 (0.834)	-0.926 (0.354)	1.116 (0.264)	-0.059 (0.953)	0.752 (0.452)
Hydropower – Solar power = 0	-0.735 (0.463)	1.982** (0.048)	0.536 (0.592)	-0.177 (0.860)	-1.401 (0.161)	-0.217 (0.828)
Hydropower – Wind power = 0	0.258 (0.796)	0.298 (0.765)	1.401 (0.161)	0.280 (0.779)	-0.736 (0.461)	-0.733 (0.464)
Biomass – Solar power = 0	-1.909* (0.056)	1.627 (0.104)	1.518 (0.129)	-1.333 (0.183)	-1.209 (0.227)	-0.921 (0.357)
Biomass – Wind power = 0	-0.898 (0.369)	0.084 (0.933)	2.266** (0.024)	-0.817 (0.414)	-0.628 (0.530)	-1.370 (0.171)
Solar power – wind power = 0	0.882 (0.378)	-1.675* (0.094)	0.979 (0.328)	0.434 (0.664)	0.576 (0.565)	-0.531 (0.595)

p-values in parentheses

Rejection of null hypothesis: ***at 1 % level   **at 5 % level   *at 10 % level

Source: OWN CALCULATIONS AND REPRESENTATION

As can be seen from Table 86, the estimated ASCs do not differ between technologies except for biomass and solar power. Hence, solar power is valued significantly different from biomass. In the previous model using generic parameter estimates, by contrast, technology-specific constants differed from each other. In the current model, differences between technologies may be captured by the technology-specific parameter estimates rather than the ASCs. Thus, the coefficient of the job creation attribute significantly differs between hydro and solar power, as well as solar and wind power. For the comparison of biomass and solar power the Wald statistic is marginally statistically significant indicating that the coefficient of jobs also differs between these two technologies, albeit only at a marginal level. The impact of CO_2 reductions on utility for biomass expansion is significantly different from the impact of reduced CO_2 emissions on utility for increased wind power. No other statistically significant differences between the coefficients on CO_2 reduction were found. Furthermore, the estimated parameters for the impact on nature and landscape do not distinguish between technologies as based on the Wald tests. The same applies to the parameters on distance and cost. Consequently, environmental impacts, distance decay and price increases are perceived equivalently across renewable technologies.

Another possibility for the investigation of people's perceptions of different renewable technologies is to compare MWTP values. For that reason, MWTP measures have been calculated on the basis of the technology-specific parameter estimates presented above. For the four choice attributes, the corresponding MWTP has been simulated for each respondent using the conditional and constrained parameter estimates. Then means, standard deviations and confidence intervals were drawn from these simulations.[245] Overall WTP for the expansion of renewable energy sources, as reflected by the positive constants, was calculated by simply dividing the appropriate ASC by the coefficient of the cost attribute.[246] The outcomes are given in Table 87. MWTP values that are not interpretable due to the sta-

---

[245] This approach is necessary due to unobserved preference heterogeneity given in the sampled population.
[246] The associated standard errors which are needed to determine confidence intervals were estimated by means of the Krinsky and Robb delta method.

tistical insignificance of underlying parameter estimates are marked in the table by an asterisk.

Table 87: MWTP across renewable technologies

Attribute	Biomass	Hydropower	Solar power	Wind power
ASC	€ 16.940 [11.471, 22.409]	€ 23.771 [17.497, 30.045]	€ 27.303 [20.199, 34.407]	€ 26.879 [18.996, 34.761]
Jobs (per 100)	€ 0.548 [0.486, 0.609]	€ 0.760 [0.677, 0.843]	€ -0.225* [-0.250, -0.200]	€ 0.765 [0.682, 0.849]
CO_2 reduction (per 10 %)	€ 1.346 [1.210, 1.483]	€ 0.827 [0.738, 0.917]	€ 1.156 [1.036, 1.276]	€ 0.731* [0.656, 0.806]
Nature (from small to strong impact)	€ -8.015 [-8.479, -7.551]	€ -6.618 [-7.033, -6.203]	€ -5.348 [-5.668, -5.028]	€ -8.499 [-9.022, -7.977]
Distance (per 5 km)	€ -0.094* [-0.104, -0.083]	€ -0.271* [-0.300, -0.242]	€ 1.025 [0.913, 1.137]	€ 0.312* [0.280, 0.345]

95 % confidence intervals in parentheses
*underlying coefficient estimate was statistically not significant

Source: OWN CALCULATIONS AND REPRESENTATION

First, people are on average willing to pay a significantly positive amount for the expansion of biomass, hydropower, solar and wind power. This can be obtained from the first row of the table below. However, since 95 % confidence intervals overlap we cannot conclude that overall WTP is higher or lower for one technology compared to another. Job creation was a statistically significant determinant of consumer utility for biomass, hydropower and wind power. Looking at MWTP, it can be shown that the creation of 100 additional jobs is valued equally for hydro and wind power. In comparison, employment effects associated with the expansion of biomass are valued significantly less as shown by the non-overlap of confidence intervals. Differences between technologies were also found with respect to the valuation of reduced CO_2 emissions. MWTP for a 10 % emission reduction obtained from an expansion of biomass is approximately € 1.3 per month; for reduced air emissions from solar power expansion MWTP, is € 1.2. These two values can be assumed to be equal from a statistical point of view since 95 % confidence intervals overlap. By contrast, the expansion of hydropower is thought to contribute less to the reduction of CO_2 emissions. This is why MWTP for a 10 % reduction is significantly lower with only € 0.8 per month. Each of the four renewable technologies causes impacts on the landscape and natural environment. These impacts are consistently valued negatively implying a disutility. In monetary terms, the disutility is highest for biomass and wind power whereas MWTP does not differ between the two technologies. Strong environmental impacts are significantly less of a concern when hydro or solar power are extended. Thereby, MWTP differs from each other, with hydropower creating a significantly higher disutility as compared to solar power. Finally, the "Not in my backyard" phenomenon was found to be a significant concern only in case of solar power expansion. Hence, people are willing to pay about € 1.0 on top of their monthly electricity bill for solar power plants to be placed 5 km further away from their home.

In the previous remarks we have seen that people's valuation of the multiple impacts of renewable energy expansion differs at least partly across technologies. Hence, it actually does matter which renewable energy is expanded since people perceive the various technologies differently. This is certainly an important result. However, it would be of further interest to find out whether the impacts of hydropower expansion are valued differently when framed in the context of other renewable energy sources. For that reason, MWTP values for hydropower resulting from the technology-specific parameter estimates of the labelled choice model have been compared with the MWTP measures obtained from the unlabelled CE presented in chapter 9.3. This comparison is given in Table 88.

The a priori expected result that hydropower is valued less when framed in the context of the expansion of other renewable energy sources was largely confirmed. First, overall WTP for hydropower expansion is € 33.1 per household and month in the unlabelled and € 23.8 in the labelled CE. This suggests that when considered in combination with other renewable energy sources hydropower expansion is valued significantly less. The opposite applies to the MWTP for job creation. Accordingly, the creation of 100 additional jobs is valued higher when hydropower expansion is framed in the context of other renewable energy sources. Hence, the involvement of other renewable energy sources leads to an appreciation of hydropower-related job creation. Looking at MWTP for reducing CO_2 emissions, it can be seen that the values differ significantly depending on whether hydropower was considered alone or in combination with other renewable energy sources. With the sole focus lying on hydropower expansion, a 10 % CO_2 emission reduction is valued with € 1.3 per month. However, if people are made aware of other renewable energy sources, the perceived contribution of hydropower expansion to the reduction of CO_2 emissions diminishes. Negative environmental impacts are considered significantly less of a concern when hydropower expansion is framed in the context of alternative renewable energy sources. Hence, if people are aware of the environmental impacts other renewable technologies cause, the impacts of hydropower on landscape and natural environment are seen less hazardous. Consequently, considering the expansion of hydropower in the broader context of renewable energy reduces the perceived energy-water trade-off. Finally, the NIMBY phenomenon occurred only with the exclusive focus on hydropower expansion, while distance between the next hydropower station and respondents' home appeared to be an insignificant determinant of utility in the labelled CE. Thus, when people are conscious of the fact that biomass, solar power plants or wind parks could be built close to their homes, the construction of a hydropower plant is seen less of a problem. This result may be attributable to the familiarity of respondents with the hydropower technology as compared to the unfamiliarity with the "new" renewable energies.

Table 88: MWTP for hydropower expansion from unlabelled and labelled CE

Attribute	Hydropower from unlabelled CE	Hydropower from labelled CE
ASC	€ 33.075 [31.721, 34.428]	€ 23.771 [17.497, 30.045]
Jobs (per 100)	€ 0.229 [0.215, 0.244]	€ 0.760 [0.677, 0.843]
CO_2 reduction (per 10 %)	€ 1.324 [1.245, 1.403]	€ 0.827 [0.738, 0.917]
Nature (from small to strong impact)	€ -13.508 [-13.907, -13.110]	€ -6.618 [-7.033, -6.203]
Distance (per 5 km)	€ 0.325 [0.304, 0.346]	€ -0.271* [-0.300, -0.242]

95 % confidence intervals in parentheses
*underlying coefficient estimate was statistically not significant

Source: OWN CALCULATIONS AND REPRESENTATION

12 Concluding remarks

Renewable energy sources play a significant role, particularly in the light of climate and energy policy objectives. These objectives are mainly determined by international agreements. Hence, the Kyoto Protocol sets binding target to reduce overall GHG emissions over the five-year period 2008-2012 by 13 % compared to the base year 1990. According to the climate and energy package of the European Union, Austria is bound to reduce its GHG emissions (beyond the emission trading system) by 16 % until 2020 compared to the year 2005. Additionally, under the EU package, Austria is committed to increase its share of renewable energy sources in gross final energy consumption to 34 % by 2020. Beside the increase of energy efficiency and long-term security of supply, the intensified use of renewable energy sources in electricity generation, the heating and the transport sector can make an important contribution to achieve the climate and energy policy targets.

Due to the natural and environmental conditions, the Austrian electricity sector is largely based on renewable energy sources. More than half (about 57 %) of gross domestic electricity generation comes from hydroelectric power. Further 10 % stem from other renewable energy sources, mainly biomass and wind power. Despite the high share of renewable energy sources, there is still enormous development potential, especially in the field of hydropower. However, the use of hydropower is associated with a conflict of interest. On the one hand, the technology is highly reliable, extremely energy-efficient and relatively cheap in production compared to other renewable energy sources. Further positive effects arising from the use of hydropower involve the emission-free generation of electricity and the associated CO_2 avoidance. In addition, hydropower investments can have positive impacts on the local economy (especially employment effects) and contribute to domestic energy security. This equally applies for the use of other renewable energy sources like biomass, wind power or photovoltaics. However, on the other hand the use of green technologies is subject to some disadvantages. Renewable energy facilities raise an issue of social acceptance. They are often seen as a blot on the landscape and a threat for the ecosystem. Important environmental concerns related to the operation of hydropower plants are the visual impact of a power plant on landscape, erosion, sedimentation and oxygen-deficiency problems due to the alteration of the water flow in the river, and correspondingly, the impacts of these changes on fish and other water-dependent wildlife. Hence, the expansion of hydropower utilisation, or in the most general sense renewable energy sources, is between the priorities of climate protection, energy generation and the preservation of nature.

The positive and negative effects associated with the intensified use of hydropower or renewable energy in general can be seen as externalities that need to be taken into account when investing in renewable technologies. However, the quantification of externalities like reduced air emissions or negative environmental impacts, is a rather difficult task since these effects are not traded in commercial markets, i.e. have no market price. For that reason, the creation of hypothetical markets is required in order to gain information on public preferences for the multiple impacts of renewable energy respectively hydropower expansion. This is usually done within the scope of direct evaluation methods or surveys. In the current investigation three different surveys have been conducted so as to evaluate the positive and negative externalities of future renewable energy investments. The main survey is related to hydropower expansion in Austria. The second survey refers to a specific hydropower project in the province of Styria and finally, hydropower expansion was considered in comparison with other renewable energy source, namely biomass, solar power (photovoltaics) and wind power.

The statistical analyses of the main hydropower sample revealed that there is a general agreement among the Austrian population upon the need to prospectively increase the use of renewable energy sources. Renewable energy use is considered as an important part of a sustainable and future-oriented energy policy. However, different renewable technologies are associated with differing degrees of popularity. Accordingly, solar power (photovoltaics) and hydropower appeared to be the most preferred renewable energy sources for future expansion followed by wind power. Biomass is, by contrast, the least preferred renewable energy source which may be attributable to the fact that the biomass technology is generally less perceived as renewable. The hydropower technology, however, has a long tradition in Austria. Hence, people have in principle a positive attitude towards hydropower use, as well as the construction of new plants along Austrian rivers. However, Austrian citizens are badly informed about the technology itself and the plans to exploit the remaining potentials, i.e. building new hydropower facilities. By means of an econometric model, the main determinants of people's attitude towards hydropower expansion have been analysed. First, public attitude depends on standard socio-economic variables. So, elderly people are generally less likely to exhibit a positive attitude towards intensified hydropower use. The same applies to higher educated people and high income households. Furthermore, people's attitude is determined by their current experience with the technology and the perceived impacts that are associated with hydropower use. For instance, people that live currently close to a hydropower station have a less positive attitude towards hydropower expansion than people living further away. By contrast, respondents who strongly perceive the positive impacts of hydropower utilisation are more likely to have a positive attitude.

Moreover, it was found that people are generally willing to pay (WTP) for green electricity from increased hydropower use. This is the result of a direct WTP question, asking people how much they would be willing to pay on top of their monthly electricity bill in order to get green electricity for their household. In relation to current electricity costs, the surcharge people are willing to accept is about 20 %. However, stated WTP depends on socio-demographic characteristics, as well as on people's individual attitude towards hydropower and their experience with the technology. Accordingly, elderly and female respondents are willing to pay less for green electricity from increased hydropower use while household income has a positive influence on stated WTP. The same applies to environmental awareness which affects stated WTP positively. These results are in line with previous research dealing with public WTP for renewable energy. Additionally, several attitudinal variables were found to have a significant impact on stated

WTP. Current experience with the hydropower technology negatively affects people's WTP for green electricity from increased hydropower use. By contrast, people's perception of the positive impacts that are associated with the intensified use of hydropower (e.g. impact on recreational activities or the satisfaction of growing electricity demand) has a positive influence on stated WTP.

Further important results of this work refer to the trade-off between economic and climate-related advantages and the negative environmental side effects future hydropower investments are associated with. For instance, people may be willing to accept the impact of hydropower plants on the ecosystem in exchange for reduced air emissions. With the help of complex choice models capturing preference heterogeneity within the sampled population, this trade-off was identified and quantified. While people exhibit a positive WTP for reduced CO_2 emissions and employment creation, they wish to be compensated for the loss of nature and landscape caused by new hydropower facilities. Hence, strong environmental impacts are associated with a considerable welfare loss, while reduced air emissions and employment affects create welfare gains. Furthermore, confirmation of the "Not in my backyard" theory was found, indicating that people are in general for an expansion of hydropower capacities, but not close to their homes. Finally, Austrians are sensitive to price increases meaning that they principally prefer an expansion of hydropower, but the electricity must be provided at a low cost.

Looking at a specific hydropower project, similar results were found. Accordingly, people are willing to pay for the provision of additional households with green electricity from the planned hydropower station and the extension of possible recreational activities. In contrast, the negative environmental side effects of the hydropower project are associated with a disutility causing a considerable welfare loss. However, in the regional case study the negative environmental impacts appeared to be less of a concern as compared to hydropower expansion in general.

The trade-off between economic and climate-related advantages and the negative environmental side effects has also been identified in the model considering different renewable technologies. Thereby, differences between technologies were found. Job creation is valued highest for hydro and wind power. In addition, biomass and solar power are thought to contribute more to the reduction of CO_2 emissions than hydro and wind power. Impacts on landscape and natural environment are caused by each renewable technology and therefore consistently valued negatively by the Austrian population. However, the disutility is highest for biomass and wind power while strong environmental impacts are significantly less of a concern when hydro or solar power are extended. Finally, the "Not in my backyard" phenomenon was found to be a significant concern only in case of solar power expansion.

Compared to the results of the choice model that focussed solely on hydropower expansion, it can be concluded that the multiple impacts of future hydropower investments are valued significantly different when considered in comparison with other renewable energy sources. Hence, job creation is valued higher when hydropower is considered in combination with other renewable technologies. In contrast, the perceived contribution of hydropower expansion to the reduction of climate-damaging CO_2 emissions diminishes if people are made aware of other renewable energy sources. An important result refers to the perceived energy-water trade-off and indicates that negative environmental impacts are considered significantly less of a concern when hydropower expansion is framed in the context of alternative renewable energy sources. Hence, hydropower is considered as the minor evil when seen in comparison with the environmental impacts that are associated with the development of other renewable energy sources such as biomass, solar or wind power.

All together, this work provides an important insight into Austrian households' perception of intensified hydropower and renewable energy use. By means of the econometric models it was possible to quantify the positive and negative externalities future renewable energy investments are associated with. These external effects need to be taken into account when energy policy decisions are to be made. Hence, this work makes an important contribution to broaden the strategic basis of decision making for the Austrian plan to expand hydropower and renewable energy capacities. The economic values gathered within the scope of this work demonstrate the importance of hydropower and renewable energy use in Austria and prioritise the political decision to open up the remaining potentials.

References

ADAMOWICZ, W., LOUVIERE, J. AND WILLIAMS, M. (1994): Combining Revealed and Stated Preference Methods for Environmental Amenities. In: Journal of Environmental Economics and Management 26, pp. 271-292.

ADAMOWICZ, W., LOUVIERE, J. AND SWAIT, J. (1998a): Introduction to Attribute-Based Stated Choice Methods. Advanis: Edmonton (Canada).

ADAMOWICZ, W., BOXALL, P., WILLIAMS, M. AND LOUVIERE, J. (1998b): Stated Preference Approaches for Measuring Passive Use Values: Choice Experiments and Contingent Valuation. In: American Journal of Agricultural Economics, Vol. 80, No. 1 (1998), pp. 64-75.

ALPIZAR, F., CARLSSON, F. AND MARTINSSON, P. (2001): Using Choice Experiments for Non-Market Valuation. Working Papers in Economics No. 52. Department of Economics: Göteborg University.

ALRIKSSON, S. AND ÖBERG, T. (2008): Conjoint Analysis for Environmental Evaluation. A review of methods and applications. In: Environmental Science and Pollution Research 15 (3), pp. 244-257.

ALVAREZ-FARIZO, B. AND HANLEY, N. (2002): Using conjoint analysis to quantify public preferences over the environmental impacts of wind farms. An example from Spain. In: Energy Policy 30 (2002), pp. 107-116.

ARROW, K., SOLOW, R., PORTNEY, P.R., LEAMER, E.E., RADNER, R. AND SCHUMAN, H. (1993): Report of the NOAA Panel on Contingent Valuation. Available under *http://www.darrp.noaa.gov/library/pdf/cvblue.pdf*. Download 31.07.2012.

BARD, J. (2006): Windkraft, Wasserkraft und Meeresenergie – Technik mit sozialer, ökologischer und ökonomischer Akzeptanz. In: Themen 2006 – Forschung und Innovation für eine nachhaltige Energieversorgung. ForschungsVerbund Sonnenenergie: Berlin, pp. 53-58.

BATEMAN, I., DAY, B.H., GEORGIOU, S. AND LAKE, I. (2006): The aggregation of environmental benefit values: welfare measures, distance decay and total WTP. Discussion Paper No. 114. Centre for Social and Economic Research on the Global Environment: Norwich.

BAUMGART, K. (2005): Bewertung landschaftsrelevanter Projekte im Schweizer Alpenraum – Die Methode der Discrete-Choice-Experimente. Geographica Bernensia: Bern.

BEN-AKIVA, M. AND LERMAN, S.R. (2000): Discrete Choice Analysis. MIT Press: Cambridge, Massachusetts.

BENNETT, J. AND BLAMEY, R. (2001): The Choice Modelling Approach to Environmental Valuation. Edward Elgar: Celtenham.

BERGMANN, A., HANLEY, N. AND WRIGHT, R. (2004): Valuing the Attributes of Renewable Energy Investments. Applied Environmental Economics Conference 2004, University: Glasgow.

BERGMANN, A., COLOMBO, S. AND HANLEY, N. (2008): Rural versus urban preferences for renewable energy developments. In: Ecological Economics 65 (2008), pp. 616-625.

BEVILLE, S. AND KERR, G. (2009): Fishing for more understanding: a mixed logit-error component model of freshwater angler site choice. University: Lincoln.

BGBL – BUNDESGESETZBLATT ÖSTERREICH (1993): Bundesgesetz über die Prüfung der Umweltverträglichkeit (Umweltverträglichkeitsprüfungsgesetz 2000 – UVP-G 2000). BGBl. Nr. 697/1993, as last amended by BGBl. I Nr. 89/2000.

BGBL – BUNDESGESETZBLATT ÖSTERREICH (2011a): Bundesgesetz zur Einhaltung von Höchstmengen von Treibhausgasemissionen und zur Erarbeitung von wirksamen Maßnahmen zum Klimaschutz (Klimaschutzgesetz – KSG). BGBl. I Nr. 106/2011.

BGBL – BUNDESGESETZBLATT ÖSTERREICH (2011b): Bundesgesetz über die Förderung der Elektrizitätserzeugung aus erneuerbaren Energieträgern (Ökostromgesetz 2012 – ÖSG 2012). BGBl. I Nr. 75/2011.

BGBL – BUNDESGESETZBLATT ÖSTERREICH (2012a): Verordnung des Bundesministers für Wirtschaft, Familie und Jugend, mit der der Förderbeitrag für Ökostrom für das Kalenderjahr 2012 bestimmt wird (Ökostromförderbeitragsverordnung 2012). BGBl. II Nr. 226/2012.

BGBL – BUNDESGESETZBLATT ÖSTERREICH (2012b): Verordnung des Bundesministers für Wirtschaft, Familie und Jugend, mit der Preise für die Abnahme elektrischer Energie aus Ökostromanlagen auf Grund von Verträgen festgesetzt werden, zu deren Abschluss die Ökostromabwicklungsstelle im Jahr 2012 verpflichtet ist (Ökostromverordnung 2012 – ÖSVO 2012). BGBl. II Nr. 471/2011.

BIGERNA, S. AND POLINORI, P. (2011): Italian consumers' willingness to pay for renewable energy sources. MPRA Paper No. 34408, University of Munich.

BLIEM, M., FRIEDL, B., BALABANOV, T. AND ZIELINSKA, I. (2011): Energie [R]evolution Österreich 2050. Institute for Advanced Studies: Vienna.

BMLFUW – BUNDESMINISTERIUM FÜR LAND- UND FORSTWIRTSCHAFT, UMWELT UND WASSERWIRTSCHAFT (2005): EU Wasserrahmenrichtlinie 2000/60/EG. Österreichischer Bericht der IST-Bestandsaufnahme. Kurz-Zusammenfassung der Ergebnisse. Vienna.

BMLFUW – BUNDESMINISTERIUM FÜR LAND- UND FORSTWIRTSCHAFT, UMWELT UND WASSERWIRTSCHAFT (2006): Eine Leitlinie für unser Wasser. Die Europäische Wasserrahmenrichtlinie. Vienna.

BMLFUW – BUNDESMINISTERIUM FÜR LAND- UND FORSTWIRTSCHAFT, UMWELT UND WASSERWIRTSCHAFT (2007a): Anpassung der Klimastrategie Österreichs zur Erreichung des Kyoto-Ziels 2008-2012. Vienna.

BMLFUW – BUNDESMINISTERIUM FÜR LAND- UND FORSTWIRTSCHAFT, UMWELT UND WASSERWIRTSCHAFT (2007b): Guter Zustand für unsere Gewässer. Die Umsetzung der europäischen Wasserrahmenrichtlinie. Vienna.

BMLFUW – BUNDESMINISTERIUM FÜR LAND- UND FORSTWIRTSCHAFT, UMWELT UND WASSERWIRTSCHAFT (2010a): EnergieStrategie Österreich. Vienna.

BMLFUW – BUNDESMINISTERIUM FÜR LAND- UND FORSTWIRTSCHAFT, UMWELT UND WASSERWIRTSCHAFT (2010b): Nationaler Gewässerbewirtschaftungsplan 2009 – NGP 2009. Vienna.

BMLFUW – BUNDESMINISTERIUM FÜR LAND- UND FORSTWIRTSCHAFT, UMWELT UND WASSERWIRTSCHAFT (2012): Österreichischer Wasserkatalog. Wasser schützen – Wasser nutzen. Kriterien zur Beurteilung einer nachhaltigen Wasserkraftnutzung. Vienna.

BMWFJ – BUNDESMINISTERIUM FÜR WIRTSCHAFT FAMILIE UND JUGEND (2010a): National Renewable Energy Action Plan 2010 for Austria under Directive 2009/28/EC of the European Parliament and of the Council. Vienna.

BMWFJ – BUNDESMINISTERIUM FÜR WIRTSCHAFT FAMILIE UND JUGEND (2010b): Energy Strategy Austria. Available under *http://www.bmwfj.gv.at/Seiten/Suchergebnisse.aspx?k=Energy%20Strategy%20Austria*. Download 18.09.2012.

BODENHÖFER, H.J., WOHLGEMUTH, N., BLIEM, M., MICHAEL, A. AND WEYERSTRASS, K. (2004): Bewertung der volkswirtschaftlichen Auswirkungen der Unterstützung von Ökostrom in Österreich. Institute for Advanced Studies Carinthia: Klagenfurt.

BÖHRINGER, C., RUTHERFORD, T.F. AND TOL, R. (2009): The EU 20/20/2020 targets: An overview of the EMF22 assessment. In: Energy Economics 31 (2009), pp. 268-273.

BOLLINO, C.A. (2009): The Willingness to Pay for Renewable Energy Sources: The Case of Italy with Socio-demographic Determinants. In: The Energy Journal, Vol. 30, No. 2, pp. 81-96.

BORCHERS, A.M., DUKE, J.M. AND PARSONS, G.R. (2007): Does willingness to pay for green energy differ by source? In: Energy Policy 35 (2007), pp. 3327-3334.

BOXALL, P.C., ADAMOWICZ, W., SWAIT, J., WILLIAMS, M. AND LOUVIERE, J. (1996): A comparison of stated preference methods for environmental valuation. In: Ecological Economics 18 (1996), pp. 242-253.

BROUWER, R., DEKKER, T., ROLFE, J. AND WINDLE, J. (2010): Choice Certainty and Consistency in Repeated Choice Experiments. In: Environmental and Resource Economics (2010) 46, pp. 93-109.

BUCHER, K. (2011): Klimakonferenz in Cancún: Ergebnisse und Ausblick. In: Die Volkswirtschaft – Das Magazin für Wirtschaftspolitik 1/2-2011, pp. 51-52.

BUNGE, T., DIRBACH, D., DREHER, B., FRITZ, K., LELL, O., RECHENBERG, B., RECHENBERG, J., SCHMITZ, E., SCHWERMER, S., STEINHAUER, M., STEUDTE, C. AND VOIGT, T. (2001): Wasserkraftanlagen als erneuerbare Energiequelle. Rechtliche und ökologische Aspekte. Umweltbundesamt: Berlin.

CAMERON, A.C. AND TRIVEDI, P.K. (2010): Microeconometrics Using Stata. Revised Edition. Stata Press: Texas.

CAPROS, P., MANTZOS, L., PAPANDREOU, V. AND TASIOS, N. (2008): Model-based Analysis of the 2008 EU Policy Package on Climate Change and Renewables. Report to the European Commission DG ENV.

CARLSEN, A.J., STRAND, J. AND WENSTOP, F. (1993): Implicit Environmental Costs in Hydroelectric Development: An Analysis of the Norwegian Master Plan for Water Resources. In: Journal of Environmental Economics and Management 25 (1993), pp. 201-211.

CARSON, R.T., FLORES, N.E. AND HANEMANN, M.W. (1998): Sequencing and Valuing Public Goods. In: Journal of Environmental Economics and Management 36 (1998), pp. 314-323.

CARSON, R.T. (1999): Contingent Valuation: A User's Guide. Department of Economics, University of California: San Diego.

CARSON, R.T. AND FLORES, N.E. (2000): Contingent Valuation: Controversies and Evidence. Department of Economics, University of California: San Diego.

CARSON, R.T. (2012): Contingent Valuation: A Practical Alternative when Prices Aren't Available. In: Journal of Economic Perspectives 26 (4), Fall 2012, pp. 27-42.

CASE, K.E., FAIR, R.C., GÄRTNER, M. AND HEATHER, K. (1999): Economics. Prentice Hall: Europe.

CHEE, Y.E. (2004): An ecological perspective on the valuation of ecosystem services. In: Biological Conservation 120 (2004), pp. 549-565.

CHRZAN, K. AND ORME, B. (2000): An Overview and Comparison of Design Strategies for Choice-Based Conjoint Analysis. Sawtooth Software Research Paper Series.

COASE, R.H. (1960): The Problem of Social Cost. In: Journal of Law and Economics, Vol. 3 (Oct. 1960), pp. 1-44.

COMMISSION OF THE EUROPEAN COMMUNITIES (2008): 20 20 by 2020. Europe's climate change opportunity. COM(2008) 30 final: Brussels.

DE JAGER, D., KLESSMANN, C., STRICKER, E., WINKEL, T., DE VISSER, E., KOPER, M., RAGWITZ, M., HELD, A., RESCH, G., BUSCH, S., PANZER, C., GAZZO, A., ROULLEAU, T., GOUSSELAND, P., HENRIET, M. AND BOUILLÉ, A. (2011): Financing Renewable Energy in the European Energy Market. ECOFYS Netherland BV: Utrecht.

DER SPIEGEL (2012): Unep-Bericht – Investitionsrekord bei erneuerbaren Energien. Available under *http://www.spiegel.de/wirtschaft/soziales/unep-bericht-investitionsrekord-bei-erneuerbaren-energien-a-838116.html*. Download 26.06.2012.

DIMITROPOULOS, A. AND KONTOLEON, A. (2009): Assessing the determinants of local acceptability of wind-farm investment: A choice experiment in the Greek Aegean Islands. In: Energy Policy 37 (2009), pp. 1842-1854.

DOBROWOLSKI, P. AND SCHLEICH, U. (2009): Zielobjekt Mur. In: Frontal 14/2009, pp. 10-14.

E3G – CHANGE AGENTS FOR SUSTAINABLE DEVELOPMENT (2009): What the EU Climate Package means for the Global Climate Deal. London.

EC – EUROPEAN COMMISSION (2008a): 20 20 by 2020. Europe's climate change opportunity. COM(2008) 30 final: Brussels.

EC – EUROPEAN COMMISSION (2008b): Proposal for a decision of the European Parliament and of the Council on the effort of Member States to reduce their greenhouse gas emissions to meet the Community's greenhouse gas emission reduction commitments up to 2020. COM(2008) 17 final: Brussels.

EC – EUROPEAN COMMISSION (2010): Emissions Trading System (EU ETS). Available under *http://ec.europa.eu/clima/policies/ets/index_en.htm*. Download 07.09.2012.

EC – EUROPEAN COMMISSION (2011a): A Roadmap for moving to a competitive low carbon economy in 2050. COM(2011) 112 final: Brussels.

EC – EUROPEAN COMMISSION (2011b): Renewable Energy: Progressing towards the 2020 target. COM(2011) 31 final: Brussels.

EC – EUROPEAN COMMISSION (2011c): Progress towards achieving the Kyoto objectives. COM(2011) 624 final: Brussels.

ECORYS – RESEARCH AND CONSULTING (2010): Assessment of non-cost barriers to renewable energy growth in EU Member States – AEON. ECORYS Netherland BV: Rotterdam.

EDER, H. (2009): Regenerative Stromerzeugung in Österreich: Möglichkeiten und Grenzen. 6. Internationale Energiewirtschaftstagung: Vienna.

EEA – EUROPEAN ENVIRONMENT AGENCY (2012): Annual European Union greenhouse gas inventory 1990-2010 and inventory report 2012. EEA Technical report No 3/2012: Copenhagen.

EHRLICH, Ü. AND REIMANN, M. (2010): Hydropower versus Non-market Values of Nature: a Contingent Valuation Study of Jägala Waterfalls, Estonia. In: International Journal of Geology, Volume 4, 2010, pp. 59-63.

EK, K. (2005): Quantifying the Preferences over the Environmental Impacts of Renewable Energy: The Case of Swedish Wind Power. University of Technology: Luleå.

ENDS – ENVIRONMENTAL DATA SERVICES (2010): Renewable Energy Europe. A special report on the National Renewable Energy Action Plans outlining goals and measures to boost renewable energy use. London.

ENERGIE-CONTROL AUSTRIA (2011a): Monitoring Report. Versorgungssicherheit Strom. Vienna.

ENERGIE-CONTROL AUSTRIA (2011b): Ökostrombericht 2011. Bericht der Energie-Control Austria gemäß § 25 Abs 1 Ökostromgesetz. Vienna.

ENERGIE-CONTROL AUSTRIA (2011c): Bestandsstatistik 2010. Available under *http://www.e-control.at/de/statistik/strom/bestandsstatistik/verteilungserzeugungsanlagen2010*. Download 09.08.2012.

ENERGIE-CONTROL AUSTRIA (2012a): Betriebsstatistik 2011. Available under *http://www.e-control.at/de/statistik/strom/betriebsstatistik/betriebsstatistik2011*. Download 09.08.2012.

ENERGIE-CONTROL AUSTRIA (2012b): Bestandsstatistik 2011. Available under *http://www.e-control.at/de/statistik/strom/bestandsstatistik/verteilungserzeugungsanlagen2011*. Download 09.08.2012.

ENERGIE-CONTROL AUSTRIA (2012c): Das Ökostrom-Fördersystem. Available under *http://www.e-control.at/de/konsumenten/oeko-energie/kosten-und-foerderungen/oekostrom-foerdersystem*. Download 01.10.2012.

ENERGIE-CONTROL AUSTRIA (2012d): Fragen und Antworten – Öko-Energie. Available under *http://www.e-control.at/de/konsumenten/oeko-energie/fragen-und-antworten*. Download 01.10.2012.

ENERGIE-CONTROL AUSTRIA (2012e): Entwicklung der Engpassleistung [in MW] jener Ökostromanlagen im Vertragsverhältnis mit Öko-BGV (bzw OeMAG) zum angegebenen Stichtag sowie Vergleich mit anerkannten Ökostromanlagen. Available under *http://www.e-control.at/de/statistik/oeko-energie/anlagenstatistik/engpassleistung-und-vertragsverhaeltnisse*. Download 20.09.2012.

ENERGIE-CONTROL AUSTRIA (2012f): Entwicklung anerkannter sonstiger Ökostromanlagen 2002-2011 (Stichtag jeweils 31.12.). Available under *http://www.e-control.at/de/statistik/oeko-energie/anlagenstatistik/anerkannte-oekostromanlagen*. Download 20.09.2012.

ENERGIE STEIERMARK (2010a): Das Projekt „Murkraftwerk Graz". Available under *http://www.e-steiermark.com/wasserkraft/murkraftwerkgraz/projekt/index.htm*. Download 10.10.2011.

ENERGIE STEIERMARK (2010b): Naherholungsgebiet. Available under *http://www.e-steiermark.com/wasserkraft/murkraftwerkgraz/Umwelt/naherholung.htm*. Download 29.10.2012.

ENERGIE STEIERMARK (2010c): Murkraftwerk Graz. Technik. Available under *http://www.e-steiermark.com/wasserkraft/murkraftwerkgraz/technik/index.htm*. Download 14.11.2012.

ENERGIE STEIERMARK (2010d): Zeitplan und UVP. Available under *http://www.e-steiermark.com/wasserkraft/murkraftwerkgraz/zeitplan/index.htm*. Download 14.11.2012.

ENERGIE STEIERMARK (2010e): Bildergalerie Wasserkraft – Murkraftwerk Graz Siegerentwurf. Available under *http://www.e-steiermark.com/e_data/energie_steiermark/news/bilder/bild.asp? Pfad=Wasserkraft - Murkarftwerk Graz Siegerentwurf*. Download 14.11.2012.

EPC – EUROPEAN PARLIAMENT AND COUNCIL (2000): Directive 2000/60/EC of the European Parliament and of the Council of 23 October 2000 establishing a framework for Community action in the field of water policy. Official Journal of the European Union: Brussels.

EPC – EUROPEAN PARLIAMENT AND COUNCIL (2001): Directive 2001/77/EC of the European Parliament and of the Council of 27 September 2011 on the promotion of electricity produced from renewable energy sources in the internal electricity market. Official Journal of the European Union: Brussels.

EPC – EUROPEAN PARLIAMENT AND COUNCIL (2003): Directive 2003/30/EC of the European Parliament and of the Council of 8 May 2003 on the promotion of the use of biofuels or other renewable fuels for transport. Official Journal of the European Union: Brussels.

EPC – EUROPEAN PARLIAMENT AND COUNCIL (2009): Directive 2009/28/EC of the European Parliament and of the Council of 23 April 2009 on the promotion of the use of energy from renewable sources and amending and subsequently repealing Directives 2001/77/EC and 2003/30/EC. Official Journal of the European Union: Brussels.

EPC – EUROPEAN PARLIAMENT AND COUNCIL (2012): Directive 2012/27/EC of the European Parliament and of the Council of 25 October 2012 on energy efficiency, amending Directives 2009/125/EC and 2010/30/EU and repealing Directives 2004/8/EC and 2006/32/EC. Official Journal of the European Union: Brussels.

EURELECTRIC (2012): CO2 emissions from different power plants in g CO2 equivalent per kWh calculated for the life cycle of the power plant. Available under *http://www.eurelectric.org/powerstats2011/Facts.asp*. Download 18.09.2012.

EUROPEAN BIOFUELS TECHNOLOGY PLATFORM (2012): Biofuel Production. Available under *http://www.biofuelstp.eu/fuelproduction.html*. Download 11.09.2012.

EUROSTAT-DATABASE (2012a): Anteil erneuerbarer Energien am Bruttoendenergieverbrauch. Available under *http://epp.eurostat.ec.europa.eu/tgm/table.do?tab=table&init=1&plugin =1&language=de&pcode=t2020_31*. Download 07.09.2012.

EUROSTAT-DATABASE (2012b): Anteil erneuerbarer Energie am Kraftstoffverbrauch des Verkehrs. Available under *http://epp.eurostat.ec.europa.eu/tgm/table.do?tab=table&init=1&plugin =1&language=de&pcode=tsdcc340*. Download 07.09.2012

EUROSTAT-DATABASE (2012c): Euro/Ecu-Wechselkurse – Jährliche Daten. Available under *http://appsso.eurostat.ec.europa.eu/nui/show.do?dataset=ert_bil_eur_a&lang=de*. Download 17.07.2012.

EUROSTAT-DATABASE (2012d): Treibhausgasemissionen. Available under *http://appsso.eurostat.ec.europa.eu/nui/show.do?dataset=env_air_gge&lang= de*. Download 11.09.2012.

EUROSTAT-DATABASE (2012e): Elektrizitätserzeugung aus erneuerbaren Energiequellen. Available under *http://epp.eurostat.ec.europa.eu/tgm/table.do?tab=table&init=1&plugin =1&language=de&pcode=tsien050*. Download 11.09.2012.

EVANS, J.R. AND MATHUR, A. (2005): The value of online surveys. In: Internet Research Vol. 15 No. 2, pp. 195-219.

FAWCETT, T. (2003). ROC Graphs: Notes and Practical Considerations for Data Mining Researchers. In: HP Labs Tech Report, No. HPL-2003-4. Paolo Alto.

FANINGER, G. (2008): Energie – Perspektiven. Energie: Gestern, Heute ... und Morgen? Berichte aus Energie- und Umweltforschung 30/2008. Bundesministerium für Verkehr, Innovation und Technologie: Vienna.

FIMERELI, E., MOURATO, S. AND PEARSON, P. (2008): Measuring preferences for low-carbon energy technologies in South-East England: the case of electricity generation. ENVECON: London.

GETZNER, M. (2012): The regional importance of free-flowing rivers for recreation. University of Technology: Vienna.

GREENE, W.H. (2002): Econometric Analysis. Fifth Edition. Prentice Hall: New Jersey.

GREENE, W.H. AND HENSHER, D.A. (2005): Heteroscedastic Control for Random Coefficients and Error Components in Mixed Logit. Working paper of the Institute of Transport and Logistics Studies: Sydney.

GREENE, W.H., HENSHER, D.A. AND ROSE, J.M. (2005): Accounting for Heterogeneity in the Variance of Unobserved Effects in Mixed Logit Models. Working paper of the Institute of Transport and Logistics Studies: Sydney.

GROOTHUIS, P.A, GROOTHUIS, J.D. AND WHITEHEAD, J.C. (2007): Green vs. Green: Measuring the Compensation Required to Site Electrical Generation Windmills in a Viewshed. Appalachian State University: Boone.

GRUBER, K.H. (2011): Herausforderungen für die Stromerzeugung durch das neue Ökostromgesetz in Österreich. In: spezial Ökoenergie, Netz und Infrastruktur '11, pp. 44-50.

HÄDER, M. AND SCHULZ, E. (2005): Beschäftigungswirkungen der Förderung erneuerbarer Energien. In: Energiewirtschaftliche Tagesfragen, 55. Jg. (2005) Heft 7, pp. 472-475.

HANEMANN, W.M. (1994): Valuing the Environment Through Contingent Valuation. In: The Journal of Economic Perspectives, Vol. 8, No. 4, pp. 19-43.

HANLEY, N., MACMILLAN, D., WRIGHT, R.E., BULLOCK, C., SIMPSON, I., PARISSON, D. AND CRABTREE, B. (1998a): Contingent Valuation Versus Choice Experiments: Estimating the Benefits of Environmentally Sensitive Areas in Scotland. In: Journal of Agricultural Economics, Vol. 49 (1), pp. 1-15.

HANLEY, N., WRIGHT, R.E. AND ADAMOWICZ, V. (1998b): Using Choice Experiments to Value the Environment. Design Issues, Current Experience and Future Prospects. In: Environmental and Resource Economics 11 (3-4), pp. 413-428.

HANNA, D., RAMSEY, E., MCBRIDE, N. AND BOND, D. (2011): Maximising the Economic Benefits of Renewable Energy. 2011 International Conference on Economics and Finance Research, Singapore, pp. 144-148.

HARRISON, G.W. (1992): Valuing Public Goods with the Contingent Valuation Method: A Critique of Kahneman and Knetsch. In: Journal of Environmental Economics and Management 23, pp. 248-257.

HARTUNG, J., ELPELT, B. AND KLÖSENER, K.-H. (1991): Statistik. Lehr- und Handbuch der angewandten Statistik. Oldenburg Verlag GmbH: München.

HAUSMAN, J.A. (1993): Contingent Valuation. A Critical Assessment. North-Holland: New York.

HAUSMAN, J. (2012): Contingent Valuation: From Dubious to Hopeless. In: Journal of Economic Perspectives 26 (4), Fall 2012, pp. 43-56.

HENSHER, D.A. AND GREENE, W.H. (2002): The Mixed Logit Model: The State of Practice. Working paper of the Institute of Transport and Logistics Studies: Sydney.

HENSHER, D.A., ROSE, J.M. AND GREENE, W.H. (2005): Applied Choice Analysis. A Primer. University Press: Cambridge, UK.

HITE, D., DUFFY, P., BRANSBY, D. AND SLATON, C. (2008): Consumer willingness-to-pay for biopower: Results from focus groups. In: Biomass and Bioenergy 32 (2008), pp. 11-17.

HOLCOMBE, R.G. (1997): A Theory of the Theory of Public Goods. In: Review of Austrian Economics 10, No. 1 (1997), pp. 1-22.

HORLACHER, H.-B. (2007): Wasserkraft – eine unverzichtbare Energiequelle. In: Wissenschaftliche Zeitschrift der Technischen Universität Dresden 56 (2007) Heft 3-4 Energie, pp. 95-99.

HUDEC, M. AND NEUMANN, C. (n.d.): Stichproben & Umfragen. Grundlagen der Stichprobenziehung. Institut für Statistik: University Vienna.

HYNES, S. AND HANLEY, N. (2006): Preservation versus development on Irish rivers: whitewater kayaking and hydro-power in Ireland. In: Land Use Policy 23 (2006), pp. 170-180.

IEA – INTERNATIONAL ENERGY AGENCY (2012): Energy Technology Perspectives 2012. Pathways to a Clean Energy System. Executive Summary. Paris.

IRENA – INTERNATIONAL RENEWABLE ENERGY AGENCY (2012): Renewable Energy Technologies: Cost Analysis Series. Hydropower. IRENA Working Paper Volume 1: Power Sector, Issue 3/5.

IVANOVA, G. (2005): Queensland Consumers' Willingness to Pay for Electricity from Renewable Energy Sources. Australia and New Zealand Society for Ecological Economics conference: Palmerston North.

JENSEN, K., MENARD, J., ENGLISH, B. AND JAKUS, P. (2004): An analysis of the residential preferences for green power – The role of bioenergy. Farm Foundation Conference on Agriculture: June 2004.

JRC – JOINT RESEARCH CENTRE OF THE EUROPEAN COMMISSION (2011): 2011 Technology Map of the European Strategic Energy Technology Plan (SET-Plan). Technology Descriptions. Publications Office of the European Union: Luxembourg.

KAHNEMAN, D. AND KNETSCH, J.L. (1992): Valuing Public Goods: The Purchase of Moral Satisfaction. In: Journal of Environmental Economics and Management 22, pp. 57-70.

KATARIA, M. (2009): Willingness to pay for environmental improvements in hydropower regulated rivers. In: Energy Economics 31 (2009), pp. 69-76.

KEITE, B. (2004): Konfliktherd kleine Wasserkraft. Ein Fisch auf dem Trockenen? In: Ökologisches Wirtschaften 5/2004, pp. 21-22.

KERR, G. AND SHARP, B. (2009): Efficiency benefits of choice model experimental design updating: a case study. Australian Agricultural and Resource Economics Society conference: Queensland.

KLEINWASSERKRAFT ÖSTERREICH (2012a): Wie macht sich der Mensch die Energie des Wassers zunutze? Available under *http://www.kleinwasserkraft.at/index.php?option=com_content&task=blogcategory&id=37&Itemid=54*. Download 20.09.2012.

KLEINWASSERKRAFT ÖSTERREICH (2012b): Kleinwasserkraft Österreich. Available under *http://www.kleinwasserkraft.at/index.php?option=com_content&task=blogcategory&id=65&Itemid=50.* Download 20.09.2012.

KLINGLMAIR, A., BLIEM, M.G., BROUWER, R. AND GRASER, L. (2012): HYDROVAL – Evaluation of Hydropower Energy Development in Austria: Exploring the Energy-Water Nexus using Public Choice Models. Final report. Research project funded by the Climate and Energy Funds of the Austrian Federal Government.

KNÖDLER, M., HIMPEL, K. AND BARBI, K. (2007): Wasser hat Energie – Wasserkraft unter der Lupe. Büro am Fluss e.V.: Plochingen.

KOHLER, U. AND KREUTER, F. (2006): Datenanalyse mit Stata. Oldenburg: Munich.

KOHL, T. (n.d.): Alternative Energien: Vergleich Solar, Wind, Wasserstoff, Wasserkraft, Geothermie, Biomasse und Gezeitenkraft hinsichtlich Ökonomie und Ökologie. Freiburg.

KOUNDOURI, P., KOUNTOURIS, Y. AND REMOUNDOU, K. (2009): Valuing a wind farm construction: A contingent valuation study in Greece. In: Energy Policy 37 (2009), pp. 1939-1944.

KRATZAT, M. AND LEHR, U. (2006): Erneuerbare Energien und Arbeitsplätze in gesamtwirtschaftlicher Betrachtung. In: Themen 2006 – Forschung und Innovation für eine nachhaltige Energieversorgung. ForschungsVerbund Sonnenenergie: Berlin, pp. 144-147.

KREWITT, W. (2002): External Costs of Energy – do the Answers Match the Questions? In: Energy Policy 20 (2002), pp. 839-848.

KRUCK, C. AND ELTROP, L. (n.d.): Stromerzeugung aus erneuerbaren Energien. Eine ökonomische und ökologische Analyse im Hinblick auf eine nachhaltige Energieversorgung in Deutschland. Zentrum für Energieforschung: Stuttgart.

KRUEGER, A.D. (2007): Valuing Public Preferences for Offshore Wind Power: A Choice Experiment Approach. University of Delaware.

KUHFELD, W.F. (1997): Efficient Experimental Designs Using Computerized Searches. Sawtooth Software Research Paper Series.

KU, S-J. AND YOO, S-H. (2010): Willingness to pay for renewable energy investment in Korea: A choice experiment study. In: Renewable and Sustainable Energy Reviews 14 (2010), pp. 2196-2201.

LADENBURG, J. AND DUBGAARD, A. (2007): Willingness to Pay for Reduced Visual Disamenities from Off-Shore Wind Farms in Denmark. In: Energy Policy 35 (2007), pp. 4059-4071.

LANCASTER, K.J. (1966): A New Approach to Consumer Theory. In: The Journal of Political Economy, Vol. 74, No. 2 (Apr. 1966), pp. 132-157.

LANCSAR, E. AND SAVAGE, E. (2004): Deriving welfare measures from discrete choice experiments: inconsistency between current methods and random utility and welfare theory. In: Health Economics Letters, John Wiley & Sons Ltd.

LAND STEIERMARK (2012a): Gemeinde- und Bezirksdaten. Available under *http://www.verwaltung.steiermark.at/cms/ziel/74836096/DE/*. Download 09.05.2012.

LAND STEIERMARK (2012b): Steiermark Wohnbevölkerung am 1.1.2012, Wanderungen 2011. In: Steirische Statistiken Heft 6/2012. Amt der Steiermärkischen Landesregierung: Graz-Burg.

LEE, J.-S. AND YOO, S.H. (2009): Measuring the environmental costs of tidal power plant construction: A choice experiment study. In: Energy Policy 37 (2009), pp. 5069-5074.

LIEBE, U. AND MEYERHOFF, J. (2005): Die monetäre Bewertung kollektiver Umweltgüter. Theoretische Grundlagen, Methoden und Probleme. Working Paper on Management in Environmental Planning 013/2005. Technische Universität: Berlin.

LONGO, A., MARKANDYA, A. AND PETRUCCI, M. (2008): The internalization of externalities in the production of electricity: Willingness to pay for the attributes of a policy for renewable energy. In: Ecological Economics 67 (2008), pp. 140-152.

LOUVIERE, J., HENSHER, D.A. AND SWAIT, J.D. (2000): Stated Choice Methods. Analysis and Applications. Cambridge University Press.

MACLEOD, M., MORAN, D. AND SPENCER, I. (2006): Counting the cost of water use in hydroelectric generation in Scotland. In: Energy Policy 34 (2006), pp. 2048-2059.

MACMILLAN, D., HANLEY, N. AND LIENHOOP, N. (2006): Contingent Valuation: Environmental polling or preference engine? In: Ecological Economics 60 (2006), pp. 299-307.

MAS-COLELL, A., WHINSTON, M. UND GREEN, J. (1995): Microeconomic Theory. First edition. Oxford University Press: New York.

MENEGAKI, A. (2008): Valuation for renewable energy: A comparative review. In: Renewable and Sustainable Energy Reviews 12 (2008), pp. 2422-2437.

MEYERHOFF, J. AND PETSCHOW, U. (1997): Umweltverträglichkeit kleiner Wasserkraftwerke. Zielkonflikte zwischen Klima- und Gewässerschutz. Institut für ökologische Wirtschaftsforschung GmbH: Berlin.

MEYERHOFF, J. AND PETSCHOW, U. (1998): Kleine Wasserkraftwerke im Spannungsfeld zwischen Klima- und Gewässerschutz. Zielkonflikte in der Umweltpolitik. In: Ökologisches Wirtschaften 1/1998, pp. 26-28.

MEYERHOFF, J., OHL, C. AND HARTJE, V. (2008): Präferenzen für die Gestaltung der Windkraft in der Landschaft – Ergebnisse einer Online-Befragung in Deutschland. Working Paper on Management in Environmental Planning 24/2008. Technische Universität: Berlin.

MEYERHOFF, J., OHL, C. AND HARTJE, V. (2010): Landscape externalities from onshore wind power. In: Energy Policy 38 (2010), pp. 82-92.

MITCHELL, R.C. AND CARSON, R.T. (1990): Using Surveys to Value Public Goods: The Contingent Valuation Method. Resources for the Future: Washington D.C.

MONTGOMERY, D.C. (1984): Design and Analysis of Experiments. Second Edition. John Wiley & Sons: Singapore.

NAVRUD, S. AND BRATEN, K.G. (2007): Consumers' Preferences for Green and Brown Electricity: a Choice Modelling Approach. In: Revue d'Économie Politique, Sep/Oct 2007, 5, pp. 795-811.

NEUBARTH, J. (n.d.): Perspektiven des Wasserkraftwerksprojekts Obere Isel aus energiewirtschaftlicher Sicht unter besonderer Berücksichtigung wirtschaftlicher Aspekte. e3 consult OG: Innsbruck.

NOMURA, N. AND AKAI, M. (2004): Willingness to pay for green electricity in Japan as estimated through contingent valuation method. In: Applied Energy 78 (2004), pp. 453-463.

OECD-DATABASE (2012): PPPs and exchange rates. Available under *http://stats.oecd.org/Index.aspx?datasetcode=SNA_TABLE4#*. Download 17.07.2012.

OECD (2012): Prices and purchasing power parities (PPP). Available under *http://www.oecd.org/about/0,3347,en_2649_34347_1_1_1_1_1,00.html*. Download 17.07.2012.

OEHLERT, G.W. (1992): A Note on the Delta Method. In: The American Statistician, Vol. 46, No. 1, pp. 27-29.

OESTERREICHS ENERGIE (2012): Zeit zum Handeln. Der Aktionsplan von Oesterreichs Energie. Vienna.

ORF (2006): Stromverbrauch deutlich über EU-Durchschnitt. Available unter *http://oesv1.orf.at/stories/95903*. Download 06.08.2012.

ORF (2012): Murkraftwerk: UVP-Bescheid ist positive. Available under *http://steiermark.orf.at/news/stories/2546964*. Download 14.09.2012.

ÖSTERREICHISCHER BIOMASSE-VERBAND (2008): 34 Prozent Erneuerbare machbar. EU-Richtlinie für erneuerbare Energien – Konsequenzen für Österreich. Vienna.

ÖSTERREICHISCHER BIOMASSE-VERBAND (2010): Nationaler Aktionsplan für erneuerbare Energie für Österreich, gemäß der Richtlinie 2009/28/EG des Europäischen Parlaments und des Rates, ausgearbeitet durch die Verbände der erneuerbaren Energien. Available under *http://www.biomasseverband.at/uploads/tx_osfopage/Kurzfassung_NAP.pdf*. Download 13.09.2012.

OTT, W., HÜRLIMANN, J. AND LEIMBACHER, J. (2008): Bewertung von Schutz-, Wiederherstellungs- und Ersatzmaßnahmen bei Wasserkraftanlagen. Bundesamt für Energie: Bern.

OWEN, A.D. (2006): Evaluating the Costs and Benefits of Renewable Energy Technologies. In: The Australian Economic Review 39 (2), pp. 207-215.

PABBRUWEE, K. (2006): The future of small hydropower within the European Union. An environmental policy study based on the European Water Framework Directive and The Renewable Energy Directive. University Groningen.

PARK, H.M. (2009): Comparing Group Means: T-tests and One-way ANOVA Using Stata, SAS, R and SPSS. Center for Statistical and Mathematical Computing: Indiana University.

PARKIN, M., POWELL, M. AND MATTHEWS, K. (2005): Economics. Sixth Edition. Pearson Education Limited: Essex.

PEARCE, D.W. AND SECCOMBE-HETT, T. (2000): Economic Valuation and Environmental Decision-Making in Europe. In: Environmental Science & Technology 2000, 34, pp. 1419-1425.

PEARCE, D.W., ÖZDEMIROGLU, E., BATEMAN, I., CARSON, R.T., DAY, B., HANEMANN, M., HANLEY, N., HETT, T., JONES-LEE, M., LOOMES, G., MOURATO, S., SUGDEN, R. AND SWANSON, J. (2002): Economic Valuation with Stated Preference Techniques. Summary Guide. Department for Transport, Local Government and the Regions: London.

PINDYCK, R.S. AND RUBINFELD, D.L. (2003): Mikroökonomie. 5., aktualisierte Auflage. Pearson Education: Munich.

PISTECKY, W. (2010): Murkraftwerk Graz. Einreichprojekt zum UVP-Verfahren. Juni 2010. Ingenieurbüro Pistecky: Vienna.

POWE, N.A., GARROD, G.D. AND MCMAHON, P.L. (2005): Mixing methods within stated preference environmental valuation: choice experiments and post-questionnaire qualitative analysis. In: Ecological Economics 52 (2005), pp. 513-526.

PÖYRY ENERGY (2008): VEÖ Wasserkraftpotentialstudie Österreich. Vienna.

RAGWITZ, M., SCHADE, W., BREITSCHOPF, B., WALZ, R., HELFRICH, N., RATHMANN, M., RESCH, G., FABER, T., PANZER, C., HAAS, R., NATHANI, C., HOLZHEY, M., ZAGAMÉ, P., FOUGEYROLLAS, A., KONSTANTINAVICIUTE, I. (2009): The impact of renewable energy policy on economic growth and employment in the European Union. Summary of the results of the Employ-RES research project. European Commission DG Energy and Transport: Brussels.

RIPL, W. (2004): Studie zur ökologischen Bewertung von kleinen Wasserkraftanlagen. Systeminstitut Aqua Terra: Berlin.

ROLFE, J., BENNETT, J. AND LOUVIERE, J. (2000): Choice modelling and its potential application to tropical rainforest preservation. In: Ecological Economics 35 (2000), pp. 289-302.

RWI – RHEINISCH-WESTFÄLISCHES INSTITUT FÜR WIRTSCHAFTSFORSCHUNG (2009): Economic impacts from the promotion of renewable energies: The German experience. Essen.

RYAN, M., BATE, A. EASTMOND, C.J. AND LUDBROOK, A. (2001): Use of discrete choice experiments to elicit preferences. In: Quality in Health Care 2001 (10), pp. 55-60.

SAMUELSON, P.A. (1954): The Pure Theory of Public Expenditure. In: The Review of Economics and Statistics, Vol. 36, No. 4 (Nov. 1954), pp. 387-389.

SAMUELSON, P.A. (1955): Diagrammatic Exposition of a Theory of Public Expenditure. In: The Review of Economics and Statistics, Vol. 37, No. 4 (Nov. 1955), pp. 350-356.

SAWTOOTH SOFTWARE (2008): The CBC Advanced Design Module (ADM) Technical Paper. Sawtooth Software Technical Paper Series.

SCHLÄPFER, F. AND ZWEIFEL, P. (2008): Nutzenmessung bei öffentlichen Gütern: Konzeptionelle und empirische Probleme in der Praxis. In: Wirtschaftsdienst 2008 (3), pp. 210-216.

SETCOM – SUSTAINABLE ENERGY IN TOURISM DOMINATED COMMUNITIES (n.d.): Erneuerbare Energieträger Vorteile/Nachteile, Verbreitung Marktreife. Agrar Plus: St. Pölten.

SLOMAN, J., WRIDE, A. AND GARRATT, D. (2012): Economics. Eights Edition. Pearson Education Limited: Essex.

SOLOMON, B.D. AND JOHNSON, N.H. (2009): Valuing climate protection through willingness to pay for biomass ethanol. In: Ecological Economics 68 (2009), pp. 2137-2144.

STANZER, G., NOVAK, S., DUMKE, H., PLHA, S., SCHAFFER, H., BREINESBERGER, J., KIRTZ, M., BIERMAYER, P. AND SPANRING, C. (2010): REGIO Energy – Regionale Szenarien erneuerbarer Energiepotenziale in den Jahren 2012/2020. Research project within the programme ENERGIE DER ZUKUNFT. Vienna/St. Pölten.

STATISTIK AUSTRIA (2011a): Statistisches Jahrbuch 2012. Vienna.

STATISTIK AUSTRIA (2011b): Demographisches Jahrbuch 2010. Vienna.

STATISTIK AUSTRIA (2011c): Demographische Indikatoren für Steiermark 1961 – 2010. Vienna.

STATISTIK AUSTRIA (2012a): Bildung in Zahlen 2010/11 – Tabellenband. Vienna.

STATISTIK AUSTRIA (2012b): Arbeitskräfterhebung 2011. Ergebnisse des Mikrozensus. Vienna.

STEPHENSON, K. AND SHABMAN, L. (2008): The Contribution of Nonmarket Valuation to Policy: The Case of Nonfederal Hydropower. Southern Agricultural Economics Association Annual Meeting: Dallas.

STERNBERG, R. (2008): Hydropower: Dimensions of social and environmental coexistence. In: Renewable and Sustainable Energy Reviews 12 (2008), pp. 1588-1621.

STIGLER, H., HUBER, C., WULZ, C. AND TODEM, C. (2005): Energiewirtschaftliche und ökonomische Bewertung potenzieller Auswirkungen der Umsetzung der EU-Wasserrahmenrichtlinie auf die Wasserkraft. Technische Universität: Graz.

STRUTZMANN, I. (2011): Die Schattenseiten der Erneuerbaren. In: Wirtschaft & Umwelt 2/2011, pp. 18-20.

SUNDQVIST, T. AND SÖDERHOLM, P. (n.d.): The Valuation of Electricity Externalities: A Critical Survey of Past Research Efforts. University of Technology: Luleå.

SUNDQVIST, T. (2002a): Quantifying Household Preferences over the Environmental Impacts of Hydropower in Sweden: A Choice Experiment Approach. University of Technology: Luleå.

SUNDQVIST, T. (2002b): Quantifying Non-Residential Preferences over the Environmental Impacts of Hydropower in Sweden: A Choice Experiment Approach. University of Technology: Luleå.

SWEAT, J. AND LOUVIERE, J. (1993): The Role of the Scale Parameter in the Estimation and Comparison of Multinomial Logit Models. In: Journal of Marketing Research 30 (3), pp. 305-314.

TICHLER, R. AND KOLLMANN, A. (2005): Volkswirtschaftliche Aspekte der Nutzung von Kleinwasserkraft in Österreich. Energie Institut of the Johannes Kepler Universität: Linz.

TRAIN, K.E. (2003): Discrete Choice Methods with Simulation. University Press: Cambridge.

TRUFFER, B. AND BRATRICH, C. (1998): Gehört die Wasserkraft in den Ökostrommix? Wissenschaftliche Grundlagen einer Zertifizierung. EAWAG: Kastanienbaum.

TRUFFER, B., MARKARD, J., BRATRICH, C. AND WEHRLI, B. (2001): Green Electricity from Alpine Hydropower Plants. In: Mountain Research and Development Vol. 21, No. 1, pp. 19-24.

UMWELTBUNDESAMT (2012a): Emissionstrends 1990-2010. Ein Überblick über die österreichischen Verursacher von Luftschadstoffen (Datenstand 2012). Vienna.

UMWELTBUNDESAMT (2012b): Klimaschutzbericht 2012. Vienna.

UMWELTDACHVERBAND (2010): Aktuelle Wasserkraftwerks Projekte der österreichischen E-Wirtschaft (in Planung). Available under
http://www.umweltdachverband.at/presse/presse-detail/?tx_ttnews%5Btt_news%5D=379&cHash=7ea3fc9a3ae47aa70cb744ddfeeb3bab.
Download 01.04.2010.

UNEP (2012): Global Trends in Renewable Energy Investment 2012. Frankfurt School of Finance and Management GmbH – UNEP Collaborating Centre for Climate & Sustainable Energy Finance: Frankfurt.

UNFCC – UNITED NATIONS FRAMEWORK CONVENTION ON CLIMATE CHANGE (2008): Kyoto Protocol Reference Manual. On Accounting of Emissions and Assigned Amount. Bonn.

UNFCC – UNITED NATIONS FRAMEWORK CONVENTION ON CLIMATE CHANGE (2009): Copenhagen Accord. Decision CP.15. Available under
http://unfccc.int/files/meetings/cop_15/application/pdf/cop15_cph_auv.pdf.
Download 07.09.2012.

UNFCCC – UNITED NATIONS FRAMEWORK CONVENTION ON CLIMATE CHANGE (2012): Background on the UNFCCC: The international response to climate change. Available under
http://unfccc.int/essential_background/items/6031.php. Download 07.09.2012.

VAN DER POL, M., SHIELL, A., AU, F., JOHNSTON, D. AND TOUGH, S. (2008): Convergent validity between a discrete choice experiment and a direct, open-ended method: Comparison of preferred attribute levels and willingness to pay estimates. In: Social Science & Medicine 67 (2008), pp. 2043-2050.

VENKATACHALAM, L. (2004): The contingent valuation method: a review. In: Environmental Impact Assessment Review 24 (2004), pp. 89-124.

VEÖ – VERBAND DER ELEKTRIZITÄTSUNTERNEHMEN ÖSTERREICHS (2008): Zukunft Wasserkraft. Masterplan zum Ausbau des Wasserkraftpotenzials. Vienna.

WHITEHEAD, J.C. (2000): A Practitioner's Primer on Contingent Valuation. Department of Economics: East Carolina University.

WISER, R.H. (2007): Using contingent valuation to explore willingness to pay for renewable energy: A comparison of collective and voluntary payment vehicles. In: Ecological Economics 62 (2007), pp. 419-432.

WOOLDRIDGE, J.M (2000): Introductory Econometrics. A Modern Approach. South Western College Publishing.

WURZEL, A. AND PETERMANN, R. (2006): Die Auswirkungen erneuerbarer Energien auf Natur und Landschaft. In: Schriftenreihe des Deutschen Rates für Landespflege 79 (2006).

YOO, S.-H. AND KWAK, S.-Y. (2009): Willingness to pay for green electricity in Korea: A contingent valuation study. In: Energy Policy 37 (2009), pp. 5408-5416.

ZARNIKAU, J. (2003): Consumer demand for "green power" and energy efficiency. In: Energy Policy 31 (2003), pp. 1661-1672.

ZORIC, J. AND HROVATIN, N. (2012): Household willingness to pay for green electricity in Slovenia. In: Energy Policy 47 (2012), pp. 180-187.

Appendix A

Tables

Table A1: Conversion of national WTP values from previous research

Attribute/WTP measure	WTP national currency	PPP (year)	WTP in USD	Exchange rate (year)	WTP in €	Annual value
BERGMANN ET AL. (2006):						
Landscape impact (from high to no impact)	8.10	0.65004 (2003)	12.46	1.1312 (2003)	11.02	44.06
Wildlife impact (from slight increase in harm to improved wildlife)	11.98		18.43		16.29	65.17
Air pollution (from slight increase to no air pollution)	14.13		21.74		19.22	76.86
LONGO ET AL. (2008):						
GHG emission reduction (by 1 %)	29.65	0.65175 (2005)	45.49	1.2441 (2005)	36.57	146.27
Electricity shortage reduction (by 1 minute)	0.36		0.55		0.44	1.78
Employment (per 1 job)	0.02		0.03		0.02	0.10
FIMERELI ET AL. (2008):						
Wind power	99.89	0.64799 (2006)	154.15	1.2556 (2006)	122.77	122.77
Biomass	58.47		90.23		71.86	71.86
Distance (per 400 m = 1 mile)	3.49		5.39		4.29	4.29
Biodiversity (from less to more)	22.80		35.19		28.02	28.02
CO_2 emission reduction (per 1 %)	1.16		1.79		1.43	1.43
NAVRUD AND BRATEN (2007):						
Wind power	1,087.00	9.78424 (2005)	111.10	1.2441 (2005)	89.30	89.30
Hydropower	-2,036.00		-208.09		-167.26	-167.26
Natural gas	-2,160.00		-220.76		-177.45	-177.45
Many small power plants (instead of few large)	-520.00		-53.15		-42.72	-42.72
More medium sized power plants (instead of few large)	-389.00		-39.76		-31.96	-31.96
KU AND YOO (2010):						
Wildlife (per 1 % improvement)	6.85	873.3189 (2006)	0.0078	1.2556 (2006)	0.0062	0.075
Air pollution (per 1 % decrease)	8.40		0.0096		0.0077	0.092
Employment (per 1 job)	10.87		0.0124		0.0099	0.119

Table A1 continued

BORCHERS ET AL. (2007):								
Generic green energy source	17.00			17.00			13.54	162.47
Solar power	21.54	1.00000	(2006)	21.54	1.2556	(2006)	17.16	205.86
Wind power	15.47			15.47			12.32	147.85
Farm methane	12.38			12.38			9.86	118.32
Biomass	10.59			10.59			8.43	101.21
ZORIC AND HROVATIN (2012):								
Mean WTP	4.18	0.68783	(2008)	6.08	1.4708	(2008)	4.13	49.58
ZOGRAFAKIS ET AL. (2010):								
Mean WTP	16.33	0.76542	(2006)	21.33	1.2556	(2006)	16.99	67.97
Median WTP	12.95			16.92			13.47	53.90
BOLLINO (2009):								
WTP – payment card	9.11	0.89566	(2006)	10.17	1.2556	(2006)	8.10	48.60
WTP – referendum	9.39			10.48			8.35	50.10
BIGERNA AND POLINORI (2011):								
Mean WTP	12.16	0.88003	(2007)	13.82	1.3705	(2007)	10.08	60.49
Median WTP	5.05			5.74			4.19	25.12
ZARNIKAU (2003):								
WTP – pre-event	7.17	1.00000	(1999)	7.17	1.0658	(1999)	6.73	80.73
WTP – post-event	6.83			6.83			6.41	76.90
MOZUMDER ET AL. (2011):								
WTP – for 10 % renewable energy	14.22	1.00000	(2011)	14.22	1.3920	(2011)	10.22	122.59
WTP – for increasing from 10 % to 20 %	5.21			5.21			3.74	44.91
IVANOVA (2005):								
WTP – voluntary	27.99	1.44128	(2004)	19.42	1.2439	(2004)	15.61	62.45
WTP – policy aim	21.95			15.23			12.24	48.97
NOMURA AND AKAI (2004):								
Median WTP	2,027.00	176.3761	(2000)	11.49	0.9236	(2000)	12.44	149.32

Table A1 continued

YOO AND KWAK (2009):						
WTP – parametric approach	1,681.00	873.3189 (2006)	1.92	1.2556 (2006)	1.53	18.40
WTP – non-parametric approach	2,072.00		2.37		1.89	22.68
EK (2005):						
Reduced noise level	0.0670		0.0068		0.0072	66.44
Mountain location (compared to on-shore)	-0.0218	9.85410 (2002)	-0.0022	0.9456 (2002)	-0.0023	-21.62
Off-shore location (compared to on-shore)	0.0347		0.0035		0.0037	34.41
Small wind parks (compared to separated located windmills)	0.0155		0.0016		0.0017	15.37
Large wind parks (compared to separated located windmills)	-0.0164		-0.0017		-0.0018	-16.26
MEYERHOFF ET AL. (2008):						
Small wind parks (compared to large)	-0.49		-0.57		-0.39	-4.65
Medium wind parks (compared to large)	0.70		0.81		0.55	6.64
Height (from 200 m to 110 m)	-1.66	0.86027 (2008)	-1.93	1.4708 (2008)	-1.31	-15.74
Environmental impacts (from medium to small)	5.11		5.94		4.04	48.46
Environmental impacts (from medium to large)	-6.72		-7.81		-5.31	-63.73
Distance (from 750 m to 1,100 m)	0.60		0.70		0.47	5.69
MEYERHOFF ET AL. (2010):						
Westsachsen						
Wind park size (from large to medium)	-0.68		-0.79		-0.54	-6.45
Red kite population (from 10 % to 5 % reduction)	2.01		2.34		1.59	19.06
Red kite population (from 10 % to 15 % reduction)	-2.49		-2.89		-1.97	-23.62
Distance (from 750 m to 1,100 m)	3.87	0.86027 (2008)	4.50	1.4708 (2008)	3.06	36.70
Distance (from 750 m to 1,500 m)	4.51		5.24		3.56	42.77
Nordhessen						
Wind park size (from large to medium)	-1.62		-1.88		-1.28	-15.36
Red kite population (from 10 % to 5 % reduction)	2.86		3.32		2.26	27.12
Red kite population (from 10 % to 15 % reduction)	-2.14		-2.49		-1.69	-20.30
Distance (from 750 m to 1,100 m)	5.22		6.07		4.13	49.51
Distance (from 750 m to 1,500 m)	5.86		6.81		4.63	55.58

Table A1 continued

LADENBURG AND DUBGAARD (2007):						
Overall model						
Distance from shore (12 km compared to 8 km)	46.00		52.87		38.58	38.58
Distance from shore (18 km compared to 8 km)	96.00	0.87006 (2007)	110.34	1.3705 (2007)	80.51	80.51
Distance from shore (50 km compared to 8 km)	122.00		140.22		102.31	102.31
Wind farms within sight						
Distance from shore (12 km compared to 8 km)	264.00		303.43		221.40	221.40
Distance from shore (18 km compared to 8 km)	407.00		467.78		341.32	341.32
Distance from shore (50 km compared to 8 km)	484.00		556.28		405.90	405.90
ALVAREZ-FARIZO AND HANLEY (2002):						
Protection of cliffs	21.52	0.76141 (1998)	28.26	1.1211 (1998)	25.21	25.21
Protection of fauna and flora	37.80		49.64		44.28	44.28
Protection of landscape	37.03		48.63		43.38	43.38
DIMITROPOULOS AND KONTOLEON (2009):	*WTA*		*WTA*		*WTA*	*WTA*
Height (from 90 m to 50 m)	−509.27	0.77091 (2007)	−660.60	1.3705 (2007)	−482.02	−482.02
Location out of a Natura 2000 area	−870.44		−1,129.11		−823.87	−823.87
Cooperation with municipalities	−1,050.89		−1,363.18		−994.66	−994.66
Wind farm size (per wind turbine)	45.84		59.46		43.38	43.38
KRUEGER (2007):						
Distance (from 0.9 to 3.6 miles)	16.96	1.00000 (2006)	16.96	1.2556 (2006)	13.51	162.09
Distance (from 0.9 to 6 miles)	23.21		23.21		18.49	221.82
Distance (from 0.9 to 9 miles)	28.17		28.17		22.44	269.23
Distance (from 0.9 to 20 miles)	37.93		37.93		30.21	362-50
KOUNDOURI ET AL. (2009):						
Mean WTP	7.73	0.77091 (2007)	10.03	1.3705 (2007)	7.32	43.90
Median WTP	9.00		11.67		8.52	51.11
GROOTHUIS ET AL. (2007):						
Median WTA	1.90	1.00000 (2005)	1.90	1.2441 (2005)	1.53	18.33

Table A1 continued

SUNDQVIST (2002a):						
Erosion and vegetation (-50 %)	0.0147	9.85410 (2002)	0.0015	0.9456 (2002)	0.0016	14.58
River adapted to all fish species	0.0166		0.0017		0.0018	16.46
SUNDQVIST (2002b):						
Erosion and vegetation (-50 %)	0.0141	9.85410 (2002)	0.0014	0.9456 (2002)	0.0015	13.98
Erosion and vegetation (-25 %)	-0.0123		-0.0012		-0.0013	-12.20
River adapted to all fish species	0.0176		0.0018		0.0019	17.45
KATARIA (2009):						
Increased fish stock (by 1 %)	15.70		1.59		1.68	1.68
Improved conditions for bird life	97.30		9.87		10.44	10.44
Species richness (moderate compared to considerably reduced)	229.00	9.85410 (2002)	23.24	0.9456 (2002)	24.58	24.58
Species richness (high compared to considerable reduced)	277.70		28.18		29.80	29.80
Vegetation and erosion (somewhat compared to considerably reduced)	413.40		41.95		44.37	44.37
Vegetation and erosion (high compared to considerably reduced)	371.20		37.67		39.84	39.84
EHRLICH AND REIMANN (2010):		0.61782 (2009)		1.3948 (2009)		
Mean WTP	17.64		28.55		20.47	20.47

Source: OECD-DATABASE, 2012, online; EUROSTAT-DATABASE, 2012c, online; OWN CALCULATIONS

Table A2: Comparison between Logit and Probit model

Variable	LOGIT coefficients	PROBIT coefficients
Dependent variable: attitude expansion		
Constant	2.952*** (0.001)	1.597*** (0.001)
Age	-0.017* (0.095)	-0.009 (0.104)
Gender	0.392 (0.199)	0.202 (0.189)
Tertiary education	-0.684** (0.033)	-0.302* (0.083)
Income	-0.596* (0.090)	-0.341* (0.069)
Hydropower	1.167*** (0.008)	0.584*** (0.006)
Positive impacts index	-1.728*** (0.000)	-0.859*** (0.000)
Negative impacts index	1.272*** (0.000)	0.630*** (0.000)
Distance	-0.935** (0.044)	-0.459* (0.055)
New plant	0.910* (0.058)	0.389 (0.101)
Recreation impact	0.800* (0.081)	0.323 (0.146)
Observations	865	865
Log likelihood	-151.539	-155.395
Wald chi-squared (prob.)	86.6 (0.000)	82.9 (0.000)
Adjusted McFadden Pseudo-R^2	0.325	0.309
AIC	325.078	332.790
BIC	377.468	385.180
Robust p-values in parentheses		
Significance: ***1 % level **5 % level *10 % level		

Source: OWN CALCULATIONS AND REPRESENTATION

Table A3: MWTP from simple calculation – hydropower

Variable	Measurement	MWTP
Hydropower	effect of the ASC	€ 33.075 [31.721, 34.428]
Jobs	per 100 jobs	€ 0.227 [-0.006, 0.460]
CO_2 reduction	per 10 % reduction	€ 0.560 [0.306, 0.814]
Impact on nature and landscape	from small to strong impact	€ -8.929 [-10.143, -7.716]
Distance	per 5 km	€ 0.280 [-0.041, 0.601]

95 % confidence intervals in parentheses

Source: OWN CALCULATIONS AND REPRESENTATION

Table A4: MWTP from simple calculation – partitioned sample hydropower

Variable	Measurement	MWTP Ebill before CE	MWTP Ebill after CE
Hydropower	effect of the ASC	€ 20.880 [17.693, 24.066]	€ 22.496 [19.596, 25.396]
Jobs	per 100 jobs	€ 0.142 [-0.200, 0.484]	€ 0.318 [0.023, 0.613]
CO_2 reduction	per 10 % reduction	€ 1.429 [1.049, 1.810]	€ 1.140 [0.805, 1.475]
Impact on nature and landscape	from small to strong impact	€ -13.520 [-15.628, -11.413]	€ -11.689 [-13.433, -9.945]
Distance	per 5 km	€ 0.020 [-0.468, 0.512]	€ 0.390 [-0.029, 0.810]

95 % confidence intervals in parentheses

Source: OWN CALCULATIONS AND REPRESENTATION

Table A5: MWTP from simple calculation – regional case study

Variable	Measurement	MWTP
Hydropower plant	effect of the ASC	€ 16.882 [11.544, 22.219]
Households	per 1,000 households	€ 0.223 [0.089, 0.356]
Impact on nature and landscape	from small to strong	€ -8.485 [-11.194, -5.777]
Recreational activities	from restricted to extended recreational activities	€ 2.677 [1.046, 4.308]

95 % confidence intervals in parentheses

Source: OWN CALCULATIONS AND REPRESENTATION

Table A6: MWTP from simple calculation – renewable energy (generic estimates)

Variable	Measurement	MWTP
Hydropower	effect of the ASC	€ 27.159 [22.947, 31.371]
Biomass	effect of the ASC	€ 21.929 [17.684, 26.175]
Solar power	effect of the ASC	€ 29.086 [24.874, 33.297]
Wind power	effect of the ASC	€ 25.582 [21.336, 29.827]
Jobs	per 100 jobs	€ 0.424 [0.111, 0.738]
CO_2 reduction	per 10 % reduction	€ 1.154 [0.761, 1.547]
Impact on nature and landscape	from small to strong impact	€ -7.142 [-8.616, -5.669]
Distance	per 5 km	€ 0.404 [-0.046, 0.854]
95 % confidence intervals in parentheses		

Source: OWN CALCULATIONS AND REPRESENTATION

Table A7: MWTP from simple calculation – renewable energy (alternative-specific estimates)

Attribute	Biomass	Hydropower	Solar power	Wind power
ASC	€ 16.940 [11.471, 22.409]	€ 23.771 [17.497, 30.045]	€ 27.303 [20.199, 34.407]	€ 26.879 [18.996, 34.761]
Jobs (per 100)	€ 0.608 [-0.113, 1.329]	€ 0.824 [-0.029, 1.678]	€ 0.261 [-0.459, 0.981]	€ 0.771 [-0.159, 1.702]
CO_2 reduction (per 10 %)	€ 1.554 [0.767, 2.341]	€ 1.229 [0.395, 2.062]	€ 0.983 [0.197, 1.768]	€ 0.452 [-0.560, 1.464]
Nature (from small to strong impact)	€ -6.376 [-10.062, -2.690]	€ -6.376 [-9.787, -2.965]	€ -6.209 [-9.751, -2.667]	€ -8.340 [-13.092, -3.588]
Distance (per 5 km)	€ -0.032 [-1.072, 1.008]	€ -0.084 [-1.164, 0.997]	€ 0.964 [-0.195, 2.122]	€ 0.569 [-0.740, 1.878]
95 % confidence intervals in parentheses				

Source: OWN CALCULATIONS AND REPRESENTATION

Appendix B

Questionnaire – Hydropower

FRAGEBOGEN – Wasserkraft

Einleitungstext

Wir führen eine Umfrage zum Thema „Wasserkraft in Österreich" durch. Wir würden Sie daher bitten, sich ca. 20 Minuten Zeit zu nehmen, um die folgenden Fragen zu diesem Thema zu beantworten. Alle Ihre Angaben sind anonym und werden streng vertraulich behandelt!

ALLGEMEINE FRAGEN ZUR WASSERKRAFT

1. **Von welchem Anbieter beziehen Sie aktuell Ihren Strom?**

    ☐ BEWAG Burgenland
    ☐ KELAG Kärnten
    ☐ Energie Klagenfurt
    ☐ EVN AG Niederösterreich
    ☐ Energie AG Oberösterreich
    ☐ Linz AG
    ☐ Salzburg AG
    ☐ Energie Steiermark
    ☐ Energie Graz
    ☐ Tiroler Wasserkraft
    ☐ Innsbrucker Kommunalbetriebe
    ☐ Vorarlberger Kraftwerke AG
    ☐ Wien Energie
    ☐ Verbund
    ☐ AAE Naturenergie
    ☐ EVN Naturkraft
    ☐ Ökostrom AG
    ☐ WEB Windenergie
    ☐ Sonstiges: ....

2. **Ist es Ihnen wichtig, dass Ihr Strom aus erneuerbaren Energiequellen wie zum Beispiel Wasserkraft, Windkraft oder Sonnenstrom (Photovoltaik) stammt?**

    ☐ Ja   ⇨ Weiter mit Frage 3!
    ☐ Nein   ⇨ Weiter mit Frage 5!

3. **Beziehen Sie Ihren Strom bewusst von einem Anbieter, der nur Strom aus erneuerbaren Energiequellen liefert?**

    ☐ Ja ⇨ Weiter mit Frage 4!
    ☐ Nein ⇨ Weiter mit Frage 5!

4. **Nehmen Sie dafür einen höheren Strompreis in Kauf?**

    ☐ Ja
    ☐ Nein
    ☐ Weiß nicht

5. **Aus welchen Energiequellen sollte Ihrer Meinung nach der in Zukunft in Österreich benötigte Strom vermehrt erzeugt werden?** *(Mehrfachnennungen möglich)*

    ☐ Erdgas
    ☐ Biomasse
    ☐ Erdöl
    ☐ Sonnenstrom (Photovoltaik)
    ☐ Kohle
    ☐ Wasserkraft
    ☐ Windkraft
    ☐ Atomenergie
    ☐ Sonstiges: ....

6. **Für wie wichtig halten Sie im Allgemeinen das Ziel, die Energiegewinnung aus erneuerbaren Energiequellen wie Wasserkraft, Windkraft oder Sonnenstrom (Photovoltaik) in Zukunft zu erhöhen?**

1 – Sehr wichtig	2 – Eher wichtig	3 – Eher unwichtig	4 – Vollkommen unwichtig
☐	☐	☐	☐

7. **Welche zwei erneuerbaren Energiequellen sollten Ihrer Meinung nach in Österreich am stärksten ausgebaut werden?** *(Bitte ordnen Sie jene zwei Energiequellen von 1 bis 2.)*

    .... Biomasse
    .... Sonnenstrom (Photovoltaik)
    .... Wasserkraft
    .... Windkraft
    .... Sonstiges: ....

8. Wie gut fühlen Sie sich im Allgemeinen über das Thema „Wasserkraft in Österreich" informiert?

1 – Sehr gut	2 – Eher gut	3 – Eher schlecht	4 – Sehr schlecht
☐	☐	☐	☐

9. Wie ist Ihre generelle Einstellung zur Wasserkraftnutzung in Österreich?

1 – Sehr positiv	2 – Eher positiv	3 – Eher negativ	4 – Sehr negativ
☐	☐	☐	☐

10. Haben Sie von dem Plan gehört, die Wasserkraft in Österreich auszubauen, das heißt neue Wasserkraftwerke zu errichten?

☐ Ja
☐ Nein

11. Wie ist Ihre generelle Einstellung zum Bau weiterer Wasserkraftwerke?

1 – Sehr positiv	2 – Eher positiv	3 – Eher negativ	4 – Sehr negativ
☐	☐	☐	☐

12. Wohnen Sie in maximal 10 km Entfernung zu einem Fließgewässer?

☐ Ja ⇨ Weiter mit Frage 13!
☐ Nein ⇨ Weiter mit Frage 14!

13. Bitte geben Sie an, wie oft Sie den folgenden Freizeitaktivitäten an dem zu Ihrem Wohnsitz nächstgelegenen Fließgewässer nachgehen.

	Häufig	Manchmal	Nie
Fischen/Angeln	☐	☐	☐
Schwimmen/Baden	☐	☐	☐
Boot fahren	☐	☐	☐
Spazieren/Wandern entlang des Ufers	☐	☐	☐
Sportliche Aktivitäten (laufen, Rad fahren etc.)	☐	☐	☐
Erholen/die Landschaft genießen	☐	☐	☐
Tierbeobachtung	☐	☐	☐
Picknick am Wasser	☐	☐	☐
Restaurant- oder Cafébesuch	☐	☐	☐
Ausflug mit der Familie	☐	☐	☐

14. **Wie viele Wasserkraftwerke gibt es Ihrer Einschätzung nach in Ihrer Umgebung?**

    ☐ Keine
    ☐ Einige
    ☐ Viele
    ☐ Weiß nicht

15. **Wie weit (Luftlinie) ist das nächste Wasserkraftwerk von Ihrem Wohnsitz entfernt?**

    Entfernung: ca. .... km

16. **Fühlen Sie sich von diesem Wasserkraftwerk positiv, negativ oder gar nicht betroffen?**

    ☐ Positiv betroffen   ⇨ Weiter mit Frage 17!
    ☐ Negativ betroffen   ⇨ Weiter mit Frage 18!
    ☐ Gar nicht betroffen   ⇨ Weiter mit Frage 19!

17. **Warum fühlen Sie sich von dem Wasserkraftwerk positiv betroffen?**
    (Bitte nur eine Möglichkeit ankreuzen!)

    ☐ Weil ich durch das Wasserkraftwerk Strom aus einer sauberen Energiequelle beziehen kann.
    ☐ Weil sich die Landschaft durch das Wasserkraftwerk zum Positiven verändert hat.
    ☐ Weil das Wasserkraftwerk bzw. der Stauraum diverse Freizeitaktivitäten ermöglicht.
    ☐ Weil ich grundsätzlich für die Nutzung der Wasserkraft bin.
    ☐ Sonstiges: ....

18. **Warum fühlen Sie sich von dem Wasserkraftwerk negativ betroffen?**
    (Bitte nur eine Möglichkeit ankreuzen!)

    ☐ Weil ich mich durch das Wasserkraftwerk bei der Ausübung meiner Freizeitaktivitäten gestört fühle.
    ☐ Weil das Wasserkraftwerk das Landschaftsbild verunstaltet.
    ☐ Weil das Wasserkraftwerk negative Auswirkungen auf die Natur (Tier- und Pflanzenwelt) hat.
    ☐ Weil ich grundsätzlich gegen die Nutzung der Wasserkraft bin.
    ☐ Sonstiges: ....

19. **Wird in der Nähe Ihres Wohnsitzes (in einem Umkreis von ca. 10 km) ein neues Wasserkraftwerk gebaut oder ist ein neues Wasserkraftwerk in Planung?**

    ☐ Ja
    ☐ Nein
    ☐ Weiß nicht

20. **Welche Auswirkung hat Ihrer Meinung nach der Bau eines Wasserkraftwerks auf die möglichen Freizeitaktivitäten (z.B. Schimmen/baden, Boot fahren)?**

- ☐ Die Möglichkeiten für Freizeitaktivitäten werden durch den Bau eines Wasserkraftwerks verbessert.
- ☐ Die Möglichkeiten für Freizeitaktivitäten werden durch den Bau eines Wasserkraftwerks verschlechtert.
- ☐ Weiß nicht.

21. **Bitte beurteilen Sie folgende Aussagen.**

	1 – Stimme voll zu	2 – Stimme eher zu	3 – Stimme eher nicht zu	4 – Stimme gar nicht zu
Die verstärkte Wasserkraftnutzung ist wichtig für die Deckung der steigenden Stromnachfrage in Österreich.	☐	☐	☐	☐
Die verstärkte Wasserkraftnutzung ist wichtig für die Reduktion von klimaschädlichen CO_2-Emissionen.	☐	☐	☐	☐
Die verstärkte Wasserkraftnutzung ist wichtig, um die Notwendigkeit von Stromimporten zu reduzieren.	☐	☐	☐	☐
Ein Wasserkraftwerk verunstaltet die Landschaft.	☐	☐	☐	☐
Ein Wasserkraftwerk gefährdet die Lebensräume von Tieren und Pflanzen.	☐	☐	☐	☐

ENTSCHEIDUNGSFRAGEN

Die folgenden Erläuterungen dienen zur Erklärung von Begriffen, die für die Beantwortung der Entscheidungsfragen benötigt werden.

Die verstärkte Nutzung erneuerbarer Energiequellen ist ein wichtiges energiepolitisches Ziel. Derzeit stammen rund 60 % des heimischen Stroms aus Wasserkraft. Trotzdem besteht noch weiteres Ausbaupotenzial. Für den Fall des Baus neuer Wasserkraftwerke, stellen Sie sich bitte vor, dass es für diesen Wasserkraftausbau verschiedene Gestaltungsmöglichkeiten gibt, die sich nach folgenden Eigenschaften unterscheiden:

Zusätzliche Arbeitsplätze:

Durch den Ausbau der Wasserkraft können in Ihrer Region zusätzliche Arbeitsplätze geschaffen werden, und zwar im Ausmaß von...			
10 Arbeitsplätzen	50 Arbeitsplätzen	100 Arbeitsplätzen	500 Arbeitsplätzen

Reduktion der CO₂-Emissionen:

Durch den Ausbau der Wasserkraft können die CO$_2$-Emissionen im Elektrizitätssektor gesenkt werden, und zwar um...			
-10%	-20%	-40%	-60%

Eingriff in Natur und Landschaftsbild: Der Bau neuer Wasserkraftwerke stellt einen Eingriff in die Natur und das Landschaftsbild dar, jedoch kann dieser Eingriff unterschiedlich stark ausfallen.

Gering	Stark
Die Kraftwerke werden so gebaut, dass sie sich gut in das Landschaftsbild einfügen. Die Lebensräume der Tiere und Pflanzen werden nur leicht beeinträchtigt.	Die Kraftwerke beeinflussen das Landschaftsbild stark. Die Lebensräume der Tiere und Pflanzen werden stark beeinträchtigt.

Entfernung zum Wohnsitz:

Ein Ausbau der Wasserkraft erfordert den Bau neuer Wasserkraftwerke. Dabei kann auch in Ihrer Umgebung ein neues Kraftwerk errichtet werden, und zwar in einer Entfernung von...			
2 km	4 km	8 km	20 km

Zusätzliche Stromkosten pro Monat:

Der Bau neuer Wasserkraftwerke ist mit Kosten verbunden, die teilweise von den Stromkunden getragen werden sollen. Monatlich erhöht sich Ihre Stromrechnung daher um...					
€ 3	€ 6	€ 9	€ 12	€ 15	€ 18

Wir würden nun gerne wissen, welche Wasserkraft-Ausbaustrategien Ihnen am meisten zusagen. Zu diesem Zweck stellen wir Ihnen nun 6 Entscheidungsfragen. Bitte betrachten Sie jede Entscheidungsfrage separat und wählen Sie jeweils die Möglichkeit aus, die Sie bevorzugen.

22. Entscheidung 1

23. Entscheidung 2

24. Entscheidung 3

25. Entscheidung 4

26. Entscheidung 5

27. Entscheidung 6

FOLGEFRAGEN

28. Wie schwierig empfanden Sie es, sich bei den vorangegangenen Entscheidungsfragen für eine der Ausbaustrategien zu entscheiden?

1 – Sehr schwierig	2 – Eher schwierig	3 – Eher leicht	4 – Sehr leicht
☐	☐	☐	☐

29. Wie wichtig waren die folgenden Eigenschaften für die Wahl einer Ausbaustrategie bei den vorangegangenen Entscheidungsfragen?

	1 – Sehr wichtig	2 – Eher wichtig	3 – Eher unwichtig	4 – Vollkommen unwichtig
Zusätzliche Arbeitsplätze	☐	☐	☐	☐
Reduktion der CO_2-Emissionen	☐	☐	☐	☐
Eingriff in Natur und Landschaftsbild	☐	☐	☐	☐
Entfernung zum Wohnsitz	☐	☐	☐	☐
Zusätzliche monatliche Zahlung	☐	☐	☐	☐

Die folgende Frage ist nur zu beantworten, wenn in allen Entscheidungen keine der beiden Ausbaustrategien gewählt wurde.

30. Warum haben Sie bei jeder Ihrer Entscheidungen keine der beiden Ausbaustrategien gewählt? *(Mehrfachnennungen möglich)*

   ☐ Ich bin strikt gegen den Ausbau der Wasserkraft.
   ☐ Ich interessiere mich nicht für die Sache.
   ☐ Der derzeitige Zustand ist bereits zufriedenstellend (keine neuen Kraftwerke nötig).
   ☐ Ich kann mir keine zusätzlichen Zahlungen leisten.
   ☐ Die zusätzlichen Zahlungen sind zu hoch.
   ☐ Ich halte andere Sachen für wichtiger.
   ☐ Sonstige Gründe: ....

PERSONENBEZOGENE FRAGESTELLUNGEN

31. Wie viele Personen leben in Ihrem Haushalt (einschließlich Ihnen selbst)?

   .... Personen

32. Wie viele Kinder leben in Ihrem Haushalt?

.... Kinder

33. Wie lässt sich Ihre derzeitige berufliche Situation beschreiben?

☐ Selbstständig beschäftigt
☐ Vollzeitbeschäftigt (mindestens 38 Stunden/Woche)
☐ Teilzeitbeschäftigt (weniger als 38 Stunden/Woche)
☐ Geringfügig beschäftigt
☐ In Ausbildung (Student/in, Schüler/in)
☐ Arbeitslos und Bezieher/in von Arbeitslosengeld
☐ Hausfrau/-mann
☐ Pensionist/in
☐ Sonstiges: ....

34. Was ist Ihre höchste abgeschlossene (formale) Schulbildung? *(Wenn Sie weiterhin in Ausbildung sind, dann geben Sie bitte den höchsten Schulabschluss vor Beginn dieser Ausbildung an.)*

☐ Höchstens Pflichtschule
☐ Lehre/Fachschule
☐ Matura
☐ Pädagogische Hochschule
☐ Universität/Fachhochschule
☐ Sonstiges: ....

35. Wie hoch ist Ihr monatliches <u>Netto-Haushaltseinkommen</u> (nach Steuern und Abgaben)?

☐ bis € 1.000
☐ € 1.001 bis € 1.500
☐ € 1.501 bis € 2.000
☐ € 2.001 bis € 2.500
☐ € 2.501 bis € 3.000
☐ € 3.001 bis € 3.500
☐ € 3.501 bis € 4.000
☐ € 4.001 bis € 4.500
☐ € 4.501 bis € 5.000
☐ mehr als € 5.000

36. Wie hoch ist derzeit Ihre <u>monatliche</u> Stromrechnung?

    ☐ bis € 20
    ☐ € 21 bis € 30
    ☐ € 31 bis € 40
    ☐ € 41 bis € 50
    ☐ € 51 bis € 60
    ☐ € 61 bis € 70
    ☐ € 71 bis € 80
    ☐ € 81 bis € 90
    ☐ € 91 bis € 100
    ☐ mehr als € 100, nämlich: ....

37. Wie genau wissen Sie über die Höhe Ihrer monatlichen Stromrechnung Bescheid?

    ☐ Ich weiß ganz genau wie hoch meine monatliche Stromrechnung ist.
    ☐ Ich kann die Höhe meiner monatlichen Stromrechnung nur grob abschätzen.

38. Wer zahlt in Ihrem Haushalt die Stromrechnung?

    ☐ Ich selbst
    ☐ Eine andere im Haushalt lebende Person
    ☐ Die Kosten werden aufgeteilt

39. Welchen Aufschlag zu Ihrer monatlichen Stromrechnung würden Sie <u>maximal</u> für den weiteren Ausbau der Wasserkraft bezahlen, damit Ihr Haushalt Ökostrom bekommt?

    ....Euro pro Haushalt und Monat

40. Spenden Sie oder irgendjemand anderer in Ihrem Haushalt für Umweltorganisationen?

    ☐ Ja
    ☐ Nein

41. Bitte geben Sie uns zum Schluss noch die Postleitzahl Ihres Wohnortes an.

    PLZ: ....

SCREENING FRAGEN MARKETAGENT

42. **Bitte nennen Sie uns Ihr Geschlecht.**

    ☐ Männlich
    ☐ Weiblich

43. **Wie alt sind Sie?**

    .... Jahre

44. **Bitte geben Sie das Bundesland an, in dem Sie Ihren Hauptwohnsitz haben.**

    ☐ Burgenland
    ☐ Kärnten
    ☐ Niederösterreich
    ☐ Oberösterreich
    ☐ Salzburg
    ☐ Steiermark
    ☐ Tirol
    ☐ Vorarlberg
    ☐ Wien

Vielen Dank für Ihre Mitarbeit!

Appendix C

Questionnaire – Regional hydropower case study

FRAGEBOGEN – Wasserkraft
Regionale Fallstudie Graz-Puntigam

Einleitungstext

Wir führen eine Umfrage zum Thema „Wasserkraft an der Mur" durch. Wir würden Sie daher bitten, sich ca. 20 Minuten Zeit zu nehmen, um die folgenden Fragen zu diesem Thema zu beantworten. Alle Ihre Angaben sind anonym und werden streng vertraulich behandelt!

ALLGEMEINE FRAGEN ZUR WASSERKRAFT

1. **Von welchem Anbieter beziehen Sie aktuell Ihren Strom?**

    ☐ BEWAG Burgenland
    ☐ KELAG Kärnten
    ☐ Energie Klagenfurt
    ☐ EVN AG Niederösterreich
    ☐ Energie AG Oberösterreich
    ☐ Linz AG
    ☐ Salzburg AG
    ☐ Energie Steiermark
    ☐ Energie Graz
    ☐ Tiroler Wasserkraft
    ☐ Innsbrucker Kommunalbetriebe
    ☐ Vorarlberger Kraftwerke AG
    ☐ Wien Energie
    ☐ Verbund
    ☐ AAE Naturenergie
    ☐ EVN Naturkraft
    ☐ Ökostrom AG
    ☐ WEB Windenergie
    ☐ Sonstiges: ....

2. **Ist es Ihnen wichtig, dass Ihr Strom aus erneuerbaren Energiequellen wie zum Beispiel Wasserkraft, Windkraft oder Sonnenstrom (Photovoltaik) stammt?**

    ☐ Ja   ⇨ Weiter mit Frage 3!
    ☐ Nein   ⇨ Weiter mit Frage 5!

3. **Beziehen Sie Ihren Strom bewusst von einem Anbieter, der nur Strom aus erneuerbaren Energiequellen liefert?**

    ☐ Ja ⇨ Weiter mit Frage 4!
    ☐ Nein ⇨ Weiter mit Frage 5!

4. **Nehmen Sie dafür einen höheren Strompreis in Kauf?**

    ☐ Ja
    ☐ Nein
    ☐ Weiß nicht

5. **Aus welchen Energiequellen sollte Ihrer Meinung nach der in Zukunft in Österreich benötigte Strom vermehrt erzeugt werden?** *(Mehrfachnennungen möglich)*

    ☐ Erdgas
    ☐ Biomasse
    ☐ Erdöl
    ☐ Sonnenstrom (Photovoltaik)
    ☐ Kohle
    ☐ Wasserkraft
    ☐ Windkraft
    ☐ Atomenergie
    ☐ Sonstiges: ....

6. **Für wie wichtig halten Sie im Allgemeinen das Ziel, die Energiegewinnung aus erneuerbaren Energiequellen wie Wasserkraft, Windkraft oder Sonnenstrom (Photovoltaik) in Zukunft zu erhöhen?**

1 – Sehr wichtig	2 – Eher wichtig	3 – Eher unwichtig	4 – Vollkommen unwichtig
☐	☐	☐	☐

7. **Welche zwei erneuerbaren Energiequellen sollten Ihrer Meinung nach in Österreich am stärksten ausgebaut werden?** *(Bitte ordnen Sie jene zwei Energiequellen von 1 bis 2.)*

    .... Biomasse
    .... Sonnenstrom (Photovoltaik)
    .... Wasserkraft
    .... Windkraft
    .... Sonstiges: ....

8. Wie gut fühlen Sie sich im Allgemeinen über das Thema „Wasserkraft in Österreich" informiert?

1 – Sehr gut	2 – Eher gut	3 – Eher schlecht	4 – Sehr schlecht
☐	☐	☐	☐

9. Wie ist Ihre generelle Einstellung zur Wasserkraftnutzung in Österreich?

1 – Sehr positiv	2 – Eher positiv	3 – Eher negativ	4 – Sehr negativ
☐	☐	☐	☐

10. Bitte beurteilen Sie folgende Aussagen.

	1 – Stimme voll zu	2 – Stimme eher zu	3 – Stimme eher nicht zu	4 – Stimme gar nicht zu
Die verstärkte Wasserkraftnutzung ist wichtig für die Deckung der steigenden Stromnachfrage in Österreich.	☐	☐	☐	☐
Die verstärkte Nutzung der Wasserkraft ist wichtig für die Reduktion von klimaschädlichen CO_2-Emissionen.	☐	☐	☐	☐
Die verstärkte Wasserkraftnutzung ist wichtig, um die Notwendigkeit von Stromimporten zu reduzieren.	☐	☐	☐	☐
Ein Wasserkraftwerk verunstaltet die Landschaft.	☐	☐	☐	☐
Ein Wasserkraftwerk gefährdet die Lebensräume von Tieren und Pflanzen.	☐	☐	☐	☐

11. Haben Sie von dem Plan gehört, die Wasserkraft an der Mur auszubauen, das heißt neue Wasserkraftwerke an der Mur zu errichten?

☐ Ja
☐ Nein

12. Wie ist Ihre generelle Einstellung zum Bau weiterer Wasserkraftwerke an der Mur?

1 – Sehr positiv	2 – Eher positiv	3 – Eher negativ	4 – Sehr negativ
☐	☐	☐	☐

13. **Bitte geben Sie an, wie oft Sie den folgenden Freizeitaktivitäten an der Mur nachgehen.**

	Häufig	Manchmal	Nie
Fischen/Angeln	☐	☐	☐
Schwimmen/Baden	☐	☐	☐
Boot fahren	☐	☐	☐
Spazieren/Wandern entlang des Ufers	☐	☐	☐
Sportliche Aktivitäten (laufen, Rad fahren etc.)	☐	☐	☐
Erholen/die Landschaft genießen	☐	☐	☐
Tierbeobachtung	☐	☐	☐
Picknick am Wasser	☐	☐	☐
Restaurant- oder Cafébesuch	☐	☐	☐
Ausflug mit der Familie	☐	☐	☐

14. **Ist Ihnen bekannt, dass in Graz-Puntigam ein neues Wasserkraftwerk gebaut werden soll?**

    ☐ Ja ⇨ Weiter mit Frage 15!
    ☐ Nein ⇨ Weiter mit Frage 19!

15. **Wie weit (Luftlinie) wäre dieses geplante Wasserkraftprojekt von Ihrem Wohnsitz entfernt?**

    Entfernung: ca. .... km

16. **Fühlen Sie sich von diesem Kraftwerksprojekt positiv, negativ oder gar nicht betroffen?**

    ☐ Positiv betroffen ⇨ Weiter mit Frage 17!
    ☐ Negativ betroffen ⇨ Weiter mit Frage 18!
    ☐ Gar nicht betroffen ⇨ Weiter mit Frage 19!

17. **Warum fühlen Sie sich von diesem Kraftwerksprojekt positiv betroffen?**
    (Bitte nur eine Möglichkeit ankreuzen!)

    ☐ Weil ich durch das Wasserkraftwerk Strom aus einer sauberen Energiequelle beziehen kann.
    ☐ Weil die Landschaft durch das Wasserkraftwerk zum Positiven verändert wird.
    ☐ Weil das Wasserkraftwerk bzw. der Stauraum diverse neue Freizeitaktivitäten ermöglichen wird.
    ☐ Weil ich grundsätzlich für die Nutzung der Wasserkraft bin.
    ☐ Sonstiges: ....

18. **Warum fühlen Sie sich von diesem Kraftwerksprojekt negativ betroffen?**
    (Bitte nur eine Möglichkeit ankreuzen!)

    ☐ Weil ich mich durch das geplante Kraftwerk bei der Ausübung meiner Freizeitaktivitäten gestört fühle.
    ☐ Weil das Kraftwerk das Landschafts- bzw. Stadtbild verunstalten wird.
    ☐ Weil das Kraftwerk negative Auswirkungen auf die Natur (Tier- und Pflanzenwelt) haben wird.
    ☐ Weil ich grundsätzlich gegen die Nutzung der Wasserkraft bin.
    ☐ Sonstiges: ....

19. **Welche Auswirkung hätte Ihrer Meinung nach der Bau dieses Wasserkraftwerks auf die möglichen Freizeitaktivitäten (z.B. Boot fahren, Radfahren, Spazieren gehen, Fischen)?**

    ☐ Die Möglichkeiten für Freizeitaktivitäten werden durch den Bau des Wasserkraftwerks verbessert.
    ☐ Die Möglichkeiten für Freizeitaktivitäten werden durch den Bau des Wasserkraftwerks verschlechtert.
    ☐ Weiß nicht.

ENTSCHEIDUNGSFRAGEN

Die folgenden Erläuterungen dienen zur Erklärung von Begriffen, die für die Beantwortung der Entscheidungsfragen benötigt werden.

Die verstärkte Nutzung erneuerbarer Energiequellen ist ein wichtiges energiepolitisches Ziel. Derzeit stammen rund 60 % des heimischen Stroms aus Wasserkraft. Trotzdem besteht noch weiteres Ausbaupotenzial und es sollen neue Wasserkraftwerke errichtet werden, darunter auch das besagte Kraftwerk Graz-Puntigam.

Stellen Sie sich vor, dass es für das geplante Wasserkraftprojekt Graz-Puntigam verschiedene Gestaltungsmöglichkeiten gibt, die sich nach folgenden Eigenschaften unterscheiden können.

Stromerzeugung:

Wenn das Kraftwerk Graz-Puntigam gebaut wird, können in Ihrer Region zusätzliche Haushalte mit Strom versorgt werden, und zwar		
5.000 Haushalte	10.000 Haushalte	15.000 Haushalte

Eingriff in Natur und Landschaftsbild: Der Bau des Kraftwerks stellt einen Eingriff in die Natur und das Landschaftsbild dar, jedoch kann dieser Eingriff unterschiedlich stark ausfallen.

Gering	Stark
Das Kraftwerk wird so gebaut, dass es sich gut in das Landschaftsbild einfügt (z.B. Staumauer zum Großteil unter Wasser). Die Lebensräume der Tiere und Pflanzen werden nur leicht beeinträchtigt.	Das Kraftwerk beeinflusst das Landschaftsbild stark (z.B. zur Gänze sichtbare Staumauer). Die Lebensräume der Tiere und Pflanzen werden stark beeinträchtigt.

Freizeitmöglichkeiten:

Erweiterte Freizeitmöglichkeiten	Eingeschränkte Freizeitmöglichkeiten
Im Zuge des Kraftwerksbaus wird ein für die Stadtbewohner nutzbarer Naherholungsraum geschaffen, der viele Möglichkeiten für die Freizeitgestaltung bietet (z.B. Radwege, Paddelschule, Mur-Schifffahrt, Café-Besuch an der Mur,...).	Die durch den Kraftwerksbau veränderte Flusslandschaft kann nicht als Naherholungsraum für die Stadtbewohner genutzt werden. Dadurch sind die Möglichkeiten der Freizeitgestaltung (z.B. Radfahren, Spazieren gehen, Boot fahren,...) eingeschränkt.

Zusätzliche Stromkosten pro Monat:

Der Bau des neuen Wasserkraftwerks ist mit Kosten verbunden, die teilweise von den Stromkunden getragen werden sollen. Monatlich erhöht sich Ihre Stromrechnung daher um...					
€ 3	€ 6	€ 9	€ 12	€ 15	€ 18

Wir würden nun gerne wissen, welche Gestaltungsmöglichkeiten des Kraftwerks Ihnen am meisten zusagen. Zu diesem Zweck stellen wir Ihnen nun 6 Entscheidungsfragen. Bitte betrachten Sie jede Entscheidungsfrage separat und wählen Sie jeweils die Möglichkeit aus, die Sie bevorzugen.

20. Entscheidung 1

21. Entscheidung 2

22. Entscheidung 3

23. Entscheidung 4

24. Entscheidung 5

25. Entscheidung 6

FOLGEFRAGEN

26. Wie schwierig empfanden Sie es, sich bei den vorangegangenen Entscheidungsfragen für eine der Möglichkeiten zu entscheiden?

1 – Sehr schwierig	2 – Eher schwierig	3 – Eher leicht	4 – Sehr leicht
☐	☐	☐	☐

27. Wie wichtig waren die folgenden Eigenschaften für die Wahl einer Möglichkeit bei den vorangegangenen Entscheidungsfragen?

	1 – Sehr wichtig	2 – Eher wichtig	3 – Eher unwichtig	4 – Vollkommen unwichtig
Anzahl der Haushalte, für die Strom erzeugt werden kann.	☐	☐	☐	☐
Eingriff in Natur und Landschaftsbild	☐	☐	☐	☐
Freizeitmöglichkeiten	☐	☐	☐	☐
Zusätzliche Stromkosten pro Monat	☐	☐	☐	☐

Die folgende Frage ist nur dann zu beantworten, wenn in allen Entscheidungen keine der beiden Möglichkeiten gewählt wurde.

28. Warum haben Sie bei jeder Ihrer Entscheidungen keine der beiden Möglichkeiten gewählt?
    (Mehrfachnennungen möglich)

    ☐ Ich bin strikt gegen den Bau des Kraftwerks.
    ☐ Ich interessiere mich nicht für die Sache.
    ☐ Der derzeitige Zustand ist bereits zufriedenstellend (kein neues Kraftwerk nötig).
    ☐ Ich kann mir keine zusätzlichen Zahlungen leisten.
    ☐ Die zusätzlichen Zahlungen sind zu hoch.
    ☐ Ich halte andere Sachen für wichtiger.
    ☐ Sonstige Gründe: ....

PERSONENBEZOGENE FRAGESTELLUNGEN

29. Wie viele Personen leben in Ihrem Haushalt (einschließlich Ihnen selbst)?

    .... Personen

30. Wie viele Kinder leben in Ihrem Haushalt?

    .... Kinder

31. Wie lässt sich Ihre derzeitige berufliche Situation beschreiben?

☐ Selbstständig beschäftigt
☐ Vollzeitbeschäftigt (mindestens 38 Stunden/Woche)
☐ Teilzeitbeschäftigt (weniger als 38 Stunden/Woche)
☐ Geringfügig beschäftigt
☐ In Ausbildung (Student/in, Schüler/in)
☐ Arbeitslos und Bezieher/in von Arbeitslosengeld
☐ Hausfrau/-mann
☐ Pensionist/in
☐ Sonstiges: ....

32. Was ist Ihre höchste abgeschlossene (formale) Schulbildung? *(Wenn Sie weiterhin in Ausbildung sind, dann geben Sie bitte den höchsten Schulabschluss vor Beginn dieser Ausbildung an.)*

☐ Höchstens Pflichtschule
☐ Lehre/Fachschule
☐ Matura
☐ Pädagogische Hochschule
☐ Universität/Fachhochschule
☐ Sonstiges: ....

33. Wie hoch ist Ihr monatliches Netto-Haushaltseinkommen (nach Steuern und Abgaben)?

☐ bis € 1.000
☐ € 1.001 bis € 1.500
☐ € 1.501 bis € 2.000
☐ € 2.001 bis € 2.500
☐ € 2.501 bis € 3.000
☐ € 3.001 bis € 3.500
☐ € 3.501 bis € 4.000
☐ € 4.001 bis € 4.500
☐ € 4.501 bis € 5.000
☐ mehr als € 5.000

34. Wie hoch ist derzeit Ihre <u>monatliche</u> Stromrechnung?

- ☐ bis € 20
- ☐ € 21 bis € 30
- ☐ € 31 bis € 40
- ☐ € 41 bis € 50
- ☐ € 51 bis € 60
- ☐ € 61 bis € 70
- ☐ € 71 bis € 80
- ☐ € 81 bis € 90
- ☐ € 91 bis € 100
- ☐ mehr als € 100, nämlich: ....

35. Wie genau wissen Sie über die Höhe Ihrer monatlichen Stromrechnung Bescheid?

- ☐ Ich weiß ganz genau wie hoch meine monatliche Stromrechnung ist.
- ☐ Ich kann die Höhe meiner monatlichen Stromrechnung nur grob abschätzen.

36. Wer zahlt in Ihrem Haushalt die Stromrechnung?

- ☐ Ich selbst
- ☐ Eine andere im Haushalt lebende Person
- ☐ Die Kosten werden aufgeteilt

37. Welchen Aufschlag zu Ihrer monatlichen Stromrechnung würden Sie <u>maximal</u> für den weiteren Ausbau der Wasserkraft bezahlen, damit Ihr Haushalt Ökostrom bekommt?

........................Euro pro Haushalt und Monat

38. Spenden Sie oder irgendjemand anderer in Ihrem Haushalt für Umweltorganisationen?

- ☐ Ja
- ☐ Nein

39. Bitte geben Sie die Postleitzahl Ihres Wohnortes an.

PLZ: ....

SCREENING FRAGEN MARKETAGENT

40. Bitte nennen Sie uns Ihr Geschlecht.

- ☐ Männlich
- ☐ Weiblich

41. Wie alt sind Sie?

……………………… Jahre

42. Bitte geben Sie das Bundesland an, in dem Sie Ihren Hauptwohnsitz haben.

- ☐ Burgenland
- ☐ Kärnten
- ☐ Niederösterreich
- ☐ Oberösterreich
- ☐ Salzburg
- ☐ Steiermark
- ☐ Tirol
- ☐ Vorarlberg
- ☐ Wien

43. In welcher der folgenden Gemeinden wohnen Sie?

- ☐ Graz
- ☐ Feldkirchen bei Graz
- ☐ Fernitz bei Graz
- ☐ Gössendorf
- ☐ Grambach
- ☐ Hausmannstätten
- ☐ Kalsdorf bei Graz
- ☐ Raaba
- ☐ Seiersberg
- ☐ Deutschfeistritz
- ☐ Eisbach
- ☐ Gratkorn
- ☐ Gratwein
- ☐ Judendorf-Straßengel
- ☐ Peggau
- ☐ Stattegg
- ☐ Weinitzen
- ☐ Hart bei Graz
- ☐ Kainbach bei Graz
- ☐ Thal

Vielen Dank für Ihre Mitarbeit!

Appendix D

Questionnaire – Renewable energy

FRAGEBOGEN – Erneuerbare Energien

Einleitungstext

Wir führen eine Umfrage zum Thema „Erneuerbare Energien in Österreich" durch. Wir würden Sie daher bitten, sich ca. 20 Minuten Zeit zu nehmen, um die folgenden Fragen zu diesem Thema zu beantworten. Alle Ihre Angaben sind anonym und werden streng vertraulich behandelt!

ALLGEMEINE FRAGEN ZU ERNEUERBAREN ENERGIEN

1. **Von welchem Anbieter beziehen Sie aktuell Ihren Strom?**

    ☐ BEWAG Burgenland
    ☐ KELAG Kärnten
    ☐ Energie Klagenfurt
    ☐ EVN AG Niederösterreich
    ☐ Energie AG Oberösterreich
    ☐ Linz AG
    ☐ Salzburg AG
    ☐ Energie Steiermark
    ☐ Energie Graz
    ☐ Tiroler Wasserkraft
    ☐ Innsbrucker Kommunalbetriebe
    ☐ Vorarlberger Kraftwerke AG
    ☐ Wien Energie
    ☐ Verbund
    ☐ AAE Naturenergie
    ☐ EVN Naturkraft
    ☐ Ökostrom AG
    ☐ WEB Windenergie
    ☐ Sonstiges: ....

2. **Ist es Ihnen wichtig, dass Ihr Strom aus erneuerbaren Energiequellen wie zum Beispiel Wasserkraft, Windkraft oder Sonnenstrom (Photovoltaik) stammt?**

    ☐ Ja   ⇨ Weiter mit Frage 3!
    ☐ Nein   ⇨ Weiter mit Frage 5!

3. **Beziehen Sie Ihren Strom bewusst von einem Anbieter, der nur Strom aus erneuerbaren Energiequellen liefert?**

    ☐ Ja ⇨ Weiter mit Frage 4!
    ☐ Nein ⇨ Weiter mit Frage 5!

4. **Nehmen Sie dafür einen höheren Strompreis in Kauf?**

    ☐ Ja
    ☐ Nein
    ☐ Weiß nicht

5. **Aus welchen Energiequellen sollte Ihrer Meinung nach der in Zukunft in Österreich benötigte Strom vermehrt erzeugt werden?** *(Mehrfachnennungen möglich)*

    ☐ Erdgas
    ☐ Biomasse
    ☐ Erdöl
    ☐ Sonnenstrom (Photovoltaik)
    ☐ Kohle
    ☐ Wasserkraft
    ☐ Windkraft
    ☐ Atomenergie
    ☐ Sonstiges: ....

6. **Für wie wichtig halten Sie im Allgemeinen das Ziel, die Energiegewinnung aus erneuerbaren Energiequellen wie Wasserkraft, Windkraft oder Sonnenstrom (Photovoltaik) in Zukunft zu erhöhen?**

1 – Sehr wichtig	2 – Eher wichtig	3 – Eher unwichtig	4 – Vollkommen unwichtig
☐	☐	☐	☐

7. **Welche zwei erneuerbaren Energiequellen sollten Ihrer Meinung nach in Österreich am stärksten ausgebaut werden?** *(Bitte ordnen Sie jene zwei Energiequellen von 1 bis 2.)*

    .... Biomasse
    .... Sonnenstrom (Photovoltaik)
    .... Wasserkraft
    .... Windkraft
    .... Sonstiges: ....

8. **Bitte beurteilen Sie folgende Aussagen.**

	1 –Stimme voll zu	2 –Stimme eher zu	3 –Stimme eher nicht zu	4 –Stimme gar nicht zu
Die verstärkte Nutzung erneuerbarer Energiequellen ist wichtig für die Deckung der steigenden Stromnachfrage in Österreich.	☐	☐	☐	☐
Die verstärkte Nutzung erneuerbarer Energiequellen ist wichtig für die Reduktion von klimaschädlichen CO_2-Emissionen.	☐	☐	☐	☐
Die verstärkte Nutzung erneuerbarer Energiequellen ist wichtig, um die Notwendigkeit von Stromimporten zu reduzieren.	☐	☐	☐	☐

ENTSCHEIDUNGSFRAGEN

Die folgenden Erläuterungen dienen zur Erklärung von Begriffen, die für die Beantwortung der Entscheidungsfragen benötigt werden.

Derzeit werden in Österreich rund 65 % des Stroms aus erneuerbaren Energiequellen (Ökostrom) erzeugt. Um die Stromerzeugung aus erneuerbaren Energiequellen weiter auszubauen, gibt es die folgenden vier Ausbaustrategien:

Ausbau BIOMASSE	Ausbau SONNENSTROM	Ausbau WASSERKRAFT	Ausbau WINDKRAFT
🍃	☀	〰	🌬

Für jede dieser Ausbaustrategien gibt es verschiedene Gestaltungsmöglichkeiten, die sich nach folgenden Eigenschaften unterscheiden:

Zusätzliche Arbeitsplätze:

Reduktion der CO$_2$-Emissionen:

Durch den Ausbau erneuerbarer Energiequellen können die CO$_2$-Emissionen im Elektrizitätssektor gesenkt werden, und zwar um...			
-10%	-20%	-40%	-60%

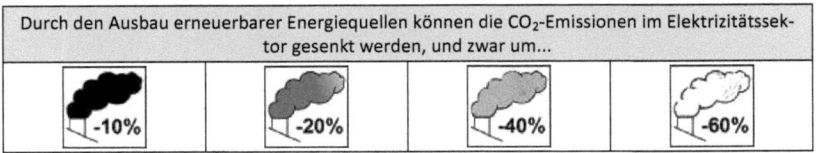

Eingriff in Natur und Landschaftsbild: Der Ausbau erneuerbarer Energiequellen stellt einen Eingriff in die Natur und das Landschaftsbild dar, jedoch kann dieser Eingriff unterschiedlich stark ausfallen.

		Wasserkraft	Windkraft/Sonnenstrom/Biomasse
	Gering	Die Kraftwerke werden so gebaut, dass sie sich gut in das Landschaftsbild einfügen. Die Lebensräume der Tiere und Pflanzen werden nur leicht beeinträchtigt.	Errichtung einzelner kleiner Anlagen mit geringem Einfluss auf das Landschaftsbild. Die Lebensräume der Tiere und Pflanzen werden nur leicht beeinträchtigt.
	Stark	Die Kraftwerke beeinflussen das Landschaftsbild stark. Die Lebensräume der Tiere und Pflanzen werden stark beeinträchtigt.	Errichtung großer Anlagen mit starkem Einfluss auf das Landschaftsbild. Die Lebensräume der Tiere und Pflanzen werden stark beeinträchtigt.

Entfernung zum Wohnsitz:

Der Ausbau erneuerbarer Energiequellen erfordert den Bau neuer Stromerzeugungsanlagen (z.B. Wasserkraftwerke, Windparks). Dabei kann auch in Ihrer Umgebung eine neue Anlage errichtet werden, und zwar in einer Entfernung von...			
2 km	4 km	8 km	20 km

Zusätzliche Stromkosten pro Monat:

Der Ausbau erneuerbarer Energiequellen ist mit Kosten verbunden, die teilweise von den Stromkunden getragen werden sollen. Monatlich erhöht sich Ihre Stromrechnung daher um...					
€ 3	€ 6	€ 9	€ 12	€ 15	€ 18

Wir würden nun gerne wissen, welche Ausbaustrategien zur Erhöhung des Anteils erneuerbarer Energiequellen Ihnen am meisten zusagen. Zu diesem Zweck stellen wir Ihnen nun 6 Entscheidungsfragen. Bitte betrachten Sie jede Entscheidungsfrage separat und wählen Sie jeweils die Möglichkeit aus, die Sie bevorzugen.

9. Entscheidung 1

10. Entscheidung 2

11. Entscheidung 3

12. Entscheidung 4

13. Entscheidung 5

14. Entscheidung 6

FOLGEFRAGEN

15. Wie schwierig empfanden Sie es, sich bei den vorangegangenen Entscheidungsfragen für eine der Ausbaustrategien zu entscheiden?

1 – Sehr schwierig	2 – Eher schwierig	3 – Eher leicht	4 – Sehr leicht
☐	☐	☐	☐

16. Wie wichtig waren die folgenden Eigenschaften für die Wahl einer Ausbaustrategie bei den vorangegangenen Entscheidungsfragen?

	1 – Sehr wichtig	2 – Eher wichtig	3 – Eher unwichtig	4 – Vollkommen unwichtig
Art der erneuerbaren Energiequelle	☐	☐	☐	☐
Zusätzliche Arbeitsplätze	☐	☐	☐	☐
Reduktion der CO_2-Emissionen	☐	☐	☐	☐
Eingriff in Natur und Landschaftsbild	☐	☐	☐	☐
Entfernung zum Wohnsitz	☐	☐	☐	☐
Zusätzliche monatliche Zahlung	☐	☐	☐	☐

Die folgende Frage ist nur zu beantworten, wenn in allen Entscheidungen keine der beiden Ausbaustrategien gewählt wurde.

17. Warum haben Sie bei jeder Ihrer Entscheidungen keine der beiden Ausbaustrategien gewählt? *(Mehrfachnennungen möglich)*

   ☐ Ich bin strikt gegen den Ausbau erneuerbarer Energiequellen.
   ☐ Ich interessiere mich nicht für die Sache.
   ☐ Der derzeitige Zustand ist bereits zufriedenstellend (kein Ausbau nötig).
   ☐ Ich kann mir keine zusätzlichen Zahlungen leisten.
   ☐ Die zusätzlichen Zahlungen sind zu hoch.
   ☐ Ich halte andere Sachen für wichtiger.
   ☐ Sonstige Gründe: ....

PERSONENBEZOGENE FRAGESTELLUNGEN

18. **Wie viele Personen leben in Ihrem Haushalt (einschließlich Ihnen selbst)?**

    .... Personen

19. **Wie viele Kinder leben in Ihrem Haushalt?**

    .... Kinder

20. **Wie lässt sich Ihre derzeitige berufliche Situation beschreiben?**

    ☐ Selbstständig beschäftigt
    ☐ Vollzeitbeschäftigt (mindestens 38 Stunden/Woche)
    ☐ Teilzeitbeschäftigt (weniger als 38 Stunden/Woche)
    ☐ Geringfügig beschäftigt
    ☐ In Ausbildung (Student/in, Schüler/in)
    ☐ Arbeitslos und Bezieher/in von Arbeitslosengeld
    ☐ Hausfrau/-mann
    ☐ Pensionist/in
    ☐ Sonstiges: ....

21. **Was ist Ihre höchste abgeschlossene (formale) Schulbildung?** *(Wenn Sie weiterhin in Ausbildung sind, dann geben Sie bitte den höchsten Schulabschluss vor Beginn dieser Ausbildung an.)*

    ☐ Höchstens Pflichtschule
    ☐ Lehre/Fachschule
    ☐ Matura
    ☐ Pädagogische Hochschule
    ☐ Universität/Fachhochschule
    ☐ Sonstiges: ....

22. **Wie hoch ist Ihr monatliches Netto-Haushaltseinkommen (nach Steuern und Abgaben)?**

    ☐ bis € 1.000
    ☐ € 1.001 bis € 1.500
    ☐ € 1.501 bis € 2.000
    ☐ € 2.001 bis € 2.500
    ☐ € 2.501 bis € 3.000
    ☐ € 3.001 bis € 3.500
    ☐ € 3.501 bis € 4.000
    ☐ € 4.001 bis € 4.500
    ☐ € 4.501 bis € 5.000
    ☐ mehr als € 5.000

23. **Wie hoch ist derzeit Ihre monatliche Stromrechnung?**

    ☐ bis € 20
    ☐ € 21 bis € 30
    ☐ € 31 bis € 40
    ☐ € 41 bis € 50
    ☐ € 51 bis € 60
    ☐ € 61 bis € 70
    ☐ € 71 bis € 80
    ☐ € 81 bis € 90
    ☐ € 91 bis € 100
    ☐ mehr als € 100, nämlich: ....

24. **Wie genau wissen Sie über die Höhe Ihrer monatlichen Stromrechnung Bescheid?**

    ☐ Ich weiß ganz genau wie hoch meine monatliche Stromrechnung ist.
    ☐ Ich kann die Höhe meiner monatlichen Stromrechnung nur grob abschätzen.

25. **Wer zahlt in Ihrem Haushalt die Stromrechnung?**

    ☐ Ich selbst
    ☐ Eine andere im Haushalt lebende Person
    ☐ Die Kosten werden aufgeteilt

26. **Welchen Aufschlag zu Ihrer monatlichen Stromrechnung würden Sie maximal für den weiteren Ausbau erneuerbarer Energiequellen bezahlen, damit Ihr Haushalt Ökostrom bekommt?**

    ....Euro pro Haushalt und Monat

27. **Spenden Sie oder irgendjemand anderer in Ihrem Haushalt für Umweltorganisationen?**

    ☐ Ja
    ☐ Nein

28. **Bitte geben Sie die Postleitzahl Ihres Wohnortes an.**

    PLZ: ....

SCREENING FRAGEN MARKETAGENT

29. **Bitte nennen Sie uns Ihr Geschlecht.**

    ☐ Männlich
    ☐ Weiblich

30. **Wie alt sind Sie?**

    .... Jahre

31. **Bitte geben Sie das Bundesland an, in dem Sie Ihren Hauptwohnsitz haben.**

    ☐ Burgenland
    ☐ Kärnten
    ☐ Niederösterreich
    ☐ Oberösterreich
    ☐ Salzburg
    ☐ Steiermark
    ☐ Tirol
    ☐ Vorarlberg
    ☐ Wien

Vielen Dank für Ihre Mitarbeit!

i want morebooks!

Buy your books fast and straightforward online - at one of world's fastest growing online book stores! Environmentally sound due to Print-on-Demand technologies.

Buy your books online at
www.get-morebooks.com

Kaufen Sie Ihre Bücher schnell und unkompliziert online – auf einer der am schnellsten wachsenden Buchhandelsplattformen weltweit! Dank Print-On-Demand umwelt- und ressourcenschonend produziert.

Bücher schneller online kaufen
www.morebooks.de

VDM Verlagsservicegesellschaft mbH
Heinrich-Böcking-Str. 6-8
D - 66121 Saarbrücken

Telefon: +49 681 3720 174
Telefax: +49 681 3720 1749

info@vdm-vsg.de
www.vdm-vsg.de

Printed by Books on Demand GmbH, Norderstedt / Germany